Undergraduate Texts in Mathematics

Editors
S. Axler
F.W. Gehring
K.A. Ribet

Springer
New York
Berlin
Heidelberg
Barcelona
Hong Kong
London
Milan
Paris
Singapore
Tokyo

Undergraduate Texts in Mathematics

Anglin: Mathematics: A Concise History and Philosophy.
Readings in Mathematics.
Anglin/Lambek: The Heritage of Thales.
Readings in Mathematics.
Apostol: Introduction to Analytic Number Theory. Second edition.
Armstrong: Basic Topology.
Armstrong: Groups and Symmetry.
Axler: Linear Algebra Done Right. Second edition.
Beardon: Limits: A New Approach to Real Analysis.
Bak/Newman: Complex Analysis. Second edition.
Banchoff/Wermer: Linear Algebra Through Geometry. Second edition.
Berberian: A First Course in Real Analysis.
Bix: Conics and Cubics: A Concrete Introduction to Algebraic Curves.
Brémaud: An Introduction to Probabilistic Modeling.
Bressoud: Factorization and Primality Testing.
Bressoud: Second Year Calculus.
Readings in Mathematics.
Brickman: Mathematical Introduction to Linear Programming and Game Theory.
Browder: Mathematical Analysis: An Introduction.
Buskes/van Rooij: Topological Spaces: From Distance to Neighborhood.
Callahan: The Geometry of Spacetime: An Introduction to Special and General Relativity.
Cederberg: A Course in Modern Geometries.
Childs: A Concrete Introduction to Higher Algebra. Second edition.
Chung: Elementary Probability Theory with Stochastic Processes. Third edition.
Cox/Little/O'Shea: Ideals, Varieties, and Algorithms. Second edition.
Croom: Basic Concepts of Algebraic Topology.
Curtis: Linear Algebra: An Introductory Approach. Fourth edition.
Devlin: The Joy of Sets: Fundamentals of Contemporary Set Theory. Second edition.
Dixmier: General Topology.
Driver: Why Math?
Ebbinghaus/Flum/Thomas: Mathematical Logic. Second edition.
Edgar: Measure, Topology, and Fractal Geometry.
Elaydi: An Introduction to Difference Equations. Second edition.
Exner: An Accompaniment to Higher Mathematics.
Exner: Inside Calculus.
Fine/Rosenberger: The Fundamental Theory of Algebra.
Fischer: Intermediate Real Analysis.
Flanigan/Kazdan: Calculus Two: Linear and Nonlinear Functions. Second edition.
Fleming: Functions of Several Variables. Second edition.
Foulds: Combinatorial Optimization for Undergraduates.
Foulds: Optimization Techniques: An Introduction.
Franklin: Methods of Mathematical Economics.
Frazier: An Introduction to Wavelets Through Linear Algebra.
Gordon: Discrete Probability.
Hairer/Wanner: Analysis by Its History.
Readings in Mathematics.
Halmos: Finite-Dimensional Vector Spaces. Second edition.
Halmos: Naive Set Theory.
Hämmerlin/Hoffmann: Numerical Mathematics.
Readings in Mathematics.
Harris/Hirst/Mossinghoff: Combinatorics and Graph Theory.
Hartshorne: Geometry: Euclid and Beyond.
Hijab: Introduction to Calculus and Classical Analysis.
Hilton/Holton/Pedersen: Mathematical Reflections: In a Room with Many Mirrors.

(continued after index)

James J. Callahan

The Geometry of Spacetime

An Introduction to Special and General Relativity

With 218 Illustrations

 Springer

James J. Callahan
Department of Mathematics
Smith College
Northampton, MA 01063-0001
USA

Editorial Board

S. Axler
Mathematics Department
San Francisco State
  University
San Francisco, CA 94132
USA

F.W. Gehring
Mathematics Department
East Hall
University of Michigan
Ann Arbor, MI 48109
USA

K.A. Ribet
Department of Mathematics
University of California
  at Berkeley
Berkeley, CA 94720-3840
USA

Mathematics Subject Classification (1991): 83-01, 83A05, 83Cxx

Library of Congress Cataloging-in-Publication Data
Callahan, James.
  The geometry of spacetime : an introduction to special and
general relativity / James J. Callahan.
    p. cm. — (Undergraduate texts in mathematics)
  Includes bibliographical references and index.
  ISBN 0-387-98641-3 (alk. paper)
   1. Relativity (Physics). 2. Space and time. 3. Algebras, Linear.
 4. Calculus of variations. 5. Mathematical physics. I. Title.
 II. Series.
 QC173.55.C36   1999
 530.11—dc21   98-48083

Printed on acid-free paper.

© 2000 Springer-Verlag New York, Inc.
All rights reserved. This work may not be translated or copied in whole or in part without the written permission of the publisher (Springer-Verlag New York, Inc., 175 Fifth Avenue, New York, NY 10010, USA), except for brief excerpts in connection with reviews or scholarly analysis. Use in connection with any form of information storage and retrieval, electronic adaptation, computer software, or by similar or dissimilar methodology now known or hereafter developed is forbidden.
The use of general descriptive names, trade names, trademarks, etc., in this publication, even if the former are not especially identified, is not to be taken as a sign that such names, as understood by the Trade Marks and Merchandise Marks Act, may accordingly be used freely by anyone.

Production managed by Terry Kornak; manufacturing supervised by Erica Bresler.
Photocomposed using the author's LaTeX files by Integre Technical Publishing Co., Inc., Albuquerque, NM.
Printed and bound by R.R. Donnelley and Sons, Harrisonburg, VA.
Printed in the United States of America.

9 8 7 6 5 4 3 2 1

ISBN 0-387-98641-3 Springer-Verlag New York Berlin Heidelberg   SPIN 10695441

To Felicity

Preface

Overview

This book is an introduction to Einstein's theories of special and general relativity. To read it, you need only a first course in linear algebra and multivariable calculus and a familiarity with the physical applications of calculus. Because the general theory is more advanced than the special, most books limit themselves to one or the other. However, I have tried to encompass both by using the geometry of spacetime as the unifying theme. Of course, we still have large mathematical bridges to cross. Special relativity is just linear algebra, but general relativity is differential geometry—specifically, the curvature of four-dimensional spacetime.

Origins of the special theory

Einstein's theory of special relativity solved a problem that was baffling physicists at the start of the twentieth century. It concerns what happens when different observers measure the speed of light. Suppose, for example, that one observer moves past a second (stationary) observer at one-tenth the speed of light. We would then expect a light beam moving in the same direction to overtake the moving observer at only nine-tenths of the speed it passes the stationary observer. In fact, careful measurements (the Michelson–Morley experiments in the 1880s) contradict this:

Light moves past all observers at the same speed, independent of their own motion.

To account for this, it was proposed that measuring rods contract slightly when they move and clocks slow down—just enough to make the velocity calculations come out right. In 1895, the Dutch physicist H. A. Lorentz even wrote down the equations that describe how lengths and times must be altered for a moving observer. But these hypotheses were *ad hoc* and just as baffling as the phenomenon they were meant to explain.

A decade later, Einstein proposed a solution that was both more radical and more satisfactory. It was built on two assumptions. The first was Galileo's *principle of relativity*:

The principle of relativity

> Two observers moving uniformly relative to one another must formulate the laws of nature in exactly the same way. In particular, no observer can distinguish between absolute rest and absolute motion by appealing to any law of nature; hence, there is no such thing as absolute motion, but only relative motion (of one observer with respect to another).

Relativity had long lain at the heart of mechanics, but Einstein made it a universal principle that applies to all physical phenomena, including electricity, magnetism, and light. His second assumption was more surprising: Rather than try to explain the invariance of the speed of light, he just accepted it as one of those laws of nature that moving observers must agree upon. From this stance, Einstein then *deduced* the transformation equations of Lorentz and the length contraction and time dilation they entailed. In this new theory of relativity (the adjective "special" came only later, when Einstein introduced a more general theory), the time coordinate is on the same relative footing as the spatial coordinates; furthermore, space and time no longer have separate existences but are fused into a single structure—spacetime. Then, in 1907, the mathematician H. Minkowski showed that Einstein's ideas could be interpreted as a new geometry of spacetime.

In Chapter 1 we review the physical problems that prompted the special theory and begin to develop the questions about coordinate transformations that lie at the heart of relativity. Because we assume that observers are in uniform relative motion, the Lorentz transformations that relate their coordinate frames are linear. Geometrically, these transformations are just like spatial rotations, except that their invariant sets are hyperbolas instead of circles. Chapter 2 describes how Einstein made Lorentz transformations the core of a comprehensive theory. We take the geometric viewpoint proposed by Minkowski and develop the Minkowski geometry of spacetime as the invariant theory of Lorentz transformations, making constant comparisons with the familiar Euclidean geometry of ordinary space as the invariant theory of rotations.

Linear spacetime geometry

We complete the study of special relativity in Chapter 3 by analyzing how objects accelerate in response to imposed forces. Motion here is still governed by Newton's laws, which carry over into spacetime in a straightforward way. We look at the geometric manifestation of acceleration as curvature—in this case, curvature of the curves that objects trace out through spacetime. The chapter also introduces the important *principle of covariance*, which says that physical laws must transform the same way as coordinates.

Special relativity is special because it restricts itself to a small class of observers—those undergoing uniform motion with no acceleration. Their coordinate frames are *inertial*; that is, Galileo's law of inertia holds in them without qualification. The law of inertia is Newton's first law of motion; it says that, in the absence of forces, a body at rest will remain at rest and one in motion will continue to move with constant velocity. The now familiar scenes of astronauts and their equipment floating freely in an orbiting spacecraft show that a frame bound to the spacecraft is inertial. Is the frame of an earthbound laboratory inertial? Objects left to themselves certainly do not float freely, but we explain the motions we see by the force of gravity.

Origins of the general theory

Our lifelong experience to the contrary notwithstanding, we must regard gravity as a rather peculiar force. It follows from Newton's second law that a given force will impart less acceleration to

Gravity and general relativity

a large mass than to a small one. But the acceleration of gravity is the same for all masses, so the gravitational force must somehow adjust itself to the mass of each object it pulls. This remarkable property makes it possible to create an artificial gravitational field in space. If we subject a spacecraft far from gravitating masses to constant linear acceleration, then objects inside will "fall down" just as they do on earth. And just as on earth, this artificial force adjusts its strength to give all objects the same downward acceleration. In fact, in any sufficiently small region of spacetime there is no way to distinguish between simple linear acceleration and gravitational acceleration caused by a massive body like the earth. This is Einstein's *principle of equivalence*; he made it the basis of a revolutionary new theory of gravity.

For Einstein, an observer in a gravitational field is simply operating in a certain kind of noninertial frame. If a physical theory is to account for gravity, he reasoned, it must allow noninertial frames on the same footing as inertial ones, and physical laws must take the same form for *all* observers. This is the familiar principle of relativity, but now it is being asserted in its most general form. A successful theory of gravity must be built on general relativity. To help us make the transition from special to general relativity, Chapter 4 considers two kinds of noninertial frames—those that rotate uniformly and those that undergo uniform linear acceleration, from the point of view of an inertial frame. We also survey Newton's theory of gravity and establish both the ordinary differential equations that tell us how a particle moves in a gravitational field and the partial differential equation that tells us how gravitating masses determine the field itself.

Gravity and special relativity are incompatible

The critical discovery in Chapter 4 is that we cannot provide a noninertial frame with the elegant and simple Minkowski geometry we find in a linear spacetime; distances are necessarily distorted. The distortions are the same sort we see in a flat map of a portion of the surface of the earth. Maps distort distances because the earth is curved, so a natural way to explain the distortions that appear when a frame contains a gravitational field is that spacetime is curved. This means that we cannot build an adequate theory of gravity out of Newtonian mechanics and special relativity, because the inertial frames of special relativity are flat.

Curvature is the key to Einstein's theory of gravity, and it is the central topic of Chapters 5 and 6. The simplest circumstance where we can see the essential nature of curvature is in the differential geometry of ordinary surfaces in three-dimensional space—the subject of Chapter 5. At each point a surface has a tangent plane, and each tangent plane has a metric—that is, a way to measure lengths and angles—induced by distance-measurement in the ambient space. With calculus techniques we can then use the metric to do geometric calculations in the surface itself. In this setting, it appears that curvature is an extrinsic feature of a surface's geometry, a manifestation of the way the surface bends in its ambient space. But this setting is both physically and psychologically unsatisfactory, because the four-dimensional spacetime in which we live does not appear to be contained in any larger space that we can perceive. Fortunately, the great nineteenth-century mathematician K. F. Gauss proved that curvature is actually an intrinsic feature of the surface, that is, it can be deduced directly from the metric without reference to the embedding. This opens the way for a more abstract theory of *intrinsic* differential geometry in which a surface patch—and, likewise, the spacetime frame of an arbitrary observer—is simply an open set provided with a suitable metric.

Chapter 6 is about the intrinsic geometry of curved spacetime. It begins with a proof of Gauss's theorem and then goes on to develop the ideas about geodesics and tensors that we need to formulate Einstein's general theory. It explores the fundamental question of relativity: If any two observers describe the same region of curved spacetime, how must their charts G and R be related? The answer is that there is a smooth map $M : G \to R$ whose differential $dM_P : TG_P \to TR_{M(P)}$ is a Lorentz map (that is, a metric-preserving linear map of the tangent spaces) at every point P of G. In other words, special relativity is general relativity "in the small." The nonlinear geometry of spacetime extends the Minkowski geometry of Chapters 2 and 3 in the same way

Intrinsic differential geometry

that the nonlinear geometry of surfaces extends Euclidean plane geometry.

Geodesics and the field equations

In Chapter 7 we take up general relativity proper. From the principle of general covariance, Einstein argues that the laws of physics should be expressed as tensor equations if they are to transform properly. Now consider a coordinate frame falling freely in a gravitational field; such a frame is inertial, so an object falling with it moves linearly and thus along a geodesic in that frame. Since geodesics are defined by tensor equations, general covariance guarantees that all observers will say that freely falling objects move on geodesics. Thus, the equations of motion in a gravitational field are the geodesic equations; moreover, the metric in any coordinate frame defines the gravitational field in that frame. The rest of the chapter is devoted to the field equations; these are derived, as they are in the Newtonian theory, from an analysis of tidal forces. Because of the connection between the field and the metric, the field equations tell us not only how the gravitational sources determine the field but how they determine the curvature of spacetime. They summarize Einstein's remarkable conclusion: Gravity is geometry.

The evidence for general relativity

In the final chapter we review the three major pieces of evidence Einstein put forward in support of his theory in the 1916 paper in which he introduced general relativity. First, Einstein demonstrated that general relativity reduces to Newtonian mechanics when the gravitational field is weak and when objects move slowly in relation to the speed of light. The second piece of evidence has to do with the assertion that gravity is curvature. If that is so, a massive object must deflect the path of anything passing it—including a beam of light. Einstein predicted that it should be possible to detect the bending of starlight by the sun during an eclipse; his predictions were fully confirmed in 1919. The third piece of evidence is the precession of the perihelion of Mercury. It was known from the 1860s that the observed value is larger than the value predicted by Newtonian theory; Einstein's theory predicted the observed value with *no* discrepancy. We follow Einstein's arguments and deduce the metric—that is, the gravitational field—associated with a spherically symmetric mass distribution. This involves solving the field equations in two settings; one is the famous Schwarzschild solution and the other is Einstein's own weak-field solution.

The road not taken

My fundamental aim has been to explore the way an individual observer views the world and how any pair of observers collaborate to gain objective knowledge of the world. In the simplest case, an observer's coordinate patch is homeomorphic to a ball in $\mathbf{R}^4$, and the tensors the observer uses to formulate physical laws are naturally expressed in terms of the coordinates in that patch. This means that it is not appropriate—at least at the introductory level—to start with a coordinate-free treatment of tensors or to assume that spacetime is a manifold with a potentially complex topology. Indeed, it is by analyzing how any pair of observers must reconcile their individual coordinate descriptions of the physical world that we can see the value and the purpose of these more sophisticated geometric ideas. To keep the text accessible to a reasonably large audience I have also avoided variational methods, even though this has meant using only analogy to justify fundamental results like the relativistic field equations.

Sources

The idea for this book originated in a series of three lectures John Milnor gave at the Universiy of Warwick in the spring of 1978. He showed that it is possible to give a unified picture of relativity in geometric terms for a mathematical audience. His approach was more advanced than the one I have taken here—he used variational methods to formulate some of his key concepts and results—but it began with a development of Minkowski geometry in parallel with Euclidean geometry that was elegant and irresistible. Nearly everything in the lectures was accessible to an undergraduate. For example, Milnor argued that when the relativistic tidal equations are expressed in terms of a Fermi coordinate frame, a symmetric 3×3 matrix appears that corresponds exactly to the matrix used to express the Newtonian tidal equations. The case for the relativistic equations is thereby made by analogy, without recourse to variational arguments.

I have used many other sources as well, but I single out four for particular mention. The first is Einstein's own papers; they appear in English translation along with other valuable papers by Lorentz and Minkowski in *The Principle of Relativity* [20]. Einstein's writing is eminently accessible, and anyone who wants a complete picture of relativity should read his 1916 paper [10]. The other three are more focused on special topics; they are the paper by F. K. Manasse and C. W. Misner on Fermi normal coordinates [21], the treatment of parallel transport in *Geometry from a Differentiable Viewpoint* by John McCleary [22], and the weak-field analysis in *Modern Geometry, Part I* by B. A. Dubrovnin et al. [7].

Since the early 1980s I have taught material in this book in an undergraduate course in either geometry or applied mathematics half a dozen times. My students have always covered Chapters 1–3 and 5 and 6 in some detail and parts of Chapters 4 and 7; we have never had the time to do Chapter 7 thoroughly or Chapter 8 at all. While the text makes progressively greater demands on the reader and the material in the later chapters is more difficult, it is no more difficult than a traditional advanced calculus course. There are points, however, where I have taken advantage of the greater emphasis on differential equations, numerical integration, and computer algebra systems found in the contemporary calculus course.

It would not have been possible for me to write this book without a sabbatical leave and also without the supportive climate over many years that enabled me to develop this material into a course; I am grateful to Smith College and to my colleagues for both. And I particularly want to thank Michael Callahan, whose modern perspective and incisive questions and comments about a number of topics sharpened my thinking about relativity.

James J. Callahan
Smith College
Northampton, MA

Contents

Preface   vii

1 Relativity Before 1905   1
    1.1   Spacetime 1
    1.2   Galilean Transformations 9
    1.3   The Michelson–Morley Experiment 15
    1.4   Maxwell's Equations 22

2 Special Relativity – Kinematics   31
    2.1   Einstein's Solution 31
    2.2   Hyperbolic Functions 43
    2.3   Minkowski Geometry 49
    2.4   Physical Consequences 73

3 Special Relativity – Kinetics   87
    3.1   Newton's Laws of Motion 87
    3.2   Curves and Curvature 108
    3.3   Accelerated Motion 123

4 Arbitrary Frames — 143
- 4.1 Uniform Rotation … 144
- 4.2 Linear Acceleration … 155
- 4.3 Newtonian Gravity … 167
- 4.4 Gravity in Special Relativity … 188

5 Surfaces and Curvature — 203
- 5.1 The Metric … 203
- 5.2 Intrinsic Geometry on the Sphere … 221
- 5.3 De Sitter Spacetime … 230
- 5.4 Curvature of a Surface … 241

6 Intrinsic Geometry — 257
- 6.1 Theorema Egregium … 257
- 6.2 Geodesics … 268
- 6.3 Curved Spacetime … 277
- 6.4 Mappings … 292
- 6.5 Tensors … 307

7 General Relativity — 329
- 7.1 The Equations of Motion … 330
- 7.2 The Vacuum Field Equations … 344
- 7.3 The Matter Field Equations … 366

8 Consequences — 385
- 8.1 The Newtonian Approximation … 386
- 8.2 Spherically Symmetric Fields … 396
- 8.3 The Bending of Light … 413
- 8.4 Perihelion Drift … 421

Bibliography — 435

Index — 439

1

CHAPTER

Relativity Before 1905

In 1905, Albert Einstein offered a revolutionary theory—special relativity—to explain some of the most troubling problems about electromagnetism and motion in the physics of the day. Soon afterwards, the mathematician Hermann Minkowski recast special relativity essentially as a new geometric structure for spacetime. The ideas of Einstein and Minkowski are the subject of the next two chapters; here we look at the physical questions that stimulated them, as well as partial solutions offered by others.

1.1 Spacetime

Spacetime itself is quite familiar: A spacetime diagram is a device you have long used to describe and analyze motion. For example, suppose a particle moves upward along the z-axis with constant velocity v meters per second. Then, if we photograph the scene using a strobe light that flashes once each second, we will see the picture on the left at the top of the next page.

Ways to describe motion

Often, though, we choose to represent the motion in a diagram like the one on the right. This is a picture of **spacetime**, because it has a coordinate axis for *time*. (It should therefore have *four*

Spacetime and worldlines

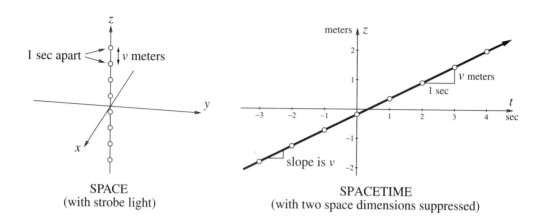

SPACE
(with strobe light)

SPACETIME
(with two space dimensions suppressed)

axes, but since x and y do not change as the particle moves, we have left out their axes to make the diagram simpler.) The motion of the particle is then represented by a straight line with slope v (or, better, v meters per second). We call this the **history**, or **worldline**, of the particle.

Spacetime is 4-dimensional

When all motion is along the z-axis, we can take spacetime to be the (t, z)-plane as we have here. When the motion ranges over a 3-dimensional region, though, spacetime is the $(1 + 3)$-dimensional (t, x, y, z)-hyperspace. Obviously, the $(1 + 1)$-dimensional spacetime is easier to visualize, and we will use it as much as we can. We call a point (t, z) or (t, x, y, z) in spacetime an **event**, because an event always happens some*where* at some *time*.

Points are events

Example: different constant velocities

Shown below is the worldline in a simple $(1 + 1)$-dimensional spacetime of a "courier" traveling along the z-axis:

E_1: left home at time t_1
E_2: arrived at z_1 at time t_2 } traveled with velocity v_1

E_3: set off for z_2 at time t_3
E_4: arrived at z_2 at time t_4 } traveled with velocity v_2

E_5: set off for home at time t_5
E_6: arrived home at time t_6 } traveled with velocity v_3

Note that the events E_2 and E_3 happen at the same place (and so do E_4 and E_5); the courier is stationary when the worldline is

§1.1 Spacetime

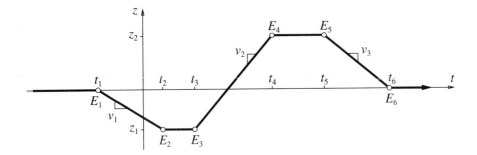

horizontal. Keep in mind that the worldline itself is *not* the path of the courier through space; on the contrary, the courier travels straight up and down the z-axis.

The next example can be found in *The Visual Display of Quantitative Information*, by Edward R. Tufte [29]. It is a picture of the schedule of trains running on the French main railway line between Paris and Lyon in the 1880s, and is thus, in effect, many instances of the previous example plotted together. The vertical axis marks distances from Paris to Lyon, and the horizontal axis is time, so it is indeed a spacetime diagram. The slanting lines are

Example: French trains

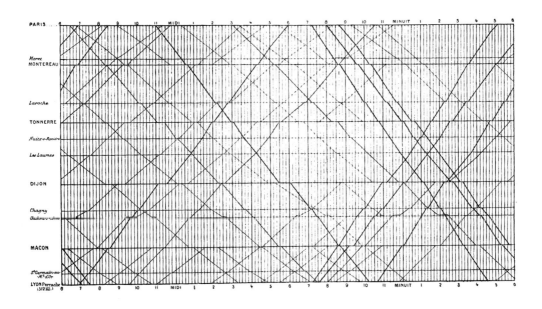

the worldlines of various trains. Those with negative slope run toward Lyon, those with positive slope toward Paris. The express trains have steeper slopes, and their horizontal segments, which represent stops along the way, are briefer and fewer in number.

Example: nonuniform velocities

An object that moves with nonuniform velocity—like a falling body—must have a *curved* worldline. The spacetime diagram below shows two balls A and B that fall to the ground ($z = 0$) from a height $z = h$. Notice that B is thrown straight up with an initial velocity $v_0 > 0$, while A is just dropped ($v_0 = 0$). The two balls are launched at the same moment ($t = 0$); A hits the ground when $t = t_A$, and B hits later, when $t = t_B$. The worldlines continue as horizontal lines because the balls are motionless once they hit the ground.

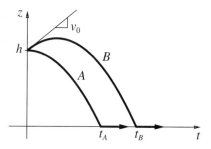

Worldlines are graphs

Once again, note that although the worldlines include parabolic arcs, the actual paths of A and B in space are vertical; A goes straight down, and B goes straight up, stops, then goes straight down. This example should be very familiar to you: Whenever $z = f(t)$ gives the position z of an object as a function of the time t, then the graph for f in the (t, z)-plane is precisely the worldline of the object. See the exercises.

Example: a photo finish

In the picture below (a photo finish of the 1997 Preakness), the winner has won "by a head." But which horse came in second? At first glance, it would appear to be the pale horse in the back, but couldn't the dark horse in the front overtake it before the two cross the finish line?

In fact, the pale horse in back *did* come in second, and this photo does prove it, because this is not an ordinary photo. An ordinary photo is a "snapshot"—a picture of space at a single

Photo courtesy of Maryland Jockey Club. Reproduced with the permission of Pimlico Racetrack.

moment in time; the camera projects a full 3-dimensional view onto a motionless piece of film. In a photo-finish camera, though, the film moves past a narrow slit that can view only the plane of the finish line. As the film is drawn past the slit at a steady rate, it records along its length the events that happen in the plane.

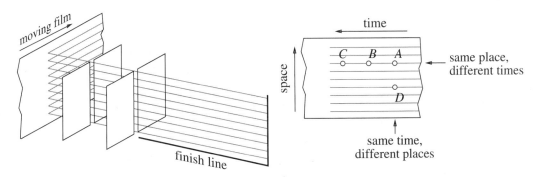

So a photo finish is a spacetime diagram. Events, like A, B, and C, that happen at the same spot on the finish line but at different times appear spread out along a horizontal line on the film, with the earliest event farthest to the right. The photo shows an entire horse because the entire horse eventually moves past the finish line. Events, like A and D, that happen at the same time but in different places on the finish line will appear on the same vertical line in the film. Thus, every vertical line is the finish line! Vertical

Spacetime with two space dimensions

lines further to the left show what happens at the finish line at successively later times. That is why we can be certain that the pale horse came in second.

If an object moves in a 2-dimensional plane, then we can draw its worldline in a $(1 + 2)$-dimensional spacetime. For example, if the object moves with uniform angular velocity around the circle $y^2 + z^2 = $ const in the (y, z)-plane, then its worldline is a helix in (t, x, y)-spacetime. The "tighter" the helix, the greater the angular velocity. Different helices of the same pitch (or "tightness") correspond to motions starting at different points on the circle.

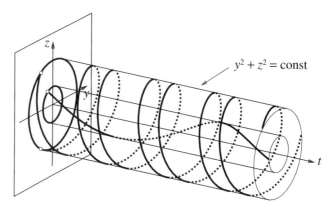

Worldlines lie on cylinders

Notice that all the helical worldlines lie on cylinders of the form $y^2 + z^2 = $ const in the $(1+2)$-dimensional (t, y, z)-spacetime. This reflects something that is true more generally: If an object moves along a curve V in the (y, z)-plane, then its worldline lies on the cylinder

$$C(V) = \{(t, y, z) : (y, z) \text{ in } V\}$$

obtained by translating that curve through (t, y, z)-space parallel to the t-axis. Thus $C(V)$ is the surface made up of all straight lines parallel to the t-axis that pass through the curve V. These lines are called the **generators** of the cylinder. We call $C(V)$ a cylinder because, if V is a circle, then $C(V)$ is the usual circular cylinder. Different parametrizations

$$V : y = f(t), \quad z = g(t)$$

of the same curve V yield different worldlines on $C(V)$.

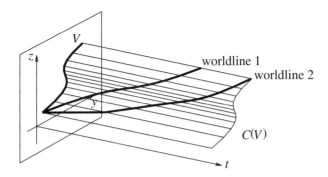

Our final spacetime pictures represent a momentary flash of light from a point source. The flash spreads out light particles, or photons, from the source with speed c ($\approx 3 \times 10^8$ m/sec). First consider only the photons that move along the z-axis; one photon travels up the z-axis with velocity c and another travels down with velocity $-c$. If we use ordinary scales, these worldlines will have enormously steep slopes. However, in the picture below we have made the slopes manageable by changing the unit of distance. Instead of using the meter, we use the *second* (or *light-second*), which is the distance light travels in one second (about 3×10^8 meters). We call this measure of distance a **geometric unit**, to distinguish it from the conventional ones.

Worldline of a photon

Geometric units: measure distance in seconds

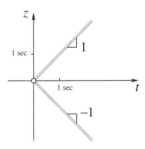

In geometric units, c is 1 second per second, or just 1, so the worldlines of the two photons have slopes ± 1. For any other motion, the numerical value of v is just a percentage of the speed of light. Typically, v is nearer to 0 than to 1, so the worldlines of ordinary motions have extremely small slopes in geometric units.

Light cones

In the full 3-dimensional space, the photons from the flash spread out in all directions and occupy the spherical shell $x^2 + y^2 + z^2 = (ct)^2$ of radius ct after t seconds have passed. In a 2-dimensional slice, the sphere becomes a circle $y^2 + z^2 = (ct)^2$, also of radius ct. If we now plot this circle in (t, y, z)-spacetime as a function of t, we get an ordinary cone. The worldlines of all the photons moving in the (y, z)-plane lie on this cone, which is called a **light cone**. Note that when we take the slice $y = 0$ to get the simple $(1+1)$-dimensional (t, z)-spacetime, we get the pair of photon worldlines we drew earlier. We also call this figure a **light cone**.

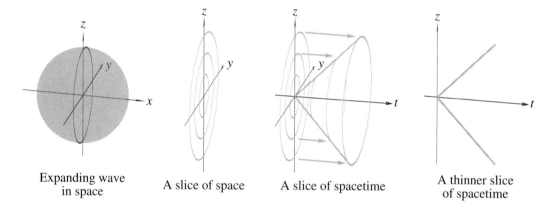

Expanding wave in space   A slice of space   A slice of spacetime   A thinner slice of spacetime

Exercises

1. (a) Sketch the worldline of an object whose position is given by $z = t^2 - t^3$, $0 \leq t \leq 1$.

   (b) How far does this object get from $z = 0$? When does that happen?

   (c) With what velocity does the object depart from $z = 0$? With what velocity does it return?

   (d) Sketch a worldline that goes from $z = 0$ to $z = 1$ and back again to $z = 0$ in such a way that both its departure and arrival velocities are 0. Give a formula $z = f(t)$ that describes this motion.

2. (a) Suppose an object falls with no air resistance, so it experiences constant acceleration $z''(t) = -g$, where g is the acceleration due to gravity ($g \approx 9.8$ m/sec^2). If its initial velocity is $z'(0) = v_0$ and its initial position is $z(0) = h$, find the formula $z(t)$ that gives its position at any time t.

   (b) Using your formula, sketch the worldlines of three objects for which v_0 takes the three values 0 m/sec, +10 m/sec, −2 m/sec. Use $h = +60$ m for all three objects.

   (c) Determine when the three objects hit the ground $z = 0$. Are they in the order you expect?

3. Two objects G and M move along the z-axis with a constant positive velocity v that is much smaller than 1 (written $v \ll 1$), separated by a distance λ; see the following figure. A signal is emitted at the event O. It travels with velocity 1 until it reaches M at the event E_1; it is then reflected back with velocity -1, reaching G at the event E_2. Express the times t_1 and t_2 in terms of λ and v.

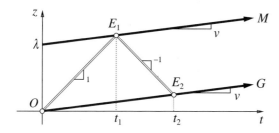

1.2 Galilean Transformations

The aim of science is to describe and interpret objective reality. But the starting point of science is individual observation, and this is inherently subjective. Nevertheless, when individual observers compare notes, they find points of common agreement, and these ultimately constitute what is physically, or objectively, real. ("Do my calculations agree with yours?" "Can I reproduce your experiment?")

The problem of objectivity

Galileo's principle of relativity

The question of objectivity has always been part of modern science. Galileo took it up in connection with motion, and was led to formulate this **principle of relativity**:

> Two observers moving uniformly relative to one another must formulate the laws of nature in exactly the same way. In particular, no observer can distinguish between absolute rest and absolute motion by appealing to any law of nature; hence, there is no such thing as *absolute* motion, but only motion in relation to an observer.

We shall read this both ways:

- Any physical law must be formulated the same way by all observers.
- Anything formulated the same way by all observers is a physical law.

As we shall see, scientists at the end of the nineteenth century were tempted to ignore Galileo's principle in order to deal with some particularly baffling problems. That didn't help, though, and when Einstein eventually solved the problems, he did so by firmly reestablishing the principle of relativity. To distinguish this from the more sweeping generalization he was to make a decade later, we call this **special relativity**.

Spacetime diagrams of two different observers

The first step to understanding what Einstein did is to make a careful analysis of the situation that Galileo addresses: two observers moving uniformly relative to one another. For us this means looking at their spacetime diagrams. We shall call these "Galilean" observers R and G; G will use Greek letters (τ, ξ, η, ζ) for coordinates, and R will use Roman (t, x, y, z). We assume that R and G approach and then move past each other with uniform velocity v. They meet at an event O, which they define to happen at time $t = \tau = 0$. In other words, they use this event to synchronize their clocks. For the sake of definiteness, we assume that corresponding axes (that is, ξ and x, η and y, ζ and z) of the two observers point in the same direction. Moreover, we assume that G moves up along R's z-axis and R moves down along G's ζ-axis. Their spacetime diagrams then look like this:

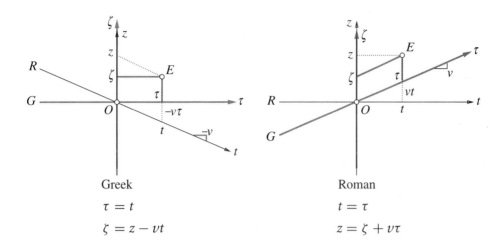

Greek
$\tau = t$
$\zeta = z - vt$

Roman
$t = \tau$
$z = \zeta + v\tau$

The equations tell us the Roman coordinates (t, z) of an event E when we know the Greek (τ, ζ), and vice versa. We note the following:

Coordinates of two observers compared

- A time axis is just the worldline of an observer.
- A right angle between the time and space axes means that the observer whose worldline is that time axis is stationary with respect to that space coordinate.
- If the angle between the time and space axes is $90° - \theta$, then that observer has velocity $v = \tan\theta$ with respect to that space coordinate.
- For a given event E, $t = \tau$ because observers measure time the same way.
- For a given event E, $z \neq \zeta$ in general because observers measure distances from themselves. The difference $z - \zeta = vt = v\tau$ is the distance between the two observers at the moment z and ζ are measured.

Notice that we make no assumption about which observer is *really* moving, or which spacetime diagram is *really* correct. According to the principle of relativity, there's simply no way to determine this. From R's point of view, only G is moving, and vice versa. The two spacetime diagrams are equally valid descriptions of reality. It is by comparing the diagrams that we discover which

All spacetime diagrams are equally valid

of their elements do reflect physical reality and which do not. Thus, time—or more precisely, the time interval between two events—*is* objectively real, because observers report the same values. By contrast, the spatial distance between two events is not. See Exercise 1.

Since we want to focus on the elements of a spacetime diagram that don't change when we go from one diagram to another, we must study the details of this transformation. The equations that convert G's coordinates to R's define a linear map $S_v : \mathbf{R}^2 \to \mathbf{R}^2$,

$$S_v : \begin{pmatrix} \tau \\ \zeta \end{pmatrix} \to \begin{pmatrix} t \\ z \end{pmatrix} = \begin{pmatrix} 1 & 0 \\ v & 1 \end{pmatrix} \begin{pmatrix} \tau \\ \zeta \end{pmatrix}.$$

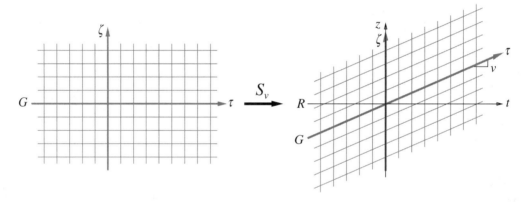

Because S_v connects the spacetime diagrams of two Galilean observers (that is, observers considered by Galileo's principle of relativity), it is commonly called a **Galilean transformation**. When we think of S_v as the matrix, we shall call it a Galilean matrix.

Geometrically, the transformation is a **shear**. It slides vertical lines up or down in proportion to their distance from the vertical axis. In fact, since the vertical line $\tau = 1$ slides by the amount v, the constant of proportionality is v and horizontal lines become parallel lines of slope v.

We can think of the Greek coordinate frame and the Roman coordinate frame as two *languages* for describing events in spacetime. In the spirit of this analogy, the matrix S_v is the *dictio-*

§1.2 Galilean Transformations

nary that translates "Greek" into "Roman". But translation dictionaries come in pairs: Besides a Greek–Roman dictionary, we need a Roman–Greek dictionary. Obviously, the inverse matrix $S_v^{-1} = S_{-v}$ plays this role.

$$\text{GREEK} \quad \underset{S_{-v}}{\overset{S_v}{\rightleftarrows}} \quad \text{ROMAN}$$

We can illustrate this with a simple example. Suppose a new particle moves uniformly with respect to R and G. Then its worldline is straight in each spacetime. If that worldline consists of the events (τ, ζ) in the Greek coordinate system, then the velocity of the particle is $\sigma = \Delta\zeta/\Delta\tau$, according to G. Since $t = \tau$ and $z = \zeta + v\tau$ (by our "translation dictionary"), R will calculate the velocity to be

$$s = \frac{\Delta z}{\Delta t} = \frac{\Delta\zeta + v\Delta\tau}{\Delta\tau} = \frac{\Delta\zeta}{\Delta\tau} + v = \sigma + v.$$

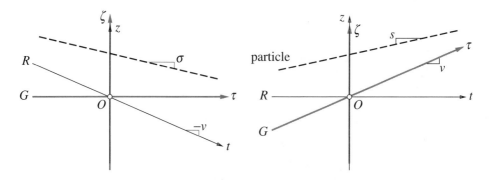

To move to the full $(1+3)$-dimensional version, we assume that the spatial axes of G and R remain parallel as G and R move. The velocity of G relative to R is a spatial vector $\mathbf{v} = (v_x, v_y, v_z)$. In the exercises you will show that the Galilean transformation is now accomplished by the 4×4 matrix

$$S_\mathbf{v} = \begin{pmatrix} 1 & 0 & 0 & 0 \\ v_x & 1 & 0 & 0 \\ v_y & 0 & 1 & 0 \\ v_z & 0 & 0 & 1 \end{pmatrix};$$

(1 + 3)-dimensional transformations

	in coordinates of:	
speed of:	GREEK	ROMAN
G	0	v
R	$-v$	0
any particle	σ	$\sigma + v$
	$s - v$	s

furthermore, $S_\mathbf{v}^{-1} = S_{-\mathbf{v}}$. In our earlier work, we assumed that motion was along the vertical axis alone and hence $v_x = v_y = 0$. In this case multiplication by the 4×4 matrix $S_\mathbf{v}$ gives $x = \xi$ and $y = \eta$. This explains why we could suppress those coordinates (and work with the simpler 2×2 matrix): They had the same values in the two frames.

Exercises

1. Suppose the events E_1 and E_2 have the coordinates $(1, 0$ and $(2, 0)$ in R. What is the spatial distance between them, according to R? What is the spatial distance, according to G?

2. Let $A(X, Y)$ denote the area of the parallelogram spanned by two vectors X and Y in the plane $\mathbf{R}^2$. If $L : \mathbf{R}^2 \to \mathbf{R}^2$ is a linear map, show that $A(L(X), L(Y)) = \det L \cdot A(X, Y)$. What does this mean when $\det L$ is negative? What does this mean when $\det L = 0$?

3. Show that every Galilean transformation $S_v : \mathbf{R}^2 \to \mathbf{R}^2$ preserves areas.

4. (a) Show that $S_v^{-1} = S_{-v}$ and $S_v S_w = S_{v+w}$ when $S_v : \mathbf{R}^2 \to \mathbf{R}^2$.

   (b) These facts imply that the set of Galilean transformations forms a group $\mathcal{G}_2$ using matrix multiplication. Show that this group is commutative, that is, that $S_v S_w = S_w S_v$. (In fact, $\mathcal{G}_2$ is isomorphic to the real numbers $\mathbf{R}$ regarded as a group using addition. Prove this if you are familiar with group theory.)

5. (a) Write the Galilean transformation $S_\mathbf{v}$ matrix when space-time has two space dimensions and relative velocity is given by the 2-dimensional vector $\mathbf{v} = (v_y, v_z)$.

   (b) Show that $S_\mathbf{v}^{-1} = S_{-\mathbf{v}}$ and $S_\mathbf{v} S_\mathbf{w} = S_{\mathbf{v}+\mathbf{w}}$. Here $\mathbf{v} + \mathbf{w}$ is ordinary vector addition.

   (c) The collection $\mathcal{G}_3$ of these matrices $S_\mathbf{v}$ is also a group. Show that $\mathcal{G}_3$ is commutative, too. (In fact, $\mathcal{G}_3$ is isomorphic to the additive group $\mathbf{R}^2$.)

6. Repeat the previous exercise with the full 4×4 Galilean transformations. ($\mathcal{G}_4$ is isomorphic to the additive group $\mathbf{R}^3$.)

1.3 The Michelson–Morley Experiment

Light doesn't transform properly under Galilean transformations. In the late nineteenth century, this problem manifested itself in several ways. We'll look at two: the Michelson–Morley experiment and Maxwell's equations. Even before we take them up in detail below, we can describe their essential implications quite simply by considering how two observers describe the motion of photons.

We assume that a momentary flash of light occurs at the event O when R and G meet, and we plot the worldlines of the photons U and D that go up and down the vertical axis. To keep the slopes reasonable, R and G will both use geometric units and measure distance in seconds (see Section 1.1). If R measures the velocity of U to be 1 and that of D to be -1, then G must measure them

Before Michelson–Morley

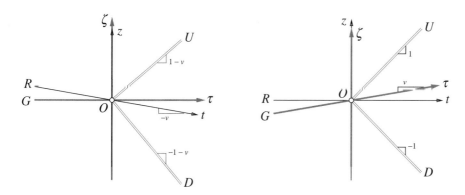

After Michelson–Morley

to be $1 - v$ and $-1 - v$, respectively.

It should, therefore, be possible to measure differences in the speed of light for two Galilean observers. But the results of the Michelson–Morley experiment imply that there are no differences! In effect, G gets the same value as R, so their spacetime diagrams look like this:

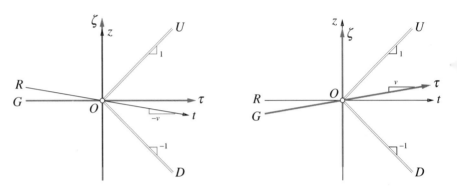

We can summarize the paradoxical conclusions of the Michelson-Morley experiment (M-M) in the following table.

speed of:		in coordinates of:	
		GREEK	ROMAN
ordinary particle		σ	$\sigma + v$
		$s - v$	s
photon	before M-M	1	$1 + v$
		$1 - v$	1
	after M-M	1	1

The Experiment

The ether

Light is considered to be a wave as well as a particle. The wave theory explains many aspects of light, such as colors, interference and diffraction patterns (like those we see on the surface of a compact disk), and refraction (the bending of light rays by lenses). But if light is a wave, there should be some medium that is doing the waving. Since light travels readily through the vacuum of empty space, that medium—which came to be called the "ether"—must permeate the entire universe. The earth must fly through it like a plane through the air. Michelson and Morley sought to measure the velocity of the earth through the ether in a series of ever more refined experiments between 1881 and 1887.

In principle, it would seem simple enough to make the measurement. Let the observer R be stationary with respect to the ether and let G be on the earth. Then measure the velocity of a beam of light emitted in the direction of G's motion with respect to R. According to the Galilean picture, the velocity of the beam will be $w = 1 - v$, where 1 is the standard speed of light in a vacuum (as measured by R) and v is the unknown velocity of G. But we have just measured w, so $v = 1 - w$.

Measuring velocity relative to the ether

This simple approach falters for two reasons. First, we have no direct awareness of the ether, so we don't know where to put R and hence which way to aim the beam of light. However, we can deal with this by sending out beams in many different directions. A more serious problem has to do with the techniques that are used to determine the speed of light. They depend on reflecting a light beam off a distant mirror and comparing the returning beam with the outgoing one. Exactly how this is done is not so important as the fact that the light ray travels at two different speeds while it is being analyzed: $1 - v$ on the outward journey and $1 + v$ on the return. Instruments can measure only the average speed for the round trip. It appears, at first glance, that the average speed is 1, and we therefore lose all evidence of the variable speed v that we are trying to determine.

In fact, the average is not 1, and the evidence does not get lost. The Michelson–Morley experiment is designed to overcome all our objections and to capture the elusive value v. The Dutch physicist H. A. Lorentz gives this contemporary (1895) description of the experiment in the opening paragraph of [19]:

> As Maxwell first remarked and as follows from a very simple calculation, the time required by a ray of light to travel from a point A to a point B and back to A must vary when the two points together undergo a displacement without

carrying the ether with them. The difference is, certainly, a magnitude of the second order; but it is sufficiently great to be detected by a sensitive interference method.

The difference is proportional to v^2

To see what Lorentz means, let us do that calculation. In the experiment, the observer G (who is moving with an unknown velocity v) sends out a light pulse that is reflected back to G by a mirror M that is firmly attached to G's frame at a fixed distance of λ light-seconds. The apparatus allows the direction from G to M to vary arbitrarily, though.

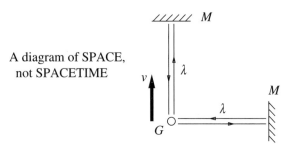

A diagram of SPACE, not SPACETIME

The exercises ask you to confirm that the travel time T for a light ray depends on the direction of the ray in the following way:

$$T_{\|}(v) = \frac{2\lambda}{1-v^2} \qquad \text{parallel to } G\text{'s motion;}$$

$$T_{\perp}(v) = \frac{2\lambda}{\sqrt{1-v^2}} \qquad \text{perpendicular to } G\text{'s motion.}$$

You can also show that the average speed of light for round trips in the two directions is not 1; instead, $c_{\|} = 1 - v^2$ and $c_{\perp} = \sqrt{1-v^2}$.

In fact, what the experiment actually measures is the *difference* between $T_{\|}$ and $T_{\perp}$. To see how this is connected to v, first look at the Taylor expansions of $T_{\|}$ and $T_{\perp}$:

$$T_{\|}(v) = 2\lambda + 2\lambda v^2 + O(v^4),$$
$$T_{\perp}(v) = 2\lambda + \lambda v^2 + O(v^4).$$

Hence the leading term in the difference is indeed proportional to v^2; this is the "magnitude of second order" that Lorentz refers to:

$$T_{\|} - T_{\perp} = \lambda v^2 + O(v^4) = v^2(\lambda + O(v^2)).$$

To get some idea how large v^2 might be, let us calculate the velocity of the earth in its annual orbit around the sun. The orbit is roughly a circle with a circumference of about 9×10^{11} meters. Since a year is about 3×10^7 seconds, the orbital speed is about 3×10^4 m/sec. In geometric units (where $c = 3 \times 10^8$ m/sec $= 1$) we get $v \approx 10^{-4}$, $v^2 \approx 10^{-8}$. Of course, if there is an "ether wind," v may be much less if the earth happens to be moving in the same direction as the "wind." However, six months later it will be heading in the opposite direction, so then v should be even larger than our estimate. In any event, the apparatus Michelson and Morley used was certainly sensitive enough to detect velocities of this order of magnitude, so a series of experiments conducted at different times and in different directions should therefore eventually reveal the motion of the earth through the ether.

Estimating v^2

But the experiment had a null outcome: it showed that $T_\parallel = T_\perp$ always. This implied that light moves past all Galilean observers at the same speed, no matter what their relative motion. Light doesn't transform properly under Galilean transformations.

The Fitzgerald Contraction Hypothesis

The experimental results contradict common sense and some very persuasive arguments. What could be wrong? Michelson himself concluded that $v = 0$ always—that is, the ether moved with the earth. Lorentz found this unsatisfactory, and in his article offered another way to remove the contradiction. His idea is more commonly know as the Fitzgerald contraction hypothesis, after the British physicist G. F. Fitzgerald, who introduced it independently.

Facing the contradiction

Noting that the travel times $T_\parallel$ and $T_\perp$ depend upon the distance λ to the mirror, Fitzgerald and Lorentz give a simple—but nonetheless astonishing—explanation for the fact that $T_\parallel$ is not larger than $T_\perp$: They say that the distance λ varies with the direction. To make this distinction, let us write

Contraction in the direction of motion

$$T_\parallel = \frac{2\lambda_\parallel}{1 - v^2}, \quad T_\perp = \frac{2\lambda_\perp}{\sqrt{1 - v^2}}.$$

Then if we set
$$\lambda_\| = \sqrt{1-v^2} \cdot \lambda_\perp,$$
we get
$$T_\| = \frac{2\lambda_\|}{1-v^2} = \frac{2\sqrt{1-v^2}\cdot\lambda_\perp}{1-v^2} = \frac{2\lambda_\perp}{\sqrt{1-v^2}} = T_\perp.$$

Since $\sqrt{1-v^2} < 1$, $\lambda_\|$ is smaller than $\lambda_\perp$. This is usually translated as saying that the apparatus (which was was supposed to keep λ fixed) must have contracted in the direction of motion—but not in any perpendicular direction.

The contraction cannot be measured directly

There is no way to measure the contraction directly, because every measuring instrument participates in the motion and hence in the contraction. In other words, a meter stick will say that λ does *not* vary. The only way to measure the contraction is indirectly, by the Michelson–Morley experiment or its equivalent.

With the Fitzgerald contraction, the transformation between coordinate frames takes a new form. The new relation (which we could call the Fitzgerald transformation) is

$$F_v : \begin{cases} t = \tau, \\ z = \sqrt{1-v^2}\,\zeta + v\tau, \end{cases} \qquad F_v = \begin{pmatrix} 1 & 0 \\ v & \sqrt{1-v^2} \end{pmatrix}.$$

To see why this is so, set $\tau = 0$ and note that when G measures a distance of $\zeta = 1$, Fitzgerald contraction implies that R will say that the distance is only $\sqrt{1-v^2} < 1$. It is not true that $F_v^{-1} = F_{-v}$ as it is for Galilean transformations; instead,

$$F_v^{-1} = \begin{pmatrix} 1 & 0 \\ \frac{-v}{\sqrt{1-v^2}} & \frac{1}{\sqrt{1-v^2}} \end{pmatrix}, \qquad F_v^{-1} : \begin{cases} \tau = t, \\ \zeta = \frac{z}{\sqrt{1-v^2}} - \frac{vt}{\sqrt{1-v^2}}. \end{cases}$$

Besides being subject to the criticism that they have been constructed after the fact merely to "preserve appearances," F_v and F_v^{-1} are not symmetric, nor do they transform the worldlines of light photons properly. You can explore this in the exercises.

Exercises

1. (a) Show that the travel time $T_\perp$ for a light ray bouncing off a mirror perpendicular to the direction of motion of an observer G is
$$T_\perp(v) = \frac{2\lambda}{\sqrt{1-v^2}}$$
when G moves with velocity v and the mirror is λ light-seconds away. What is the average velocity of light in a perpendicular direction?

   (b) Show that the travel time $T_\parallel$ when the light ray moves in the same direction as G is
$$T_\parallel(v) = \frac{2\lambda}{1-v^2}.$$
What is the average velocity of light in the direction of G's motion?

   (c) Determine the third-order Taylor expansions of $T_\perp(v)$ and $T_\parallel(v)$ and show thereby that
$$T_\parallel - T_\perp = v^2\left(\lambda + O(v^2)\right).$$

2. (a) The linear map $C_v : \mathbf{R}^2 \to \mathbf{R}^2$ that performs simple compression by the factor $\sqrt{1-v^2}$ in the direction of motion is
$$C_v = \begin{pmatrix} 1 & 0 \\ 0 & \sqrt{1-v^2} \end{pmatrix}.$$
Indicate the effect of C_v by sketching the image of a grid of unit squares in the target; describe the image grid in words.

   (b) Let F_v be the "Fitzgerald transformation" defined in the text. Show that $F_v = S_v \circ C_v \stackrel{?}{=} C_v \circ S_v$, where S_v is the Galilean shear of velocity v.

3. Sketch the image of a grid of unit squares under the linear map F_v. How is this image related to the images produced by C_v and S_v? In particular, how are the images of *vertical* lines

under F_v and S_v related, and how are the images of *horizontal* lines related?

4. (a) Consider upward and downward moving photons in G whose worldlines have equations $\zeta = \pm \tau$ and hence slopes ± 1. Express the slopes $m_\pm$ of the image worldlines in R under the map $F_v : G \to R$ as functions of the relative velocity v.

   (b) Show that $m_+ > +1$ and $m_- > -1$ whenever $0 < v < 1$. For which v does m_+ attain its maximum value?

5. Explain why F_v cannot be a valid description of the relation between the coordinates of G and R.

1.4 Maxwell's Equations

Electric and magnetic fields satisfy Maxwell's equations

To explain why one magnet can act upon another "at a distance"—that is, without touching it or anything connected to it—we say that the magnet is surrounded by a force field; it is this magnetic field that acts on the other magnet. Electric charges also act on one another at a distance, and we explain this by the intervention of an electric field. These fields permeate space and vary from place to place and from moment to moment. But not all variations are possible; the functions defining the fields must satisfy the partial differential equations given by James Clerk Maxwell in 1864.

Light is electromagnetic

One consequence of these "field equations" is that a disturbance in the field (caused, for example, by a vibrating electric charge) will propagate through space and time in a recognizable way and with a definite velocity that depends on the medium. In other words, there are electromagnetic waves. Moreover, the velocity of an electromagnetic wave turns out to be the same as the velocity of light. The natural conclusion is that light must be electromagnetic in nature and that the ether must carry the electric and magnetic fields. This leads us back to the earth's motion, but now the question is, How does that motion affect Maxwell's equations?

The Equations

Maxwell's equations (in R's coordinate frame) concern:

- **electric forces**, described by an electric vector **E** varying over space and time:

$$\mathbf{E}(t, x, y, z) = (\underbrace{E_1(t, x, y, z)}_{x \text{ component}}, \underbrace{E_2(t, x, y, z)}_{y \text{ component}}, \underbrace{E_3(t, x, y, z)}_{z \text{ component}})$$

- **magnetic forces**, likewise described by a varying vector:

$$\mathbf{H}(t, x, y, z) = (H_1(t, x, y, z), H_2(t, x, y, z), H_3(t, x, y, z))$$

- a scalar function $\rho(t, x, y, z)$, **electric charge density**
- a vector function $\mathbf{J}(t, x, y, z)$, **electric current density**

These functions are not independent of one another, but must together satisfy the following partial differential equations—Maxwell's equations—which summarize all the observed facts about electricity and magnetism. We write them here in geometric units, where $c = 1$.

Maxwell's equations

$$\nabla \cdot \mathbf{E} = \rho, \qquad\qquad \nabla \cdot \mathbf{H} = 0,$$
$$\nabla \times \mathbf{E} = -\frac{\partial \mathbf{H}}{\partial t}, \qquad \nabla \times \mathbf{H} = \frac{\partial \mathbf{E}}{\partial t} + \mathbf{J},$$
$$\nabla = \left(\frac{\partial}{\partial x}, \frac{\partial}{\partial y}, \frac{\partial}{\partial z}\right).$$

Everything is written in R's frame; how would it look in G's frame? Rather than answer this question in full, we focus on one important consequence arrived at by the following series of steps.

Change to G's frame

1. Assume that we are in empty space, so $\rho = 0$ and $\mathbf{J} = \mathbf{0}$.
2. Consider the following derivation (where subscripts denote partial derivatives):

$$\mathbf{E}_{tt} = \frac{\partial}{\partial t}(\nabla \times \mathbf{H}) \qquad \left(\nabla \times \mathbf{H} = \frac{\partial \mathbf{E}}{\partial t} + \mathbf{J}, \text{ but } \mathbf{J} = \mathbf{0}\right)$$
$$= \nabla \times \mathbf{H}_t$$

$$= -\nabla \times (\nabla \times \mathbf{E}) \qquad\qquad \left(\nabla \times \mathbf{E} = -\frac{\partial \mathbf{H}}{\partial t}\right)$$

$$= -\nabla(\nabla \cdot \mathbf{E}) + \nabla^2 \mathbf{E} \qquad\qquad \text{(calculus identity)}$$

$$= \nabla^2 \mathbf{E} \qquad\qquad (\nabla \cdot \mathbf{E} = \rho = 0)$$

$$= \left(\frac{\partial^2}{\partial x^2} + \frac{\partial^2}{\partial y^2} + \frac{\partial^2}{\partial z^2}\right)\mathbf{E}.$$

3. The second-order partial differential equation $\mathbf{E}_{tt} = \nabla^2 \mathbf{E}$ is called the **wave equation**; it must hold for each component $E = E_i$ of the electric vector $\mathbf{E}$:

$$E_{tt} = E_{xx} + E_{yy} + E_{zz}.$$

4. Assume that the electric field depends only on t and z: $E = E(t, z)$. For example, a plane wave parallel to the (x, y)-plane will satisfy this condition. The wave equation for each component then takes the simple form

$$E_{tt} = E_{zz}, \qquad \text{or} \qquad E_{tt} - E_{zz} = 0.$$

We now ask, What does this simple wave equation become in G's frame of reference, when G moves along R's z-axis with velocity v?

The Galilean transformation of $E_{tt} - E_{zz}$

The appropriate Galilean transformation is $S_v : t = \tau,\ z = \zeta + v\tau$. The following equations convert the components of the electric vector to G's coordinates:

$$E(t, z) = E(\tau, \zeta + v\tau) = \mathcal{E}(\tau, \zeta).$$

In other words, we use the "dictionary" S_v to convert E (in R's frame) to $\mathcal{E}$ (in G's frame) by composing E with S_v:

$$\text{GREEK} \underset{S_v^{-1}}{\overset{S_v}{\rightleftarrows}} \text{ROMAN}$$

$$\mathcal{E} \searrow \qquad \swarrow E$$

$$R$$

Using the multivariable chain rule, we could then express the derivatives of $\mathcal{E}$ with respect to τ and ζ in terms of the derivatives of E with respect to t and z.

But this is the wrong way round: To see how $E_{tt} - E_{zz}$ transforms, we want the derivatives of E with respect to t and z in terms of the derivatives of $\mathcal{E}$ with respect to τ and ζ. The solution is to use the inverse "dictionary" S_v^{-1}: $\tau = t$, $\zeta = z - vt$:

$$\mathcal{E}(\tau, \zeta) = \mathcal{E}(t, z - vt) = E(t, z).$$

Now the chain rule works the right way. Here is one derivative calculated in detail:

$$E_t = \frac{\partial E}{\partial t} = \frac{\partial \mathcal{E}}{\partial \tau}\frac{\partial \tau}{\partial t} + \frac{\partial \mathcal{E}}{\partial \zeta}\frac{\partial \zeta}{\partial t} = \mathcal{E}_\tau - v\mathcal{E}_\zeta.$$

The others are

$$E_z = \mathcal{E}_\zeta, \qquad E_{tt} = \mathcal{E}_{\tau\tau} - 2v\mathcal{E}_{\tau\zeta} + v^2\mathcal{E}_{\zeta\zeta}, \qquad E_{zz} = \mathcal{E}_{\zeta\zeta}.$$

Thus in G's coordinate frame the simple wave equation $E_{tt} - E_{zz} = 0$ transforms into

$$\mathcal{E}_{\tau\tau} - \mathcal{E}_{\zeta\zeta} + v^2\mathcal{E}_{\zeta\zeta} - 2v\mathcal{E}_{\tau\zeta} = 0.$$

The two extra terms mean that the wave equation has a different form in G's frame (if $v \neq 0$, which we are certainly assuming).

According to Galileo's principle of relativity, this shouldn't happen: Observers in uniform motion relative to each other must formulate and express physical laws in the same way. Maxwell's equations and the wave equation are physical laws, so they can't have different forms for different observers.

Does Fitzgerald contraction help? If we replace $\zeta = z - vt$ (which comes from S_v^{-1}) by

The "Fitzgerald" transformation of $E_{tt} - E_{zz}$

$$\zeta = \frac{z}{\sqrt{1 - v^2}} - \frac{vt}{\sqrt{1 - v^2}}$$

from F_v^{-1}, then

$$E(t, z) = \mathcal{E}\left(t, \frac{z}{\sqrt{1 - v^2}} - \frac{vt}{\sqrt{1 - v^2}}\right)$$

and

$$E_t = \mathcal{E}_\tau - \frac{v}{\sqrt{1-v^2}}\mathcal{E}_\zeta, \qquad E_z = \frac{1}{\sqrt{1-v^2}}\mathcal{E}_\zeta,$$

$$E_{tt} = \mathcal{E}_{\tau\tau} - \frac{2v}{\sqrt{1-v^2}}\mathcal{E}_{\tau\zeta} + \frac{v^2}{1-v^2}\mathcal{E}_{\zeta\zeta}, \qquad E_{zz} = \frac{1}{1-v^2}\mathcal{E}_{\zeta\zeta}.$$

Therefore,

$$E_{tt} - E_{zz} = \mathcal{E}_{\tau\tau} - \frac{2v}{\sqrt{1-v^2}}\mathcal{E}_{\tau\zeta} + \frac{v^2-1}{1-v^2}\mathcal{E}_{\zeta\zeta}$$

$$= \mathcal{E}_{\tau\tau} - \mathcal{E}_{\zeta\zeta} - \frac{2v}{\sqrt{1-v^2}}\mathcal{E}_{\tau\zeta};$$

the Fitzgerald transformation got rid of one of the two extra terms!

To get rid of the other, Lorentz proposed a further alteration of the Galilean transformation that has come to be known as the Lorentz transformation.

The Lorentz Transformation

Since a modification of the equation connecting space coordinates got rid of one of the extra terms in the transformed wave equation, perhaps a similar modification on the time coordinates will get rid of the other. Here is the matrix for Lorentz's proposed transformation with the Fitzgerald matrix for comparison:

$$L_v^{-1} = \begin{pmatrix} \frac{1}{\sqrt{1-v^2}} & \frac{-v}{\sqrt{1-v^2}} \\ \frac{-v}{\sqrt{1-v^2}} & \frac{1}{\sqrt{1-v^2}} \end{pmatrix}, \qquad F_v^{-1} = \begin{pmatrix} 1 & 0 \\ \frac{-v}{\sqrt{1-v^2}} & \frac{1}{\sqrt{1-v^2}} \end{pmatrix}.$$

Even before we see whether it solves our problem, we can see that the Lorentz matrix has an aesthetic appeal that the Fitzgerald matrix lacks: It is a symmetric matrix—a reflection of the symmetric treatment of time and space variables that was designed into the transformation. Furthermore, $L_v^{-1} = L_{-v}$, the same simple relation enjoyed by the Galilean transformations.

The Lorentz transformation does indeed make the wave equation transform properly. If we write the transformation in the form

The Lorentz transformation of $E_{tt} - E_{zz}$

§1.4 Maxwell's Equations

$$\tau = \frac{z - vt}{\sqrt{1-v^2}}, \qquad \zeta = \frac{z - vt}{\sqrt{1-v^2}},$$

then

$$E(t, z) = \mathcal{E}\left(\frac{z - vt}{\sqrt{1-v^2}}, \frac{z - vt}{\sqrt{1-v^2}}\right)$$

and

$$E_t = \frac{\mathcal{E}_\tau - v\mathcal{E}_\zeta}{\sqrt{1-v^2}}, \qquad E_z = \frac{-v\mathcal{E}_\tau + \mathcal{E}_\zeta}{\sqrt{1-v^2}},$$

$$E_{tt} = \frac{\mathcal{E}_{\tau\tau} - 2v\mathcal{E}_{\tau\zeta} + v^2\mathcal{E}_{\zeta\zeta}}{1-v^2}, \qquad E_{zz} = \frac{v^2\mathcal{E}_{\tau\tau} - 2v\mathcal{E}_{\tau\zeta} + \mathcal{E}_{\zeta\zeta}}{1-v^2}.$$

Thus

$$E_{tt} - E_{zz} = \frac{(1-v^2)\mathcal{E}_{\tau\tau} + (v^2-1)\mathcal{E}_{\zeta\zeta}}{1-v^2} = \mathcal{E}_{\tau\tau} - \mathcal{E}_{\zeta\zeta}.$$

Lorentz introduced these ideas in the 1895 article [19] that we cited in the previous section. He called the new expression for τ—certainly an odd mixture of R's space and time coordinates—a "local time," but he never gave it a good a priori physical explanation. When Einstein eventually did offer an explanation a decade later, Lorentz never completely accepted it! (See Pais [26], pages 166 ff.).

Exercises

1. (a) Show that $\mathbf{A} \times (\mathbf{B} \times \mathbf{C}) = (\mathbf{A} \cdot \mathbf{C})\mathbf{B} - (\mathbf{A} \cdot \mathbf{B})\mathbf{C}$, where $\mathbf{A}$, $\mathbf{B}$, and $\mathbf{C}$ are any vectors in $\mathbf{R}^3$.
   (b) Deduce the corollary $\nabla \times (\nabla \times \mathbf{F}) = \nabla(\nabla \cdot \mathbf{F}) - \nabla^2 \mathbf{F}$, where $\mathbf{F}$ is any smooth vector function. (Note: Since the vector ∇ is an *operator*, we must write it to the left of the scalar $\nabla \cdot \mathbf{F}$.)

2. Show that $\nabla \cdot (\nabla \times \mathbf{F}) = 0$ for any smooth vector function $\mathbf{F}$.

3. From Maxwell's equations deduce the *conservation of charge*:
$$\frac{\partial \rho}{\partial t} = -\nabla \cdot \mathbf{J}.$$

4. Assume that $\mathcal{E}(\tau, \zeta) = \mathcal{E}(t, z - vt) = E(t, z)$ and verify that

$$E_z = \mathcal{E}_\zeta, \qquad E_{tt} = \mathcal{E}_{\tau\tau} - 2v\mathcal{E}_{\tau\zeta} + v^2\mathcal{E}_{\zeta\zeta}, \qquad E_{zz} = \mathcal{E}_{\zeta\zeta}.$$

5. Assume that $\mathcal{E}$ and E are related by the Lorentz transformation

$$E(t, z) = \mathcal{E}\left(\frac{t - vz}{\sqrt{1 - v^2}}, \frac{z - vt}{\sqrt{1 - v^2}}\right)$$

and verify that

$$E_t = \frac{\mathcal{E}_\tau - v\mathcal{E}_\zeta}{\sqrt{1 - v^2}}, \qquad E_z = \frac{-v\mathcal{E}_\tau + \mathcal{E}_\zeta}{\sqrt{1 - v^2}},$$

$$E_{tt} = \frac{\mathcal{E}_{\tau\tau} - 2v\mathcal{E}_{\tau\zeta} + v^2\mathcal{E}_{\zeta\zeta}}{1 - v^2}, \qquad E_{zz} = \frac{v^2\mathcal{E}_{\tau\tau} - 2v\mathcal{E}_{\tau\zeta} + \mathcal{E}_{\zeta\zeta}}{1 - v^2}.$$

6. (a) When we revert to conventional units, the simple wave equation derived from Maxwell's equations has the form $E_{tt} = c^2 E_{zz}$. Show that the function

$$E(t, z) = f(z - ct) + g(z + ct)$$

solves the wave equation, for any functions $f(u)$, $g(u)$. Note that f and g are functions of a *single* variable.

(b) Suppose $w = f(u)$ has a spike at the origin (a "soliton"), as shown in the graph below. Let $E(t, z) = f(z - ct)$. Show that the graph of $w = E(t_0, z)$, for a fixed t_0, is the translate of $w = E(0, z)$ by the amount ct_0. In particular, the translation puts the spike at $z = ct_0$. Use the graphs to explain that $E(t, z)$ represents a spike traveling in the direction of positive z with velocity c.

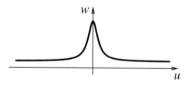

(c) Give the corresponding interpretation for $E(t, z) = f(z + ct)$.

Further Reading for Chapter 1

The biography of Einstein by Pais [26] deals extensively with Einstein's scientific work and places him in the scientific context of his times. In his definitive 1905 paper on special relativity [11], Einstein himself discusses the physical problems he sought to address. The *Feynman Lectures* [13] give clear discussions of all the physical issues on which special relativity is based.

2 | Special Relativity – Kinematics

CHAPTER

Material objects have mass and move in response to forces; **kinematics** is that part of the study of motion that does not take force and mass into account. Since kinematics is the simplest part of dynamics, and since some of the most basic ideas of special relativity are kinematic in nature, we look at them first.

2.1 Einstein's Solution

According to the Michelson–Morley experiment and to Maxwell's equations, the speed of light is constant, independent of the motion of the observer measuring it. Instead of considering this a troubling contradiction, Einstein argued from Galileo's principle of relativity that it must simply be a law of physics. He then considered what this new law must imply.

Accept the constancy of the speed of light as a law of physics

In geometric terms, the contradiction arises because we use the Galilean transformation S_v, which alters the light cone. Einstein reasoned that the way out is to construct a new transformation $B_v : G \to R$ that *preserves* the light cone. (Henceforward, we shall use G and R as a shorthand for the spacetimes of G and R, respectively.) How should B_v be defined?

B_v is linear

- **The transformation is linear**. Suppose an object A has a straight worldline in G's spacetime. Then A has a constant velocity with respect to G. But then A will have a constant velocity with respect to R, because R and G are in uniform motion with respect to one another. Hence A must have a straight worldline in R's frame, too. So B_v must map straight lines to straight lines. Since we implicitly assume that B_v is invertible and carries the origin to the origin, it follows that B_v is linear. (While this result is part of linear algebra, it is not part of the usual first linear algebra course. The exercises will guide you through a proof.)

$B_v^{-1} = B_{-v}$

- **The transformation has the same form for all pairs of observers**. Galilean relativity stresses that there should be no essential distinction between uniformly moving observers. Therefore, G and R should use the same transformation when they convert from their own spacetime to the other's. When G's velocity with respect to R is v, R's velocity with respect to G is $-v$. Therefore, since $B_v : G \to R$, we must have $B_{-v} : R \to G$. But the map $R \to G$ must at the same time be the *inverse* of the map $G \to R$, so $B_{-v} = B_v^{-1}$.

The Graphical Solution

Map the light cone to itself

Assume, as we have before, that R sets off a spark of light at the event O. Then one photon travels up the vertical axis; another travels down. In G's spacetime, below, we see the situation at time $\tau = \tau_0$: G is at G_0, one photon is at U, and one photon is at D. Because the speed of light is 1, U and D are both τ_0 light-seconds from G_0.

Slide points along the cone

Consider now what the map B_v has to accomplish. In R, the images of U and D must lie on the light cone, and G_0 must lie on G's worldline. Because B_v is linear and D, G_0, and U are equally spaced along a line in G, their images must be equally spaced along another line in R. This is where the problem lies: If we map D, G_0, and U to a *vertical* line in R (which is what the Galilean transformation does), the three points are no longer equally spaced; G_0 will be closer to the upward light cone. But this is easy to fix: Just rotate the line clockwise around G_0, letting

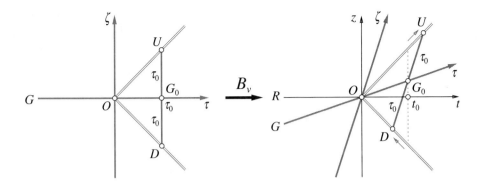

D and U slide along the light cone. The lower segment will get shorter and the upper one longer; at some point the two will be the same length.

We have suddenly stumbled upon the heart of Einstein's revolution: From G's point of view, D, G_0, and U happen at the same time, because they lie on a vertical line ($\tau = \tau_0$). But from R's point of view, they happen at *different* times: first D, then G_0, then U. The transformation B_v preserves the light cone, but at the cost of losing simultaneity. We do not even know whether $t_0 = \tau_0$, that is, whether R and G agree about the time of the event G_0. (As we shall see, they don't!)

Simultaneity is lost

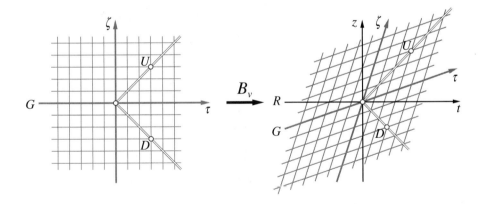

Even without the quantitative details we have a good qualitative picture. Because B_v is linear, it maps the ζ-axis and all the vertical lines $\tau = $ const to lines that are parallel to the line

Map is like a "collapsing crate"

containing D and U. It is not a shear; it is more like the partial collapse of a grid of interlocking cardboard spacers that separate the bottles in a case of wine. Notice that, in this collapse, grid points that lie along the $\pm 45°$ lines (the light cone) are shifted in or out but stay at $\pm 45°$: The light cone is preserved.

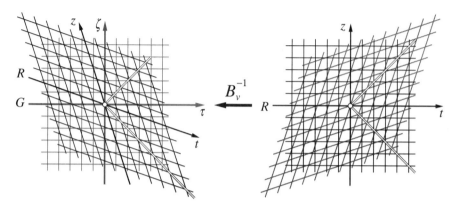

The inverse map

The inverse map, shown above going right to left, must undo the effect of B_v. It must take the collapsed grid in R (shown in light gray) and map it back to an orthogonal grid in G. In doing so, the orthogonal grid in R (shown in black) will map to a grid in G that is collapsed in the opposite direction. In fact, this is the same as B_{-v}. To see why, look back at the figure showing B_v and change v to $-v$. The time axis will then slope downward rather than upward, and the space axis will tilt backward rather than forward—exactly like B_v^{-1}.

Connecting B_v and B_{-v}

There is one more connection between B_v and B_{-v} that we need when we compute the coefficients of B_v. Suppose G and R flip their vertical axes:

$$F : \begin{cases} \tau_1 = \tau, & t_1 = t, \\ \zeta_1 = -\zeta, & z_1 = -z. \end{cases}$$

In terms of the new coordinates, the velocity of G relative to R is now $\Delta z_1/\Delta t_1 = -\Delta z/\Delta t = -v$. Thus, if G_1 and R_1 denote the "flipped" versions of G and R, then the map $G_1 \to R_1$ must be given by B_{-v} when the map $G \to R$ is given by B_v. That is what the following diagram says.

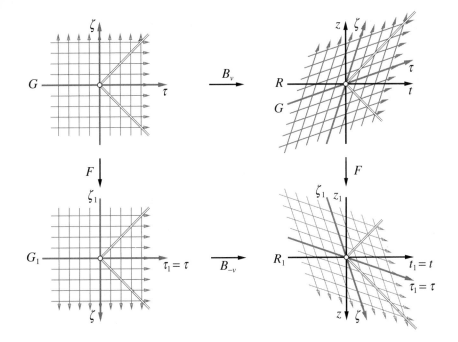

Start at G in the upper left; then a vertical flip followed by B_{-v} produces the same result as B_v followed by a vertical flip: $B_{-v}F = FB_v$. We solve for B_{-v},

$$B_{-v} = FB_vF^{-1} = FB_vF,$$

to get the relation we need. The last equality holds because F is its own inverse. Because the equation $B_{-v}F = FB_v$ tells us how F commutes with the Bs, the diagram above (and more typically its abstract form, shown in the margin) is called a **commutative diagram**.

Commutative diagrams

Eigenvectors and Eigenvalues

Because each part of the light cone is invariant, the vectors

$$U = \begin{pmatrix} 1 \\ 1 \end{pmatrix} \quad \text{and} \quad D = \begin{pmatrix} 1 \\ -1 \end{pmatrix}$$

Vectors that B_v simply stretches

(and any scalar multiples of them) that lie on the light cone are special: B_v merely expands or contracts them without altering their direction. That is, there are positive numbers λ_U and λ_D for which $B_v U = \lambda_U U$ and $B_v D = \lambda_D D$. A square grid in G parallel to U and D (rather than to the coordinate axes) will be stretched by the factor λ_U in the direction of U and compressed by the factor λ_D in the direction of D. The sides of the grid will remain parallel to the U and D vectors.

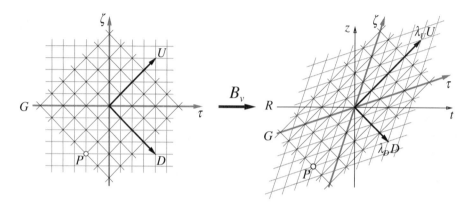

Linearity guarantees that B_v acts in a consistent way on the two grids. For example, consider the point P in G. On the upright gray grid, P is 2 grid units west of the origin and 4 units south. Its image is in the same relative position in the gray image grid. On the diagonal black grid, it is 1 grid unit southeast of the origin and 3 units southwest. Its image in these coordinates is in the same relative position in the black image grid.

We can therefore use either grid to define B_v. However, B_v has a simple and elegant geometric description if we use the (U, D)-grid: Stretch by the factor λ_U in the U direction and by λ_D in the D direction. Here "stretch by λ" is meant to include the possibility of a contraction when $|\lambda| \leq 1$, a flip when $\lambda < 0$, and a complete collapse when $\lambda = 0$.

In fact, the directions of the vectors U and D and the corresponding stretch factors λ_U and λ_D completely characterize B_v. For this reason U and D are called characteristic vectors, or **eigenvectors**, and λ_U and λ_D are called the corresponding characteristic values, or **eigenvalues**. (The names come from German, where *eigen* means "of one's own." In French, the same idea is

Simple description of B_v

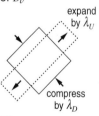

Eigenvectors and eigenvalues

conveyed by the word *propre*, so characteristic values and vectors are also called proper values and vectors.)

Definition 2.1 *Let M be an $n \times n$ matrix; A column vector X is an **eigenvector** of M with corresponding **eigenvalue** λ if $X \neq 0$ and $MX = \lambda X$.*

Note that any nonzero scalar multiple of an eigenvector is also an eigenvector with the same eigenvalue. The eigenvalues of M are the roots of the function $p(\lambda) = \det(M - \lambda I)$; $p(\lambda)$ is a polynomial of degree n, so M has exactly n eigenvalues (counting repeated roots as many times as they appear). Since the polynomial may have complex roots, M may have complex eigenvalues—even when all of its coefficients are real. However, it can be shown that the eigenvalues of a symmetric matrix (one equal to its own transpose: $M^t = M$) are real, and the eigenvectors corresponding to different eigenvalues are orthogonal to one another.

An $n \times n$ matrix has n eigenvalues

Suppose $\{X_1, X_2, \ldots, X_n\}$ is a basis for $\mathbf{R}^n$ consisting of eigenvectors of M, and suppose the corresponding eigenvalues $\lambda_1, \lambda_2, \ldots, \lambda_n$ are all real. Then, in terms of this basis, the action of $M : \mathbf{R}^n \to \mathbf{R}^n$ becomes transparently simple:

Using eigenvectors to describe the action of M

$$Y = a_1 X_1 + \cdots + a_n X_n \quad \Longrightarrow \quad MY = \lambda_1 a_1 X_1 + \cdots + \lambda_n a_n X_n.$$

Thus, if we build a grid parallel to the eigenvectors, M will simply stretch the grid in different directions by the value of the corresponding eigenvalue. Conversely, this action defines M because $\{X_1, X_2, \ldots, X_n\}$ is a basis, so the eigenvalues and the directions of the eigenvectors characterize M completely.

Unfortunately, not every matrix has a basis of eigenvectors. Furthermore, some or all of the eigenvalues may be complex; in these cases the geometric description is more complicated, but eigenvectors and eigenvalues still provide the clearest path to a geometric understanding of the action of the matrix. The exercises explore some of these issues for 2×2 matrices. The following is a result we need immediately to compute B_v; it concerns a general 2×2 matrix, its determinant, and its trace:

$$M = \begin{pmatrix} a & b \\ c & d \end{pmatrix}, \quad \det(M) = ad - bc, \quad \operatorname{tr}(M) = a + d.$$

Proposition 2.1 *If λ_1 and λ_2 are the eigenvalues of a 2×2 matrix M, then $\det(M) = \lambda_1 \lambda_2$ and $\operatorname{tr}(M) = \lambda_1 + \lambda_2$.*

The Computational Solution

Theorem 2.1 *B_v is the Lorentz matrix: $B_v = \dfrac{1}{\sqrt{1-v^2}} \begin{pmatrix} 1 & v \\ v & 1 \end{pmatrix}$.*

PROOF: Let $B_v = \begin{pmatrix} a & b \\ c & d \end{pmatrix}$; we must find a, b, c, and d in terms of v.

We start by noting that G's worldline in R—which is the image of the τ-axis—must have slope $v = \Delta z / \Delta t$. But

$$\begin{pmatrix} a & b \\ c & d \end{pmatrix} \begin{pmatrix} \tau \\ 0 \end{pmatrix} = \begin{pmatrix} \tau a \\ \tau c \end{pmatrix} = \begin{pmatrix} t \\ z \end{pmatrix},$$

so $v = c/a$, or $c = av$. Thus $a \neq 0$, for otherwise G's velocity would be $\pm\infty$.

Condition: the light cone is invariant

The next step is to use the invariance of the light cone, that is, to use the fact that

$$U = \begin{pmatrix} 1 \\ 1 \end{pmatrix} \quad \text{and} \quad D = \begin{pmatrix} 1 \\ -1 \end{pmatrix}$$

are eigenvectors of B_v. The equation $B_v U = \lambda_U U$ becomes

$$B_v U = \begin{pmatrix} a & b \\ c & d \end{pmatrix} \begin{pmatrix} 1 \\ 1 \end{pmatrix} = \begin{pmatrix} a+b \\ c+d \end{pmatrix} = \lambda_U \begin{pmatrix} 1 \\ 1 \end{pmatrix}.$$

The last equality requires that $c + d = a + b = \lambda_U$. Similarly,

$$B_v D = \begin{pmatrix} a & b \\ c & d \end{pmatrix} \begin{pmatrix} 1 \\ -1 \end{pmatrix} = \begin{pmatrix} a-b \\ c-d \end{pmatrix} = \lambda_D \begin{pmatrix} 1 \\ -1 \end{pmatrix}$$

requires that $-(c - d) = a - b = \lambda_D$. This gives us two equations

$$a + b = c + d, \qquad a - b = -c + d$$

that then imply $d = a$, $c = b$. Since we already know that $c = av$, at this stage we have

$$B_v = \begin{pmatrix} a & b \\ b & a \end{pmatrix} = \begin{pmatrix} a & av \\ av & a \end{pmatrix} = a \begin{pmatrix} 1 & v \\ v & 1 \end{pmatrix}.$$

§2.1 Einstein's Solution

To determine the remaining coefficient, a, we use the condition $B_v^{-1} = B_{-v} = FB_vF$. First rewrite this as $B_v = FB_v^{-1}F$ and then calculate the determinant:

$$\det(B_v) = \det(FB_v^{-1}F) = \det(F)\det(B_v^{-1})\det(F)$$
$$= -1 \cdot \det(B_v^{-1}) \cdot -1 = \det(B_v^{-1}) = \det(B_v)^{-1}.$$

Condition: B_v "flipped" is B_{-v}

Thus $\det^2(B_v) = 1$, implying $\det(B_v) = \pm 1$. But $\det(B_v) = \lambda_U \lambda_D > 0$, so $\det(B_v) = +1$, and hence

$$\det(B_v) = a^2 - b^2 = a^2 - (av)^2 = a^2(1 - v^2) = 1.$$

Condition: eigenvalues are positive

We solve this for a (and choose the positive square root because $2a = \text{tr}(B_v) = \lambda_U + \lambda_D > 0$):

$$a = \frac{1}{\sqrt{1 - v^2}}.$$

END OF PROOF

For future reference we note this alternative expression for B_v:

$$B_v = \begin{pmatrix} a & b \\ b & a \end{pmatrix}, \quad a^2 - b^2 = 1, \quad a > 0.$$

The matrix B_v is the Lorentz matrix for the transformation $G \to R$. However, instead of creating it after the fact to fit the facts (as Lorentz had done), Einstein has deduced it from a simple hypothesis: The speed of light is constant for all Galilean observers.

In the setting Einstein created, the Lorentz transformation must replace the classical Galilean transformation. The Galilean transformation has a clear and relatively simple geometric interpretation—as a shear. Is there something similar for the Lorentz transformation?

In the next section we shall introduce the hyperbolic functions in order to recast the Lorentz matrix in a more geometric form. We will then be able to see a striking and valuable similarity with ordinary Euclidean geometry.

Exercises

1. Suppose that $L : \mathbf{R}^2 \to \mathbf{R}^2$ is a linear map. Using the fact that L is *additive* ($L(X + Y) = L(X) + L(Y)$ for all vectors X and Y) and *homogeneous* ($L(rX) = rL(X)$ for all vectors X and real numbers r), show the following:

   - L maps straight lines to straight lines.
   - L maps parallel lines to parallel lines.
   - L maps equally spaced points on a line to equally spaced points on the image (but the spacing may be different).

2. The purpose of this exercise is to prove that if $M : \mathbf{R}^2 \to \mathbf{R}^2$ is continuous and invertible, maps straight lines to straight lines, and maps the origin to the origin, then M is linear—that is, it is *additive* (for all vectors X and Y, $M(X+Y) = M(X)+M(Y)$) and *homogeneous* (for all vectors X and real numbers r, $M(rX) = rM(X)$). Homogeneity is the more complicated of the two to prove.

   (a) Show that M maps parallel lines to parallel lines. (Hint: If the lines α and β are parallel but $M(\alpha)$ and $M(\beta)$ intersect, then M is not invertible at the intersection point.)

   (b) Use part (a) to show that M is additive. (Consider the image of the parallelogram spanned by any pair of vectors X and Y; your proof should also make explicit use of the fact that M maps the origin to the origin.)

   (c) Let the vector X be given and suppose $Y_j \to X$ as $j \to \infty$. Then $M(Y_j) \to M(X)$; why? Explain why this implies $M(X + Y_j) \to M(2X)$ and, using the fact that M is additive, $M(X + Y_j) \to 2M(X)$ as well. Conclude $M(2X) = 2M(X)$.

   (d) Prove, by induction on the positive integer k, that $M((k-1)X + Y_j) \to M(kX)$ and $M((k-1)X + Y_j) \to kM(X)$. Conclude $M(kX) = kM(X)$ for any positive integer k.

   (e) Use the fact that M is additive to show $M(-kX) = -M(kX)$ and hence that $M(-kX) = -kM(X)$ for any negative integer $-k$. Thus $M(pX) = pM(X)$ for *any* integer p.

(f) Now prove that $M(qX) = qM(X)$ for any rational number $q = p/n$. Suggestion: let $X = nZ$ and consider $nM(pZ) = pM(nZ)$.

(g) Now let r be any real number and q_j a sequence of rational numbers converging to r. Use $M(q_j X) = q_j M(X)$ to prove $M(rX) = rM(X)$. This shows that M is homogeneous and hence linear.

3. Ignoring the results of Theorem 2.1 and using only the condition $B_{-v} = FB_v F$, prove that a is an even function of v, while b is an odd function. In other words, if we write

$$B_v = \begin{pmatrix} a(v) & b(v) \\ b(v) & a(v) \end{pmatrix}$$

to indicate that a and b are functions of v, then $a(-v) = a(v)$ and $b(-v) = -b(v)$.

4. **Eigenvalues and eigenvectors**. Suppose

$$M = \begin{pmatrix} a & b \\ c & d \end{pmatrix}, \quad X = \begin{pmatrix} x \\ y \end{pmatrix} \neq 0.$$

Then X is an eigenvector of M with eigenvalue λ if $MX = \lambda X$. (We require $X \neq 0$ because $X = 0$ satisfies the equation for every λ.)

(a) Using the fact that the equation $(M - \lambda I)X = 0$ has a nonzero solution, deduce that $\det(M - \lambda I) = 0$ and that λ is a root of the equation

$$\lambda^2 - \operatorname{tr}(M)\lambda + \det(M) = 0.$$

Conclude that M has two eigenvalues λ_1, λ_2; they may be equal or they may be complex.

(b) Prove that $\det(M) = \lambda_1 \lambda_2$ and $\operatorname{tr}(M) = \lambda_1 + \lambda_2$.

(c) Suppose X_1 and X_2 are eigenvectors corresponding to λ_1 and λ_2, and $\lambda_1 \neq \lambda_2$. Prove that X_1 and X_2 are linearly independent.

(d) Suppose M is symmetric; that is, $c = b$. Prove that its eigenvalues must be real. Prove that the eigenvectors X_1,

X_2 corresponding to λ_1 and λ_2 are orthogonal (that is, $X_1 \cdot X_2 = 0$) if $\lambda_1 \neq \lambda_2$. Prove that *every* nonzero vector is an eigenvector if $\lambda_1 = \lambda_2 = \lambda$, and show that this implies $M = \lambda I$.

5. Determine the eigenvalues and eigenvectors of the flip $F : \mathbf{R}^2 \to \mathbf{R}^2$ defined in the text.

6. Suppose M is a 2×2 symmetric matrix whose eigenvalues are equal. Prove that M must be a multiple of the identity matrix.

7. Suppose M is a 2×2 invertible matrix. Show that the eigenvalues of M^{-1} are the inverses of the eigenvalues of M; in particular, the eigenvalues of M must be nonzero. How are the eigenvectors of M^{-1} and M related?

8. (a) Determine the eigenvalues and eigenvectors of the following matrices and verify that the eigenvectors are orthogonal when the matrix is symmetric:

$$\begin{pmatrix} 6 & 2 \\ 2 & 3 \end{pmatrix}, \begin{pmatrix} 6 & -2 \\ -2 & 3 \end{pmatrix}, \begin{pmatrix} 4 & 4 \\ 2 & -3 \end{pmatrix}, \begin{pmatrix} 2 & 1 \\ 0 & 1 \end{pmatrix}, \begin{pmatrix} 2 & 1 \\ 1 & 1 \end{pmatrix}.$$

(b) Consider each matrix M in part (a) as a linear map $M : \mathbf{R}^2 \to \mathbf{R}^2$. Sketch the image under M of a grid of unit squares.

(c) For each matrix M in part (a), construct a grid in $\mathbf{R}^2$ based on the eigenvectors of M and sketch the image of that grid under M. Compare your results with the grids for the map B_v sketched in the text.

9. Determine the eigenvalues and eigenvectors of the circular and hyperbolic **rotation** matrices

$$R_\theta = \begin{pmatrix} \cos\theta & -\sin\theta \\ \sin\theta & \cos\theta \end{pmatrix}, \quad H_u = \begin{pmatrix} \cosh u & \sinh u \\ \sinh u & \cosh u \end{pmatrix}.$$

(These linear maps are discussed in the following sections of this chapter.)

2.2 Hyperbolic Functions

Definitions

$$\sinh u = \frac{e^u - e^{-u}}{2}, \quad \tanh u = \frac{\sinh u}{\cosh u}, \quad \operatorname{sech} u = \frac{1}{\cosh u},$$

$$\cosh u = \frac{e^u + e^{-u}}{2}, \quad \coth u = \frac{\cosh u}{\sinh u}, \quad \operatorname{csch} u = \frac{1}{\sinh u}.$$

The functions sinh and tanh are often pronounced "cinch" and "tanch"; cosh and sech are pronounced as they are spelled; and coth and csch are usually just called the hyperbolic cotangent and the hyperbolic cosecant.

Pronunciation

The two functions on the left satisfy the fundamental identity

Why hyperbolic?

$$\cosh^2 u - \sinh^2 u = 1 \quad \text{for all } u.$$

It follows that the point $(x, y) = (\cosh u, \sinh u)$ lies on the unit hyperbola $x^2 - y^2 = 1$ (on the branch where $x > 0$). This is analogous to the way the point $(x, y) = (\cos u, \sin u)$ is found on the unit circle $x^2 + y^2 = 1$ because of the familiar trigonometric identity

$$\cos^2 u + \sin^2 u = 1 \quad \text{for all } u.$$

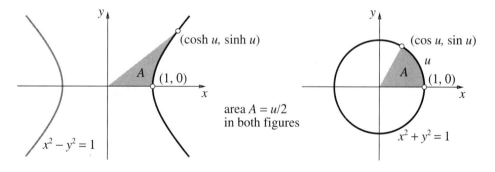

area $A = u/2$ in both figures

But the point $(\cos u, \sin u)$ doesn't merely lie on the circle; it is exactly u radians around the circle from the point $(1, 0)$. This means that we can use the circle $x^2 + y^2 = 1$ to *define* $\cos u$ and $\sin u$—and thus call them *circular* functions. Can we do the same for $\cosh u$ and $\sinh u$? Unfortunately, the distance from

Why hyperbolic functions? (cosh u, sinh u) to (1, 0) along the hyperbola is *not* equal to u—at least not if we use the Euclidean arc length along the hyperbola.

However, if we switch to areas, then the analogy works perfectly. The circular arc from (1, 0) to (cos u, sin u) cuts off a sector A whose area is exactly $u/2$ square units. (If u is negative, the sector runs clockwise from (1, 0).) As it happens, the hyperbolic arc from (1, 0) to (cosh u, sinh u) *also* cuts off a sector A of area $u/2$ square units. See the exercises. In this way we can use the hyperbola $x^2 - y^2 = 1$ to give a new definition of cosh u and sinh u—explaining why they are called hyperbolic functions.

Hyperbolic identities As the original definitions suggest, there are indeed extensive parallels between the hyperbolic and the circular functions. The following table is a list of identities you should verify, noting whether and how each identity differs from its circular analogue.

Addition Formulas

$$\sinh(u \pm v) = \sinh u \cosh v \pm \cosh u \sinh v,$$
$$\cosh(u \pm v) = \cosh u \cosh v \pm \sinh u \sinh v,$$
$$\tanh(u \pm v) = \frac{\tanh u \pm \tanh v}{1 \pm \tanh u \tanh v},$$
$$\coth(u \pm v) = \frac{1 \pm \coth u \coth v}{\coth u \pm \coth v}.$$

Double and Half Angles

$$\sinh(2u) = 2 \sinh u \cosh u, \qquad \cosh(2u) = \cosh^2 u + \sinh^2 u,$$

$$\sinh\left(\frac{u}{2}\right) = \sqrt{\frac{\cosh u - 1}{2}}, \qquad \cosh\left(\frac{u}{2}\right) = \sqrt{\frac{\cosh u + 1}{2}}.$$

Algebraic Relations

$$\cosh^2 u - \sinh^2 u = 1, \qquad 1 - \tanh^2 u = \operatorname{sech}^2 u, \qquad \coth^2 u - 1 = \operatorname{csch}^2 u.$$

Calculus

$$\frac{d}{du} \sinh u = \cosh u, \qquad \frac{d}{du} \tanh u = \operatorname{sech}^2 u, \qquad \frac{d}{du} \operatorname{sech} u = - \operatorname{sech} u \tanh u,$$

$$\frac{d}{du} \cosh u = \sinh u, \qquad \frac{d}{du} \coth u = - \operatorname{csch}^2 u, \qquad \frac{d}{du} \operatorname{csch} u = - \operatorname{csch} u \coth u,$$

§2.2 Hyperbolic Functions

$$\int \sinh^2(u)\,du = \frac{\sinh 2u}{4} - \frac{u}{2}, \qquad \int \cosh^2(u)\,du = \frac{\sinh 2u}{4} + \frac{u}{2}.$$

The graphs of $w = \cosh u$ and $w = \sinh u$ are shown with the exponential functions $w = \frac{1}{2}e^u$, $w = \frac{1}{2}e^{-u}$, and $w = -\frac{1}{2}e^{-u}$. Note that $\cosh u$ is the sum of the first and the second of these, while $\sinh u$ is the sum of the first and the third.

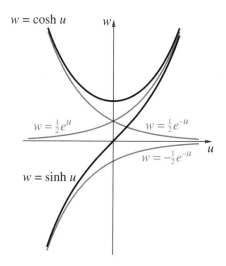

Notice that $w = \sinh u$ is one-to-one and maps the u-axis onto the w-axis. This means that given any real number w_0, there is a unique u_0 for which $w_0 = \sinh u_0$. We say that the hyperbolic sine is invertible and we write the inverse as $\sinh^{-1}$, so $u_0 = \sinh^{-1}(w_0)$.

The hyperbolic sine is invertible

Proposition 2.2 *If (x, y) is any point on the hyperbola $x^2 - y^2 = 1$ and $x > 0$, then there is a unique u for which*

$$x = \cosh u, \qquad y = \sinh u.$$

PROOF: Take $u = \sinh^{-1}(y)$. The result also follows from the geometric connection between $\cosh u$ and $\sinh u$: Draw the hyperbolic sector A defined by the three points $(0, 0)$, $(1, 0)$, and (x, y); then $u = 2 \cdot \text{area}\, A$. END OF PROOF

This proposition gives us a useful parametrization of that branch of the hyperbola $x^2 - y^2 = 1$ where $x > 0$. In fact, this is just what we need for the Lorentz matrix.

Parametrizing a branch of a hyperbola

Corollary 2.1 *We can write the Lorentz matrix B_v in the alternative form*

$$H_u = \begin{pmatrix} \cosh u & \sinh u \\ \sinh u & \cosh u \end{pmatrix},$$

where $u = \tanh^{-1} v$, or $v = \tanh u$.

PROOF: While computing the formula for B_v we derived the alternative form

$$B_v = \begin{pmatrix} a & b \\ b & a \end{pmatrix}, \qquad a^2 - b^2 = 1, \quad a > 0.$$

The proposition provides a unique u for which $a = \cosh u$, $b = \sinh u$. Futhermore, $v = b/a = \sinh u / \cosh u = \tanh u$.

END OF PROOF

Boosts and hyperbolic rotations

We now have two ways to describe the Lorentz transformation—physically and geometrically. When we use matrix B_v, we call the transformation a **boost** (or a boost by velocity v). This is a physical description; it is expressed in terms of the physical velocity parameter v. By contrast, the matrix H_u gives us a geometric description; in the next section we shall see that u can be thought of as a "hyperbolic angle" and H_u as a **hyperbolic rotation**. There are direct analogues with ordinary Euclidean angles and rotations. The two forms of the Lorentz transformation are related as follows:

$$H_u = B_{\tanh u} \quad \text{and} \quad B_v = H_{\tanh^{-1} v}.$$

Exercises

1. Deduce from the definitions of $\sinh u$ and $\cosh u$ that $\cosh^2 u - \sinh^2 u = 1$ for all u. Do the same with the other two algebraic identities.

2. Verify all the addition formulas for the hyperbolic functions and note whether and how they differ from the corresponding addition formulas for the circular functions. Do the same for the double and half angle formulas and the calculus identities.

§2.2 Hyperbolic Functions

3. Prove that $\cosh u \pm \sinh u = e^{\pm u}$.

4. Show that $H_{u_1} \cdot H_{u_1} = H_{u_1+u_2} = H_{u_2} \cdot H_{u_1}$, where

$$H_u = \begin{pmatrix} \cosh u & \sinh u \\ \sinh u & \cosh u \end{pmatrix}.$$

5. (a) Suppose $v = \tanh u$; show that

$$\sinh u = \frac{v}{\sqrt{1-v^2}}, \qquad \cosh u = \frac{1}{\sqrt{1-v^2}}.$$

   (b) Conclude that $H_u = \dfrac{1}{\sqrt{1-v^2}} \begin{pmatrix} 1 & v \\ v & 1 \end{pmatrix} = B_v$.

   (c) Using this expression for B_v, calculate $B_{v_1} \cdot B_{v_2}$ and show that there is some v_3 for which $B_{v_3} = B_{v_1} \cdot B_{v_2} = B_{v_2} \cdot B_{v_1}$. Express v_3 in terms of v_1 and v_2; prove that $|v_3| < 1$ when $|v_1| < 1$ and $|v_2| < 1$.

6. (a) Sketch on the same axes the graphs of $w = \tanh u$, $w = \coth u$, $w = \operatorname{sech} u$, and $w = \operatorname{csch} u$.

   (b) Your sketches should show that $0 < \operatorname{sech} u \le 1$ and $-1 < \tanh u < 1$. Prove these statements analytically.

7. (a) Using the parametrization $x = \cosh u$, $y = \sinh u$ of $x^2 - y^2 = 1$, $x > 0$, as a model, construct parametrizations of these curves:

$$x^2 - y^2 = 1, \ x < 0; \quad y^2 - x^2 = 1, \ y > 0; \quad y^2 - x^2 = 1, \ y < 0.$$

   (b) Sketch each curve and mark on it the location of the points where $u = -1, 0, 1$. Put an arrow on each curve that shows the direction in which u increases.

8. Make a careful sketch of the parametrized curve $P_u = (\cosh u, \sinh u)$ with $-1 \le u \le 2$. On your sketch draw the three line segments that connect the pairs of points $\{P_{-1}, P_2\}$, $\{P_{-1/2}, P_{3/2}\}$, $\{P_0, P_1\}$. These lines should be parallel and parallel to the tangent to the curve at $u = \frac{1}{2}$; are they?

9. Give a simple argument that explains why the circular sector A at the beginning of the section has area $u/2$, given that the circle has radius 1 and the defining arc has length u.

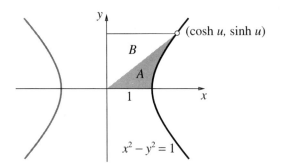

10. Prove that area $A = u/2$ in the hyperbolic sector above. Here is one approach you can take.

    (a) Explain why the right branch of the hyperbola is the graph $x = \sqrt{1 + y^2}$. Then show that
    $$\text{area}(A + B) = \int_0^{\sinh u} \sqrt{1 + y^2}\, dy = \frac{u}{2} + \frac{\sinh u \cosh u}{2}.$$
    (One way to evaluate the integral is via the substitution $y = \sinh s$.)

    (b) Find area B and subtract it from area$(A + B)$.

11. Find the area of the *circular* sector by an integral argument similar to the one you used for the hyperbolic sector in the previous exercise. Note the similarities.

12. (a) Obtain the Taylor series for e^u, $\cosh u$, and $\sinh u$.

    (b) Obtain the third-order Taylor polynomial for $\tanh u$.

13. Calculate the derivatives of the three inverse functions $\sinh^{-1} u = \text{arcsinh}\, u$, $\cosh^{-1} u = \text{arccosh}\, u$, and $\tanh^{-1} u = \text{arctanh}\, u$.

14. (a) Show that $\arcsin(\tanh u) = \arctan(\sinh u)$.

    (b) Show that $\tanh^{-1} w = \text{arctanh}\, w = \dfrac{1}{2} \ln\left(\dfrac{1 + w}{1 - w}\right)$.

15. Pronunciation lesson (from Bill Watson); identify the cuisine:

    (a) $\dfrac{e^{2r} + 1}{2e^r}$ (b) $\dfrac{2e}{e^2 + 1}$

2.3 Minkowski Geometry

Rotations and the Euclidean Norm

In analytic plane geometry, the basic tasks are measuring lengths and angles; everything else depends on them. The tool for these tasks is an **inner product** in $\mathbf{R}^2$. The standard Euclidean inner product is

$$X_1 \cdot X_2 = X_1^t X_2 = (x_1, y_1) \begin{pmatrix} x_2 \\ y_2 \end{pmatrix} = x_1 x_2 + y_1 y_2.$$

The norm and inner product

Here we write an element X of $\mathbf{R}^2$ in the usual way as a column vector; the transpose X^t is the corresponding row vector. By using the transpose of the vector on the left, we make the inner product into an ordinary matrix multiplication. This will be useful later on. From the inner product we get the length, or Euclidean **norm**, of a vector,

$$\|X\| = \sqrt{X \cdot X} = \sqrt{x^2 + y^2}.$$

We can now make the basic measurements. For two points, or vectors, X_1 and X_2 in $\mathbf{R}^2$, they are

distance from X_1 to $X_2 = \|X_2 - X_1\| = \sqrt{(x_2 - x_1)^2 + (y_2 - y_1)^2}$,

angle between X_1 and $X_2 = \alpha = \arccos\left(\dfrac{X_1 \cdot X_2}{\|X_1\| \|X_2\|}\right).$

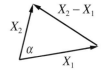

It is a fundamental geometric principle that lengths and angles do not change when the plane undergoes a rotation. Counterclockwise rotation by θ radians is given by the matrix

$$R_\theta = \begin{pmatrix} \cos\theta & -\sin\theta \\ \sin\theta & \cos\theta \end{pmatrix}.$$

(See the exercises.) We prove that rotations, as defined here, do indeed preserve lengths and angles by proving that they preserve the norm and inner product. In fact, it is sufficient to prove just the second, because the norm is defined in terms of the inner product. However, before giving that proof we look first at a direct computational proof for the norm.

Rotations preserve lengths and angles

Proposition 2.3 $\|R_\theta X\| = \|X\|$ *for every X in $\mathbf{R}^2$.*

PROOF: Let $\begin{pmatrix} p \\ q \end{pmatrix} = \begin{pmatrix} \cos\theta & -\sin\theta \\ \sin\theta & \cos\theta \end{pmatrix} \begin{pmatrix} x \\ y \end{pmatrix}$. We must show that $p^2 + q^2 = x^2 + y^2$:

$$\begin{aligned} p^2 + q^2 &= (x\cos\theta - y\sin\theta)^2 + (x\sin\theta + y\cos\theta)^2 \\ &= x^2\cos^2\theta - 2xy\cos\theta\sin\theta + y^2\sin^2\theta \\ &\quad + x^2\sin^2\theta + 2xy\sin\theta\cos\theta + y^2\cos^2\theta \\ &= x^2(\cos^2\theta + \sin^2\theta) + y^2(\sin^2\theta + \cos^2\theta) \\ &= x^2 + y^2. \end{aligned}$$

END OF PROOF

Corollary 2.2 *The rotation R_θ maps each circle $x^2 + y^2 = r^2$ to itself.*

Proposition 2.4 $R_\theta X_1 \cdot R_\theta X_2 = X_1 \cdot X_2$.

PROOF: Write the inner product as a matrix multiplication:

$$\begin{aligned} R_\theta X_1 \cdot R_\theta X_2 &= (R_\theta X_1)^t R_\theta X_2 = X_1^t R_\theta^t R_\theta X_2 \\ &= X_1^t R_{-\theta} R_\theta X_2 = X_1^t I X_2 = X_1 \cdot X_2 \end{aligned}$$

See the exercises to verify that $R_\theta^t = R_\theta^{-1} = R_{-\theta}$. END OF PROOF

Corollary 2.3 *The distance from $R_\theta X_1$ to $R_\theta X_2$ is the same as the distance from X_1 to X_2. The angle between $R_\theta X_1$ and $R_\theta X_2$ is the same as the angle between X_1 and X_2.*

PROOF: The distance from $R_\theta X_1$ to $R_\theta X_2$ is

$$\|R_\theta X_2 - R_\theta X_1\| = \|R_\theta(X_2 - X_1)\| = \|X_2 - X_1\|,$$

the distance from X_1 to X_2. The angle between $R_\theta X_1$ and $R_\theta X_2$ is

$$\arccos\left(\frac{R_\theta X_1 \cdot R_\theta X_2}{\|R_\theta X_1\| \|R_\theta X_2\|}\right) = \arccos\left(\frac{X_1 \cdot X_2}{\|X_1\| \|X_2\|}\right),$$

the angle between X_1 and X_2. END OF PROOF

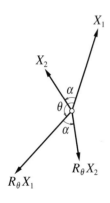

Absolute quantities have geometric meaning

Thus, even though the coordinates of a point change under rotations, its norm does not. In other words, coordinates are *relative* to a coordinate frame, but the norm is *absolute*: It has the same value in all frames (obtained from one another by rotation). Because the norm is absolute, it has a true geometric meaning. Likewise, the inner product of two vectors is absolute, not relative, so it also has a true geometric meaning.

One practical consequence of an absolute norm is that we can now calibrate coordinate frames—that is, we can transfer units of measurement from one frame to another. In the figure below, points on the different axes that intersect the same circle must be the same number of units from the origin. Furthermore, a 90° rotation even allows us to calibrate the x- and y-axes of the *same* frame.

Calibration

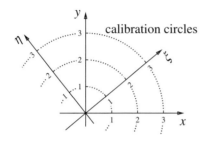

Proposition 2.5 *The product of two rotations is another rotation; specifically, $R_\theta \cdot R_\varphi = R_{\theta+\varphi}$. In other words, to multiply two rotations, add their angles. Furthermore, $R_0 = I$, the identity matrix, and $R_\theta^{-1} = R_{-\theta}$.*

PROOF: Exercises.

This proposition says that the set $O_+(2, \mathbf{R})$ of rotations is a group. This group is commutative, essentially because addition of angles is commutative:

$$R_\theta R_\varphi = R_{\theta+\varphi} = R_{\varphi+\theta} = R_\varphi R_\theta.$$

The Minkowski Norm

The Lorentz transformation H_u is designed to preserve the light cone, which is the set $t^2 - z^2 = 0$. In fact, H_u does much more;

H_u preserves every $t^2 - z^2 = k$

according to the following theorem, it preserves each of the hyperbolas $t^2 - z^2 = k$, for all k. (The common asymptote of these hyperbolas is the light cone.)

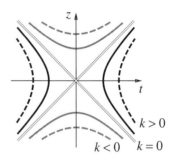

Theorem 2.2 *If $\tau^2 - \zeta^2 = k$ and*

$$\begin{pmatrix} t \\ z \end{pmatrix} = H_u \begin{pmatrix} \tau \\ \zeta \end{pmatrix} = \begin{pmatrix} \cosh u & \sinh u \\ \sinh u & \cosh u \end{pmatrix} \begin{pmatrix} \tau \\ \zeta \end{pmatrix},$$

then $t^2 - z^2 = k$. The parameter u can have any real value.

PROOF: We calculate

$$\begin{aligned} t^2 - z^2 &= (\tau \cosh u + \zeta \sinh u)^2 - (\tau \sinh u + \zeta \cosh u)^2 \\ &= \tau^2 \cosh^2 u + 2\tau\zeta \cosh u \sinh u + \zeta^2 \sinh^2 u \\ &\quad - \tau^2 \sinh^2 u - 2\tau\zeta \sinh u \cosh u - \zeta^2 \cosh^2 u \\ &= \tau^2(\cosh^2 u - \sinh^2 u) - \zeta^2(\cosh^2 u - \sinh^2 u) \\ &= \tau^2 - \zeta^2. \end{aligned}$$

END OF PROOF

$t^2 - z^2$ corresponds to $x^2 + y^2$

This proof is a direct translation of the proof that a rotation preserves the norm: $\|R_\theta X\| = \|X\|$ (Proposition 2.3). All that is involved is converting the circular identities to their hyperbolic counterparts. Since $t^2 - z^2$ corresponds to the norm $\|X\|^2 = x^2 + y^2$ that we use for measuring distance in the Euclidean plane, the correspondence suggests that we should treat $t^2 - z^2$ as a norm for measuring "distance" in spacetime. This is the approach that Hermann Minkowski took in 1907.

The light cone separates spacetime

The approach is complicated by the fact that $t^2 - z^2$ can be negative as well as positive or zero. To see what this implies, let us work with events $E = (t, x, y, z)$ in the full $(1+3)$-dimensional spacetime. We define

$$Q(E) = t^2 - x^2 - y^2 - z^2,$$

or $Q(E) = t^2 - z^2$ if $E = (t, z)$. The set $Q(E) = 0$ is a cone that separates spacetime into two regions: Q is positive in one and negative in the other.

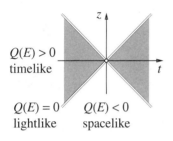
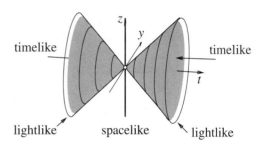

We see that $Q(E) > 0$ for all points inside the cone (shown shaded in the figure above). Since the image of any observer's time axis under a Lorentz map H_u will lie in the interior of the cone, we say that events E in the interior are **timelike**. The exterior of the cone (where $Q(E) < 0$) will contain the images of other observers' space axes, so we say that events E in this region are **spacelike**. Finally, we say that events on the light cone itself are **lightlike**.

Timelike, spacelike, and lightlike events

Definition 2.2 *The **Minkowski norm** of an event $E = (t, x, y, z)$ in spacetime is*

$$\|E\| = \begin{cases} \sqrt{Q(E)} & \text{if E is \textbf{timelike}: } Q(E) > 0; \\ \sqrt{-Q(E)} & \text{if E is \textbf{spacelike}: } Q(E) < 0; \\ 0 & \text{if E is \textbf{lightlike}: } Q(E) = 0. \end{cases}$$

Since we are using geometric units, in which length has the dimensions of time, the Minkowski norm also has the dimensions of time. Thus, for example, if $E = (5 \text{ sec}, 3 \text{ sec})$, then $\|E\| = 4$ sec. This is not an error: Minkowski geometry has a 3-5-4 triangle where Euclidean geometry has a 3-4-5 triangle.

The Minkowski norm is a time

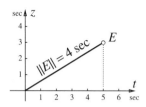

A nonintuitive length and "unit circle"

When we use this norm to measure lengths in spacetime, we get results that are, at first, surprising and nonintuitive. For a start, the "unit circle" in this geometry is the pair of hyperbolas $t^2 - z^2 = \pm 1$. Thus, all the vectors below are *unit* vectors—timelike on the left, spacelike on the right. Of course, if we were to measure them instead with the Euclidean norm (which our eyes do instinctively), then they get longer the nearer they are to the asymptotes $z = \pm t$.

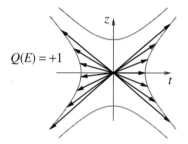
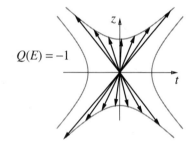

Calibration

We can use the Minkowski norm to calibrate spacetime coordinate frames the same way we used the ordinary norm to calibrate Euclidean frames. Thus, in the figure below, the spacing of units along the τ- and ζ-axes must be exactly as they appear, even though it looks too large to our Euclidean eyes.

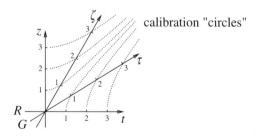

Since the physical distinction between past and future is important, we now refine our partition of spacetime to take this into account.

Past and future in spacetime

Definition 2.3 *Spacetime consists of the following six mutually exclusive sets of events, or vectors, $E = (t, x, y, z)$ (or just $E = (t, z)$):*

$\mathcal{T}_+$: *the future timelike set $Q(E) > 0$, $t > 0$;*
$\mathcal{T}_-$: *the past timelike set $Q(E) > 0$, $t < 0$;*
$\mathcal{S}$: *the spacelike set $Q(E) < 0$;*
$\mathcal{L}_+$: *the future lightlike set $Q(E) = 0$, $t > 0$;*
$\mathcal{L}_-$: *the past lightlike set $Q(E) = 0$, $t < 0$;*
$\mathcal{O}$: *the origin.*

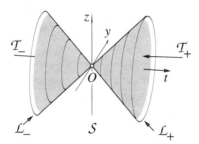

Theorem 2.3 *Each H_u maps each of the six regions of spacetime to itself.*

PROOF: Since Theorem 2.2 says that $Q(H_u(E)) = Q(E)$, the only thing left to prove is that the future and past sets are individually preserved. So suppose E is in $\mathcal{T}_+$:

$$E = \begin{pmatrix} \tau \\ \zeta \end{pmatrix}, \quad \tau > 0; \quad H_u(E) = \begin{pmatrix} \cosh u & \sinh u \\ \sinh u & \cosh u \end{pmatrix} \begin{pmatrix} \tau \\ \zeta \end{pmatrix} = \begin{pmatrix} t \\ z \end{pmatrix}.$$

We must show that $t > 0$. Since E is in $\mathcal{T}_+$, $\tau > 0$ and $-\tau < \zeta < \tau$. Therefore,

$$t = \tau \cosh u + \zeta \sinh u > \tau \cosh u - \tau \sinh u = \tau e^{-u} > 0.$$

The other three results are obtained in a similar way. END OF PROOF

The interval between events

Definition 2.4 *The **separation** between two the events E_1 and E_2 in spacetime is the vector $E_2 - E_1$, and the **interval** between them is $\|E_2 - E_1\|$. We say that the interval is timelike, spacelike, or lightlike if the corresponding separation is timelike, spacelike, or lightlike, respectively.*

The interval, like the norm, is measured in seconds. Note that the interval between two events does not have to be of the same type as the events themselves. For example, in the figure on the left below, the interval between S_1 and S_2 is timelike, and the interval between T_1 and T_2 is spacelike.

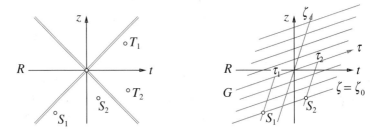

The norm and the interval have physical meaning

Theorem 2.2 implies that $\|H_u E\| = \|E\|$ for every event E and every real number u. Hence the Minkowski norm of an event, and the interval between two events, are *absolute* quantities: All uniformly moving observers assign them the same values. Thus, by Galileo's principle of relativity, they are objectively real—they have a physical meaning that all observers will agree on.

A timelike interval measures ordinary time

To see what this means concretely, consider the following. In the figure on the right above, the interval between S_1 and S_2 is timelike, so the separation $S_2 - S_1$ has slope v with $|v| < 1$. Since $|v| < 1$, we can introduce an observer G who travels with velocity v relative to R. Then S_1 and S_2 will lie on a line $\zeta = \zeta_0$ parallel to G's time axis. From G's point of view, S_1 and S_2 happen at the same place but at different times τ_1 and τ_2. According to G, what separates S_1 and S_2 is a pure time interval of duration $\tau_2 - \tau_1$; the separation vector has the simple form $S_2 - S_1 = (\tau_2 - \tau_1, 0)$.

Of course, R considers that S_1 and S_2 happen at different places z_1 and z_2 as well as different times t_1 and t_2; $S_2 - S_1 = (t_2 - t_1, z_2 - z_1)$. Nevertheless, by Theorem 2.2 the norms $\|S_2 - S_1\|_R$ and

$\|S_2 - S_1\|_G$ must be equal. Thus

$$\sqrt{(t_2-t_1)^2 - (z_2-z_1)^2} = \|S_2-S_1\|_R = \|S_2-S_1\|_G = \tau_2 - \tau_1.$$

In other words, when $S_2 - S_1$ is timelike, every observer's calculation of the interval $\|S_2 - S_1\|$ yields the ordinary time interval between S_1 and S_2 as calculated by an observer "traveling with" those events. In this sense the interval between S_1 and S_2 is objective.

In a similar way, if the interval between two events is spacelike, there will be an observer who says that they happen at the same time but at different places. For that observer, the separation is purely spatial. Consequently, the interval between those events, as calculated by any observer whomsoever, will give that spatial separation—the ordinary distance—as measured by an observer for whom the events happen simultaneously.

A spacelike interval measures ordinary distance

Hyperbolic Angles and Rotations

In the analogy we are developing between R_θ and H_u, u corresponds to the Euclidean angle θ, so we have to be able to interpret u somehow as a "hyperbolic angle" and H_u as a rotation through that angle. The connection between the hyperbolic functions and the unit hyperbola will give us just what we need. But before we pursue this, we have to reckon with the fact that H_u acts separately on each of the six distinct regions of spacetime. Since worldlines of different observers correspond to future timelike vectors, we focus on the future timelike set $\mathcal{T}_+$, which we will henceforth describe more briefly as the **future set** $\mathcal{F}$.

Focus on the future set $\mathcal{F}$

Theorem 2.4 *For every real number u, H_u maps $\mathcal{F}$ one-to-one onto itself.*

PROOF: We already know that H_u maps $\mathcal{F}$ into itself. But so does its inverse $H_u^{-1} = H_{-u}$. It follows that H_u is one-to-one and onto.
END OF PROOF

Proposition 2.6 *The set $\mathcal{F}$ is closed under additon and multiplication by positive scalars.*

PROOF: The second statement is clear. To prove the first, suppose

$$E_1 = \begin{pmatrix} t_1 \\ z_1 \end{pmatrix} \quad \text{and} \quad E_2 = \begin{pmatrix} t_2 \\ z_2 \end{pmatrix}$$

are future vectors. Then

$$0 < t_1, \quad -t_1 < z_1 < t_1,$$
$$0 < t_2, \quad -t_2 < z_2 < t_2.$$

Therefore, $0 < t_1 + t_2$ and $-(t_1 + t_2) < z_1 + z_2 < t_1 + t_2$, so the sum $E_1 + E_2$ is also a future vector. END OF PROOF

Future rays

The map R_θ changes the angle of any ray through the origin in the Euclidean plane by θ radians. We will therefore be interested in what H_u does to a **future ray**, which we define to be a ray through the spacetime origin that lies in the future set $\mathcal{F}$.

Definition 2.5 *Two future rays ρ_1 and ρ_2 determine the **hyperbolic angle** $\angle \rho_1 \rho_2$ from ρ_1 to ρ_2. If ρ_1 and ρ_2 intersect the unit hyperbola $t^2 - z^2 = 1$ at the points $(\cosh \alpha_1, \sinh \alpha_1)$ and $(\cosh \alpha_2, \sinh \alpha_2)$, respectively, then $\angle \rho_1 \rho_2$ measures $\alpha_2 - \alpha_1$ hyperbolic radians.*

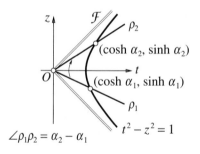

Proposition 2.2 guarantees that α_1 and α_2 are uniquely defined. Notice that the hyperbolic angle has a sign, and $\angle \rho_2 \rho_1 = -\angle \rho_1 \rho_2$. When confusion is unlikely, we drop the adjective *hyperbolic*.

Theorem 2.5 *Let ρ be a future ray; then the hyperbolic angle from ρ to $H_u(\rho)$ measures u hyperbolic radians.*

§2.3 Minkowski Geometry

PROOF: Suppose ρ intersects the unit hyperbola in the point $(\cosh\alpha, \sinh\alpha)$. Then the ray $H_u(\rho)$ contains the point

$$H_u \begin{pmatrix} \cosh\alpha \\ \sinh\alpha \end{pmatrix} = \begin{pmatrix} \cosh u & \sinh u \\ \sinh u & \cosh u \end{pmatrix} \begin{pmatrix} \cosh\alpha \\ \sinh\alpha \end{pmatrix} = \begin{pmatrix} \cosh(u+\alpha) \\ \sinh(u+\alpha) \end{pmatrix}$$

that lies on the unit hyperbola. (We used the addition formulas for the hyperbolic sine and cosine). By definition, the hyperbolic angle from ρ to $H_u(\rho)$ is $u + \alpha - \alpha = u$. END OF PROOF

Corollary 2.4 *If the hyperbolic angle from ρ_1 to ρ_2 is β, then $\rho_2 = H_\beta(\rho_1)$.*

PROOF: This is the converse of the theorem; it is true because there is a unique ray that lies β hyperbolic radians from ρ_1. END OF PROOF

Corollary 2.5 *H_u preserves the hyperbolic angle between future rays.*

PROOF: Suppose the hyperbolic angle from ρ_1 to ρ_2 is β. Then, by the previous corollary, $\rho_2 = H_\beta(\rho_1)$. But then

$$H_u(\rho_2) = H_u(H_\beta(\rho_1)) = H_{u+\beta}(\rho_1) = H_\beta(H_u(\rho_1)).$$

By Theorem 2.5, the hyperbolic angle from $H_u(\rho_1)$ to $H_u(\rho_2)$ is β. END OF PROOF

Corollary 2.6 *If $\angle\rho_1\rho_2 = \beta$, then the signed area of the sector on the unit hyperbola cut off from ρ_1 to ρ_2 is $\beta/2$.*

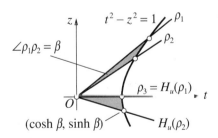

PROOF: Let ρ_3 be the positive t-axis; choose u such that $H_u(\rho_1) = \rho_3$. Then $\angle H_u(\rho_1)H_u(\rho_2)$ is also β, by the previous corollary.

Therefore, $H_u(\rho_2)$ intersects the unit hyperbola in the point $(\cosh \beta, \sinh \beta)$. By Exercise 10 of Section 2.2, the signed area of the new sector is $\beta/2$. But $\det(H_u) = 1$, so H_u is an area- and orientation-preserving map. Thus the two sectors have the same signed area, namely $\beta/2$. END OF PROOF

H_u is a hyperbolic rotation

With these results we have the complete analogy between R_θ and H_u; they explain the statement "H_u rotates points in the future set $\mathcal{F}$ by the hyperbolic angle u." The last corollary even gives us an alternative definition of the measure of the hyperbolic angle as twice the ordinary signed area of the sector on the unit hyperbola cut off by the two rays.

To prove the corollaries, we used the following theorem; it states that hyperbolic rotations form a commutative group with the same properties as the group of circular rotations.

Hyperbolic rotations form a group

Theorem 2.6 *The product of two hyperbolic rotations is another hyperbolic rotation; specifically, $H_u \cdot H_w = H_{u+w}$. In other words, to multiply two rotations, add their angles. Furthermore, $H_0 = I$, the identity map, $H_u^{-1} = H_{-u}$, and $H_u H_w = H_w H_u$.*

The Minkowski Inner Product

The standard Euclidean inner product of two vectors X_1 and X_2 is

$$X_1 \cdot X_2 = X_1^t I X_2 = (x_1, y_1) \begin{pmatrix} 1 & 0 \\ 0 & 1 \end{pmatrix} \begin{pmatrix} x_2 \\ y_2 \end{pmatrix} = x_1 x_2 + y_1 y_2.$$

Since $X \cdot X = x^2 + y^2 = \|X\|^2$, the analogy between the Euclidean and the Minkowski norms suggests that we should define the **Minkowski inner product** of two events E_1 and E_2 to be

$$E_1 \cdot E_2 = E_1^t J_{1,1} E_2 = (t_1, z_1) \begin{pmatrix} 1 & 0 \\ 0 & -1 \end{pmatrix} \begin{pmatrix} t_2 \\ z_2 \end{pmatrix} = t_1 t_2 - z_1 z_2.$$

The matrices $J_{p,q}$

By writing the inner products this way, we see that they obey all the usual rules of matrix multiplication. We also see that the difference between the Euclidean and Minkowski inner products lies only in the defining matrix—the identity matrix I in one case and the matrix $J_{1,1}$ in the other. To define the Minkowski inner product in the full $(1 + 3)$-dimensional spacetime, just use the matrix

$$J_{1,3} = \begin{pmatrix} 1 & 0 & 0 & 0 \\ 0 & -1 & 0 & 0 \\ 0 & 0 & -1 & 0 \\ 0 & 0 & 0 & -1 \end{pmatrix}.$$

Then $E \cdot E = E^t J_{1,3} E = t^2 - x^2 - y^2 - z^2 = Q(E)$, so

$$E \cdot E = \begin{cases} -\|E\|^2 & \text{if } E \text{ is spacelike,} \\ \|E\|^2 & \text{otherwise.} \end{cases}$$

Theorem 2.7 *Hyperbolic rotations preserve the Minkowski inner product.*

PROOF: $H_u(E_1) \cdot H_u(E_2) = (H_u E_1)^t J_{1,1} H_u E_2 = E_1^t H_u^t J_{1,1} H_u E_2$

$$= E_1^t \begin{pmatrix} \cosh u & \sinh u \\ \sinh u & \cosh u \end{pmatrix} \begin{pmatrix} 1 & 0 \\ 0 & -1 \end{pmatrix} \begin{pmatrix} \cosh u & \sinh u \\ \sinh u & \cosh u \end{pmatrix} E_2$$

$$= E_1^t \begin{pmatrix} \cosh u & -\sinh u \\ \sinh u & -\cosh u \end{pmatrix} \begin{pmatrix} \cosh u & \sinh u \\ \sinh u & \cosh u \end{pmatrix} E_2$$

$$= E_1^t \begin{pmatrix} 1 & 0 \\ 0 & -1 \end{pmatrix} E_2$$

$$= E_1 \cdot E_2. \hspace{2cm} \text{END OF PROOF}$$

If E_1 and E_2 are future vectors that lie on the future rays ρ_1 and ρ_2, respectively, we define $\angle E_1 E_2 = \angle \rho_1 \rho_2$. The next result is analogous to $X_1 \cdot X_2 = \|X_1\| \|X_2\| \cos \theta$ for two vectors X_1 and X_2 in the Euclidean plane that subtend the angle θ.

The angle between future vectors

Theorem 2.8 *If E_1 and E_2 are arbitrary future vectors and $\angle E_1 E_2 = \beta$, then $E_1 \cdot E_2 = \|E_1\| \|E_2\| \cosh \beta$.*

PROOF: We prove the result first for the *unit* vectors in the directions of E_1 and E_2; these are $U_i = E_i / \|E_i\|$, $i = 1, 2$. The vector U_i lies on the same ray as E_i, so $\angle U_1 U_2 = \angle E_1 E_2 = \beta$. Because U_1 is a unit vector, we can choose H_u such that

$$H_u(U_1) = \begin{pmatrix} 1 \\ 0 \end{pmatrix}.$$

Then
$$H_u(U_2) = \begin{pmatrix} \cosh \beta \\ \sinh \beta \end{pmatrix}$$
by the proof of Corollary 2.6. Consequently,
$$H_u(U_1) \cdot H_u(U_2) = \begin{pmatrix} 1 \\ 0 \end{pmatrix} \cdot \begin{pmatrix} \cosh \beta \\ \sinh \beta \end{pmatrix} = \cosh \beta.$$
Since H_u preserves the Minkowski inner product, $U_1 \cdot U_2 = \cosh \beta$. Finally,
$$E_1 \cdot E_2 = (\|E_1\| U_1) \cdot (\|E_2\| U_2) = \|E_1\| \|E_2\| (U_1 \cdot U_2)$$
$$= \|E_1\| \|E_2\| \cosh \beta. \qquad \text{END OF PROOF}$$

The first of the two following corollaries of Theorem 2.8 is a direct analogue of the Euclidean law of cosines—but notice the difference in sign. The second is particularly intriguing, as it goes completely in the opposite direction from its Euclidean analogue; it can be paraphrased as saying "a straight line is the *longest* distance between two points."

Corollary 2.7 *The law of hyperbolic cosines.* If E_1 and E_2 are future vectors and $\angle E_1 E_2 = \beta$, then
$$\|E_1 + E_2\|^2 = \|E_1\|^2 + \|E_2\|^2 + 2\|E_1\| \|E_1\| \cosh \beta.$$

PROOF: We know that $E_1 + E_2$ is a future vector. Therefore, since $\angle E_1 E_2 = \beta$, we have
$$\|E_1 + E_2\|^2 = (E_1 + E_2) \cdot (E_1 + E_2) = (E_1 \cdot E_1) + 2(E_1 \cdot E_2) + (E_2 \cdot E_2)$$
$$= \|E_1\|^2 + 2\|E_1\| \|E_2\| \cosh \beta + \|E_2\|^2. \qquad \text{END OF PROOF}$$

Corollary 2.8 *The reverse triangle inequality.* If E_1 and E_2 are future vectors, then $\|E_1 + E_2\| \geq \|E_1\| + \|E_2\|$.

PROOF: Suppose $\angle E_1 E_2 = \beta$; since $\cosh \beta \geq 1$, we have
$$\|E_1 + E_2\| = \sqrt{\|E_1\|^2 + 2\|E_1\| \|E_2\| \cosh \beta + \|E_2\|^2}$$
$$\geq \sqrt{\|E_1\|^2 + 2\|E_1\| \|E_2\| + \|E_2\|^2}$$
$$= \|E_1\| + \|E_2\|. \qquad \text{END OF PROOF}$$

The Pythagorean Theorem

There is no reasonable way to talk about the hyperbolic angle between a timelike and a spacelike vector; for example, no hyperbolic rotation can map one to the other. However, by analogy with Euclidean geometry we can say that two vectors E_1 and E_2 are **perpendicular**, or **orthogonal**, and write $E_1 \perp E_2$, if $E_1 \cdot E_2 = 0$. Apart from lightlike vectors, which are orthogonal to themselves, orthogonal vectors must be of different types—one timelike and one spacelike. Since we use the same units on the space and time axes (so the light cone has slope ± 1), orthogonal vectors in Minkowski geometry make equal Euclidean angles with the light cone. In Euclidean geometry, the slopes of perpendicular lines are m and $-1/m$; in Minkowski geometry, they are m and $+1/m$.

Orthogonal vectors

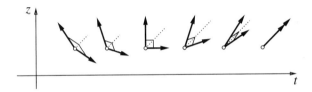

Suppose the timelike vector E_{ti} and the spacelike vector E_{sp} are orthogonal: $E_{ti} \perp E_{sp}$. Then these vectors form the sides of two different right triangles; the hypotenuse of one is $E_{ti} + E_{sp}$ and of the other is $E_{ti} - E_{sp}$. The hypotenuses can be of any type—timelike, spacelike, or lightlike—but the two possibilities are always of the same type.

Right triangles

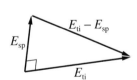 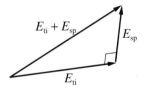

Proposition 2.7 *If $E_{ti} \perp E_{sp}$ where E_{ti} is timelike and E_{sp} is spacelike, then the vectors $E_{ti} \pm E_{sp}$ are always of the same type.*

PROOF: Since E_{ti} is timelike, we can choose a hyperbolic rotation H_u that puts $H_u(E_{ti})$ on R's time axis, so $H_u(E_{ti}) = (a, 0)^t$ for some

$a \neq 0$. Since H_u preserves the Minkowski inner product, $H_u(E_{\text{ti}}) \cdot H_u(E_{\text{sp}}) = 0$, so $H_u(E_{\text{sp}}) = (0, b)^t$ for some b. Therefore,

$$H_u(E_{\text{ti}} \pm E_{\text{sp}}) = H_u(E_{\text{ti}}) \pm H_u(E_{\text{sp}}) = \begin{pmatrix} a \\ \pm b \end{pmatrix}$$

and

$$Q(E_{\text{ti}} \pm E_{\text{sp}}) = Q(H_u(E_{\text{ti}} \pm E_{\text{sp}})) = a^2 - b^2,$$

so $E_{\text{ti}} \pm E_{\text{sp}}$ are of the same type. END OF PROOF

Theorem 2.9 (The Pythagorean theorem) *Suppose $E_{\text{ti}} \perp E_{\text{sp}}$ where E_{ti} is timelike and E_{sp} is spacelike. Then*

$$\pm \|E_{\text{ti}} \pm E_{\text{sp}}\|^2 = \|E_{\text{ti}}\|^2 - \|E_{\text{sp}}\|^2,$$

where we take the leading sign to be a minus if the vectors $E_{\text{ti}} \pm E_{\text{sp}}$ are spacelike and a plus otherwise.

PROOF: Recall that $E \cdot E = \pm \|E\|^2$ and we use the minus sign if and only if E is spacelike. Therefore,

$$\begin{aligned}
\pm \|E_{\text{ti}} \pm E_{\text{sp}}\|^2 &= (E_{\text{ti}} \pm E_{\text{sp}}) \cdot (E_{\text{ti}} \pm E_{\text{sp}}) \\
&= (E_{\text{ti}} \cdot E_{\text{ti}}) \pm 2(E_{\text{ti}} \cdot E_{\text{sp}}) + (E_{\text{sp}} \cdot E_{\text{sp}}) \\
&= \|E_{\text{ti}}\|^2 - \|E_{\text{sp}}\|^2.
\end{aligned}$$

END OF PROOF

Congruence

Rigid motions

In Euclidean geometry, rotation is a *rigid motion*: When we rotate the plane about the origin, the rotated image of a figure is congruent to the original. We declare the same to be true in Minkowski geometry: The image of a figure in the future set $\mathcal{F}$ under a hyperbolic rotation is congruent to the original. While it is not immediately obvious to our Euclidean eyes, all the rectangles on the right in the figure below are congruent to one another. They are, in Minkowski geometry, the "same" rectangle.

Reflections

There are two other kinds of rigid motion: reflection and translation. Every reflection in the Euclidean plane can be shown to be equal to the product of an ordinary rotation and the particular reflection across the *x*-axis.

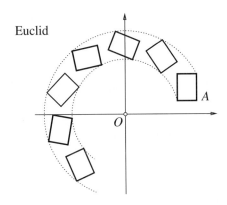
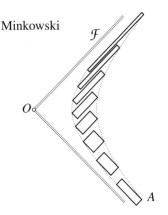

Proposition 2.8 *If $F_\theta : \mathbf{R}^2 \to \mathbf{R}^2$ is reflection across the line that makes an angle of θ with the x-axis, then $F_\theta = R_{2\theta} F_0$.*

Proposition 2.9 *Each F_θ preserves the Euclidean inner product: $F_\theta^t I F_\theta = I$. Conversely, if a matrix M preserves the Euclidean inner product, it is either a rotation or a reflection.*

Proofs are in the exercises. How might we define a reflection in Minkowski geometry? The Euclidean reflection F_θ fixes points on the line that makes an angle of θ with the x-axis and flips the orthogonal line on itself. Since we have a notion of orthogonality in the Minkowski plane, we can use the same idea to define a hyperbolic reflection.

Definition 2.6 *Let λ_u be the line through the origin that makes an angle of u hyperbolic radians with the t-axis, and let $\lambda_u^\perp$ be the orthogonal line. The **hyperbolic reflection across** λ_u is the linear map K_u that fixes points on the line λ_u and flips the line $\lambda_u^\perp$ on itself.*

That is, $K_u(x) = -x$ for every vector x in $\lambda_u^\perp$.

We can translate the definition into the language of eigenvalues and eigenvectors: K_u is the linear map whose (unit) eigenvectors are

$$\begin{pmatrix} \cosh u \\ \sinh u \end{pmatrix} \quad \text{and} \quad \begin{pmatrix} \sinh u \\ \cosh u \end{pmatrix},$$

and its corresponding eigenvalues are $+1$ and -1. Thus $\det K_u = +1 \cdot -1 = -1$. The analogy with Euclidean reflections suggests the following theorem.

Theorem 2.10 $K_u = H_{2u}K_0$.

PROOF: Consider the matrix

$$M = H_{2u}K_0 = \begin{pmatrix} \cosh 2u & \sinh 2u \\ \sinh 2u & \cosh 2u \end{pmatrix} \begin{pmatrix} 1 & 0 \\ 0 & -1 \end{pmatrix} = \begin{pmatrix} \cosh 2u & -\sinh 2u \\ \sinh 2u & -\cosh 2u \end{pmatrix}.$$

We show that $M = K_u$ by showing that M has the same eigenvectors and eigenvalues as K_u. The addition formulas for the hyperbolic sine and cosine give

$$M\begin{pmatrix} \cosh u \\ \sinh u \end{pmatrix} = \begin{pmatrix} \cosh 2u \cosh u - \sinh 2u \sinh u \\ \sinh 2u \cosh u - \cosh 2u \sinh u \end{pmatrix} = \begin{pmatrix} \cosh(2u - u) \\ \sinh(2u - u) \end{pmatrix},$$

so

$$M\begin{pmatrix} \cosh u \\ \sinh u \end{pmatrix} = +1 \cdot \begin{pmatrix} \cosh u \\ \sinh u \end{pmatrix}.$$

In a similar way,

$$M\begin{pmatrix} \sinh u \\ \cosh u \end{pmatrix} = \begin{pmatrix} \cosh 2u \sinh u - \sinh 2u \cosh u \\ \sinh 2u \sinh u - \cosh 2u \cosh u \end{pmatrix} = \begin{pmatrix} -\sinh u \\ -\cosh u \end{pmatrix},$$

so

$$M\begin{pmatrix} \sinh u \\ \cosh u \end{pmatrix} = -1 \cdot \begin{pmatrix} \sinh u \\ \cosh u \end{pmatrix}. \qquad \text{END OF PROOF}$$

Corollary 2.9 *Each K_u preserves the Minkowski inner product: $K_u^t J_{1,1} K_u = J_{1,1}$. Conversely, if M preserves the Minkowski inner product and maps $\mathcal{F}$ to itself, it is either a hyperbolic rotation or a hyperbolic reflection.*

PROOF: The first statement follows from the fact that K_u can be written as a product of matrices that individually preserve the Minkowski norm. To prove the second, let

$$M = \begin{pmatrix} a & b \\ c & d \end{pmatrix}.$$

Since M maps $\mathcal{F}$ to itself, the vector

$$\begin{pmatrix} a & b \\ c & d \end{pmatrix}\begin{pmatrix} 1 \\ 0 \end{pmatrix} = \begin{pmatrix} a \\ c \end{pmatrix}$$

must lie in $\mathcal{F}$; in particular, $a > 0$. The condition $M^t J_{1,1} M = J_{1,1}$ translates to

$$\begin{pmatrix} a & c \\ b & d \end{pmatrix} \begin{pmatrix} 1 & 0 \\ 0 & -1 \end{pmatrix} \begin{pmatrix} a & b \\ c & d \end{pmatrix} = \begin{pmatrix} a^2 - c^2 & ab - cd \\ ab - cd & b^2 - d^2 \end{pmatrix} = \begin{pmatrix} 1 & 0 \\ 0 & -1 \end{pmatrix}.$$

The conditions $a^2 - c^2 = 1$ and $b^2 - d^2 = -1$ imply that the columns of M,

$$C_1 = \begin{pmatrix} a \\ c \end{pmatrix} = \begin{pmatrix} \cosh u \\ \sinh u \end{pmatrix}, \qquad C_2 = \begin{pmatrix} b \\ d \end{pmatrix} = \pm \begin{pmatrix} \sinh v \\ \cosh v \end{pmatrix},$$

lie on the timelike and spacelike unit circles, respectively, and thus can be parametrized as shown, for appropriate hyperbolic angles u and v. Since $a > 0$, there is only one choice for the sign of C_1, but there is no similar restriction on the sign of C_2.

Finally, the equation $ab - cd = 0$ implies $C_1 \perp C_2$ and translates to

$$\cosh u \sinh v - \sinh u \cosh v = \sinh(v - u) = 0.$$

This implies $v - u = \sinh^{-1}(0) = 0$, so M is either

$$\begin{pmatrix} \cosh u & \sinh u \\ \sinh u & \cosh u \end{pmatrix} = H_u \qquad \text{or} \qquad \begin{pmatrix} \cosh u & -\sinh u \\ \sinh u & -\cosh u \end{pmatrix} = K_{u/2}.$$

END OF PROOF

Clearly, hyperbolic reflections help make Minkowski geometry a full and complete theory. But what role do they play in the physics of spacetime? The answer has to do with an assumption we first made in Section 1.2 and have carried with us. There we assumed that two Galilean observers will orient their spatial axes the same way: G's positive ζ-axis will point the same way as R's positive z-axis. This assumption kept our calculations simple; it wasn't essential. If we abandon it now and allow G and R to give their spatial axes opposite orientations, then we will need a hyperbolic reflection to map G's spacetime to R's.

The physical need for hyperbolic reflections

To see why this is true, suppose X is an object that is stationary with respect to G and is located at $\zeta = +1$. Then the worldline of X lies parallel to the τ-axis in the view of both G and R, but R will put the worldline *below* the τ-axis. If the slope of the τ-axis in R is $v = \tanh u$, then we can map G to R by first reflecting G on

Reversing spatial orientation

itself by K_0 and then performing the hyperbolic rotation H_u. The result is the hyperbolic reflection $H_u K_0 = K_{u/2}$.

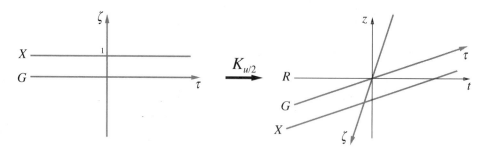

Enlarging the Lorentz group

By assuming that observers gave the same orientation to their spatial axes, we were inadvertently overlooking certain genuine Lorentz transformations. Now that we allow observers to orient their spatial axes arbitrarily, we must enlarge the class of Lorentz transformations to include hyperbolic reflections. Of course, we continue to assume that all observers see time flowing in the same direction, so we will still require Lorentz transformations to preserve the future set $\mathcal{F}$.

Definition 2.7 *The **orthochronous Lorentz group** $\mathcal{L}$ is the set of all linear maps $L : \mathbf{R}^2 \to \mathbf{R}^2$ that preserve the Minkowski inner product and the future set $\mathcal{F}$. The group operation is composition of maps or, equivalently, multiplication of matrices.*

The adjective *orthochronous* here means roughly "does the right thing to time." For us, it means that the image of the positive time axis will lie in $\mathcal{F}$. Since we shall never have a reason to use maps that do otherwise, we shall usually drop the adjective and refer to $\mathcal{L}$ simply as the Lorentz group.

Multiplication rules in $\mathcal{L}$

The definition includes the assertion that $\mathcal{L}$ is a group, which means that the product of two Lorentz maps is another Lorentz map. We already know that the product of two rotations is a rotation. In the exercises you are asked to verify that the product of two reflections is *also* a rotation and the product of a reflection and a rotation is another reflection. In fact, you can reduce any product of rotations and reflections to a single map. If the original product involves an even number of reflections, the result is a rotation; otherwise, the result is a single reflection. One way to demonstrate this is to commute the factors in a product until all the reflections are together on one end. The commutation rules in $\mathcal{L}$ are given in the following commutative diagrams, in which the sources and targets are represented simply by dots in order to focus attention on the maps themselves.

$$\begin{array}{ccc} \cdot \xrightarrow{H_{-u}} \cdot & \cdot \xrightarrow{K_{2v-u}} \cdot & \cdot \xrightarrow{K_{v-u}} \cdot \\ K_v \downarrow \quad \downarrow K_v & K_v \downarrow \quad \downarrow K_v & H_u \downarrow \quad \downarrow H_u \\ \cdot \xrightarrow{H_u} \cdot & \cdot \xrightarrow{K_u} \cdot & \cdot \xrightarrow{K_v} \cdot \end{array}$$

For example, the first diagram asserts that $H_u K_v = K_v H_{-u}$. This can be written in the alternative form

$$K_v^{-1} H_u K_v = K_v H_u K_v = H_{-u},$$

and when we take $v = 0$, this becomes $H_{-u} = K_0 H_u K_0$, which is identical to the result $B_{-v} = F B_v F$ extracted from the first commutative diagram in Section 2.1.

Let us derive the assertion in the second diagram, which can be expressed as $K_{2v-u} = K_v K_u K_v$. Since $K_p = H_{2p} K_0$, we have

$$K_v K_u K_v = H_{2v} \underbrace{K_0 H_{2u} K_0}_{H_{-2u}} H_{2v} K_0 = H_{2v} H_{-2u} H_{2v} K_0 = H_{2(2v-u)} K_0 = K_{2v-u}.$$

Derivations of the other diagrams are in the exercises.

The last kind of rigid motion is translation: $T(X) = X + C$ for some fixed vector C. This has the same effect in Minkowski and Euclidean geometry: It shifts the origin.

Translations

Definition 2.8 *A **rigid motion** in the Minkowski plane is an inhomogeneous Lorentz map, that is, a Lorentz map L from $\mathcal{L}$ followed by a translation. $M(E) = LX + C$.*

Definition 2.9 *Suppose A and B are two geometric figures in the Minkowski plane; A **is congruent to** B, $A \sim B$, if there is a rigid motion M that maps A onto B: $M(A) = B$.*

Proposition 2.10 *The interval between two events E_1 and E_2 is preserved by any rigid motion M.*

PROOF: It is enough to show that $Q(M(E_2) - M(E_1)) = Q(E_2 - E_1)$. But

$$M(E_2) - M(E_1) = LE_2 + C - (LE_1 + C) = L(E_2 - E_1),$$

and since L preserves the Minkowski inner product,

$$Q(M(E_2) - M(E_1)) = Q(L(E_2 - E_1)) = L(E_2 - E_1) \cdot L(E_2 - E_1)$$
$$= (E_2 - E_1) \cdot (E_2 - E_1) = Q(E_2 - E_1). \qquad \text{END OF PROOF}$$

Proposition 2.11 *Congruent figures have the same (Euclidean) area, up to sign.*

PROOF: Suppose $A \sim B$ and $M(A) = L(A) + C = B$. Then area$(B) = $ area$(L(A) + C) = $ area$(L(A))$ because translation preserves area. Therefore,

$$\text{area}(B) = \text{area}(L(A)) = \det(L) \cdot \text{area}(A) = \pm 1 \cdot \text{area}(A),$$

so the result follows. END OF PROOF

Exercises

In these exercises, $R_\theta : \mathbf{R}^2 \to \mathbf{R}^2$ is the linear map defined by the matrix multiplication

$$R_\theta \begin{pmatrix} x \\ y \end{pmatrix} = \begin{pmatrix} \cos\theta & -\sin\theta \\ \sin\theta & \cos\theta \end{pmatrix} \begin{pmatrix} x \\ y \end{pmatrix}.$$

1. Show that R_θ rotates the plane $\mathbf{R}^2$ by θ radians. (Suggestion: Since a linear map is completely determined by its action on a basis, just show that the basis vectors $(1, 0)$ and $(0, 1)$ are rotated θ radians.)

2. Exercise 1 immediately implies that the inverse of R_θ is $R_{-\theta}$; why? Prove that $R_\theta^{-1} = R_{-\theta}$ also by direct computation.

3. Exercise 1 also implies that $R_\theta R_\varphi = R_{\theta + \varphi}$; why? Give a second proof by direct computation using the addition formulas for the sine and the cosine functions. This result implies that the set $O_+(2, \mathbf{R})$ of rotation matrices forms a **group**.

4. Prove that a hyperbolic rotation H_u preserves each of the sets $\mathcal{T}_-$, $\mathcal{L}_+$, and $\mathcal{L}_-$.

5. Suppose E_1 and E_2 are future vectors and $\angle E_1 E_2 = \beta$. Then $E_1 \cdot E_2 = \|E_1\|\|E_2\| \cosh \beta$, and since $\cosh \beta \geq 1$, it follows that $E_1 \cdot E_2 \geq \|E_1\|\|E_2\|$. This is the **reverse Cauchy–Schwarz inequality**. Prove it from first principles, using only the definitions of the inner product and the norm in a $(1+1)$-dimensional spacetime.

6. (a) Explain why the reflection $F_\theta : \mathbf{R}^2 \to \mathbf{R}^2$ across the line through the origin that makes an angle θ with the positive x-axis has the following eigenvalues and eigenvectors:
$$\lambda_1 = +1, \ X_1 = \begin{pmatrix} \cos \theta \\ \sin \theta \end{pmatrix}, \quad \lambda_2 = -1, \ X_2 = \begin{pmatrix} -\sin \theta \\ \cos \theta \end{pmatrix}.$$
   (b) Show that $R_{2\theta} F_0$ has the same eigenvalues and eigenvectors, and deduce that $F_\theta = R_{2\theta} F_0$. Write the matrix representation of F_θ.
   (c) Verify that $F_\theta^t I F_\theta = F_\theta^t F_\theta = I$.

7. Show that the product of two reflections is a rotation: $F_\alpha F_\beta = R_{2(\alpha - \beta)}$.

8. The aim of this exercise is to prove that if the matrix $M = \begin{pmatrix} a & b \\ c & d \end{pmatrix}$ preserves the Euclidean inner product, $M^t M = I$, then M is either a rotation or a reflection.
   (a) Show that $M^t M = I$ implies that the columns of M are vectors that lie on the unit circle and are orthogonal to each other.
   (b) Deduce that there is a θ for which
$$\begin{pmatrix} a \\ c \end{pmatrix} = \begin{pmatrix} \cos \theta \\ \sin \theta \end{pmatrix}, \quad \begin{pmatrix} b \\ d \end{pmatrix} = \pm \begin{pmatrix} -\sin \theta \\ \cos \theta \end{pmatrix}.$$
   Explain why this implies that M is either a rotation or a reflection. Note: The **orthogonal group** $O(2, \mathbf{R})$ is the set of all 2×2 matrices M that preserve the Euclidean inner product $M^t M = I$. This exercise shows that every *orthogonal* matrix is either a rotation or a reflection.

9. Let H_u be the hyperbolic rotation through the hyperbolic angle u, and let K_v be the hyperbolic reflection across the line λ_v that makes an angle of v hyperbolic radians with the positive t-axis. Prove the following commutation relations:

$$H_{-u} = K_0 H_u K_0, \qquad H_{-u} = K_v H_u K_v, \qquad K_{v-u} = H_{-u} K_v H_u.$$

10. (a) Prove that the product of any two hyperbolic reflections is a hyperbolic rotation: $K_{v_1} K_{v_2} = H_u$. Express u in terms of v_1 and v_2.

    (b) Prove that the product of a hyperbolic rotation and a hyperbolic reflection is another hyperbolic reflection: $H_{u_1} K_{u_2} = K_v$. Express v in terms of u_1 and u_2.

11. Construct a *nonorthochronous* Lorentz map—that is, a linear map M that preserves the Minkowski inner product but does not preserve the future set $\mathcal{F}$.

12. (a) Consider the parametrized curve $P_u = (\cosh u, \sinh u)$ and the three line segments that connect the pairs of points $\{P_{-3/2}, P_{3/2}\}$, $\{P_{-1}, P_1\}$, and $\{P_{-1/2}, P_{1/2}\}$. Explain why these are parallel, and parallel to the tangent to the curve at the point P_0.

    (b) Now consider the three line segments that connect the pairs of points $\{P_{-1}, P_2\}$, $\{P_{-1/2}, P_{3/2}\}$, $\{P_0, P_1\}$. Prove that these are also parallel, and parallel to the tangent to the curve at $P_{1/2}$. Suggestion: Consider the hyperbolic rotation $H_{1/2}$.

13. (a) Let A be a rectangle in $\mathcal{F}$ whose sides are parallel to the lines $z = \pm t$, and suppose the Euclidean lengths of the sides are a and b, as shown in the figure below.

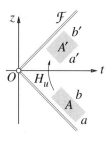

Let $H_u : \mathcal{F} \to \mathcal{F}$ be hyperbolic rotation by the hyperbolic angle u. Show that $A' = H_u(A)$ is another rectangle whose sides are likewise parallel to the lines $z = \pm t$, and their Euclidean lengths are

$$a' = e^u a, \qquad b' = e^{-u} b.$$

(b) Describe the image of a Euclidean circle S of radius r in $\mathcal{F}$ under the hyperbolic rotation H_u. Along the way you will need the result of the following exercise. To describe the ellipse $H_u(S)$ you will need to give the lengths and directions of its semimajor and semiminor axes.

14. Show that the image of a circle under an invertible linear map $L : \mathbf{R}^2 \to \mathbf{R}^2$ is an ellipse. Show, furthermore, that if the matrix of L is symmetric, then the major and minor axes of the image ellipse are in the directions of the eigenvectors of that matrix. If the eigenvalues are $\lambda_1 \leq \lambda_2$ and the circle has radius r, then the semimajor and semiminor axes of the image ellipse are $\lambda_2 r$ and $\lambda_1 r$.

2.4 Physical Consequences

Length and Time

On the face of it, length, time, and mass are independent quantities. However, once all Galilean observers agree that a certain ratio of length to time—namely c, the speed of light—is a physical constant, that fixed ratio ties one of those units to the other. So one of the first consequences of Einstein's relativity theory is that length and time are no longer independent. We can, for example, express time in meters. However, we choose to make time primary and express length in seconds. We take time as the primary unit because it is the natural one to use along an observer's worldline: For that observer, all events on the worldline differ only in the *time* they happen, not the *place*.

We could set the ratio c to any value, but the most practical choice is $c = 1$. Since

Calibrating length to time

$$c = 2.9979246 \times 10^8 \, \frac{\text{meters}}{\text{second}}$$

in conventional units, the calibration sets 2.9979246×10^8 meters equal to 1 second, or 1 meter equal to 3.3356409×10^{-9} seconds. In visual terms, we make the light cone have slope ± 1 and use it to map units along the time axis to any space axis.

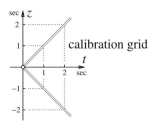

"Geometrized" units

By applying the same idea to other physical constants we can "geometrize" additional units. For example, the constant G that appears in Newton's law of universal gravitation has the value

$$G = 6.67 \times 10^{-11} \, \frac{\text{m}^3}{\text{kg sec}^2}$$

in conventional units. Therefore, by setting $G = 1$ we can calibrate mass to time:

$$1 \text{ kg} = 6.67 \times 10^{-11} \times \frac{(3.3356409 \times 10^{-9} \text{ sec})^3}{1 \text{ sec}^2} = 2.48 \times 10^{-36} \text{ sec}.$$

When we use geometric units, velocity becomes dimensionless, energy and momentum both have the dimensions of mass and thus of time, and acceleration has the dimensions of time^{-1}.

Velocity Limit

The time axis must be inside the light cone

No Galilean observer G can travel faster than the speed of light relative to another Galilean observer R. This is an assumption we have already used frequently—for example, to derive the form of the Lorentz transformation—but we haven't proven that it must be true. However, it follows directly from the constancy of the

§2.4 Physical Consequences

speed of light. In G's own coordinate frame, the time axis (which is that observer's worldline) must be the central axis of symmetry of the light cone. In particular, G's worldline must lie *inside* the light cone, and this must still be true when we map $G \to R$. But if the velocity v of G relative to R were greater than c, then G's worldline would have a slope greater than the slope of the light cone and thus would lie outside the light cone. This is impossible.

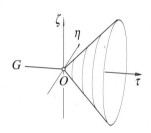
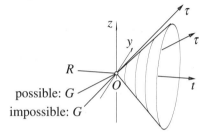

Simultaneity

Proposition 2.12 *Given any spacelike event E whatsoever, there is an observer who says that E and O are simultaneous.*

PROOF: Because E is spacelike, there is a real number k for which $E = (t, z) = (k \sinh u, k \cosh u)$ in R's coordinate frame. Let G have velocity $v = \tanh u$ relative to R. Then E will lie on G's ζ-axis; G will say that E and O both happen at time $\tau = 0$.   END OF PROOF

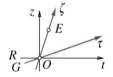

Since Galilean observers need not agree that two events happened at the same time, the principle of relativity implies that the notion of *simultaneity* is not physically meaningful. In effect, we replace it with the constancy of the speed of light.

The Objective Future

According to R, the events that happen after O lie in the half-plane $t > 0$; according to G, they lie in the half-plane $\tau > 0$. But these are different sets, so R and G disagree about what constitutes the future. However, they *do* agree about the events in the intersection, which we can call the *common future* of R and G.

The common future

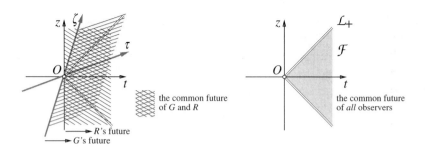

the common future of G and R

R's future
G's future

the common future of all observers

The common future of all observers

When a third observer comes in, the common future grows smaller, and when we consider all possible Galilean observers, *their* common future reduces to the union of $\mathcal{F}$ and the future light cone $\mathcal{L}_+$. This is because no spacelike event E can be in the intersection, because we saw already that there is *some* observer who considers E to have happened at the same time as O. We call the common future of all observers the *objective future* of O. If P is any event whatsoever, we can translate the origin to P and define, in the same way, the **objective future** of P. We can define the objective past of P in a similar way.

Causality

Definition 2.10 *The **causal future** of an event A is the set of all events that A can influence. The **causal past** of A is the set of all events that can influence A.*

The principle of causality

We start with the principle that causes happen before effects: If the event A causes, or influences, the event B, then A must happen before B. Since we want this principle of causality to be a physical law, A must happen before B *for all Galilean observers*. Thus B must be in the objective future of A, and the causal future of A is identical with the objective future of A.

Three causal regions

Likewise, A must be in the objective past of B. Each event P thus divides spacetime into three regions: Besides the causal future and causal past of P, there is a region consisting of events that can neither influence nor be influenced by P. In fact, these are the events Q for which the separation $Q - P$ is spacelike.

§2.4 Physical Consequences

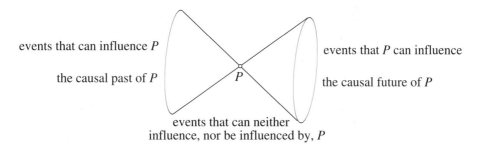

All the events that P can influence lie on worldlines that have slope v where $|v| \leq 1$: Causality cannot travel faster than the speed of light. Thus, immediate action at a distance is impossible.

Rigidity

One of the intriguing consequences of the causal structure of spacetime is that no physical object is completely rigid. Here is a specific example to illustrate. Suppose R takes a rod lying on the positive z-axis between $z = 0$ and $z = 1$ and hits the end at the origin with a hammer at time $t = 0$ in such a way as to send the rod up the z-axis with velocity v. In the figure below, the worldlines of five equally-spaced points on the rod are plotted. The plot on the left shows what happens if the whole rod were to move rigidly—that is, without altering the distances between points. In this scenario, the far end moves immediately, but the principle of causality rules out immediate action at a distance.

No object is completely rigid

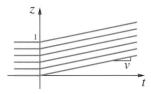 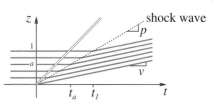

The picture on the right is better; it takes causality into account. Notice that the point on the rod at $z = a$ does not begin to move at $t = 0$; its worldline stays horizontal until some later time $t = t_a$. The simplest assumption we can make is that t_a is proportional to a: $t_a = ka$, or $a = (1/k)t_a = pt_a$. The line $z = pt$ in

A shock wave travels through the rod

spacetime marks a "shock wave" that propagates through the rod with velocity p, where $|v| < |p| < 1$, imparting motion to points in the rod as it propagates. The entire rod is in motion only after the wave reaches its far end, which happens at time $t = t_1$.

This is not the whole story. Notice that the worldlines on the right are closer together than those on the left: The shock wave has compressed the rod. If the rod is elastic, it will rebound and expand back to its original size. This will involve further shocks traveling through the rod—none of which we describe here.

Addition of Velocities

Can galaxies move faster than the speed of light?

Distant galaxies are traveling away from the earth at enormous speeds, the farthest ones at speeds close to the speed of light. Consider two galaxies at opposite ends of the sky each moving away from the earth at two-thirds the speed of light. Ordinary reasoning (as framed by Galilean transformations) says that an observer on one of the galaxies would see the other galaxy moving at four-thirds the speed of light, which is impossible.

Combining boosts

Let us consider this in general terms. Suppose G has velocity v_1 relative to R, and R has velocity v_2 relative to a third observer C. Then we can map one frame to another by appropriate *velocity boosts* (that is, Lorentz transformations expressed directly in terms of the velocity parameter):

$$B_{v_1} : G \to R, \qquad B_{v_2} : R \to C.$$

We want to know the velocity v associated with the boost $B_v = B_{v_2} B_{v_1}$ that maps G directly to C: $B_v : G \to C$.

The addition formula

Since the rule for combining hyperbolic rotations is simpler, we convert the boosts to hyperbolic rotations by the relation $H_u = B_v$, where $v = \tanh u$ and $u = \tanh^{-1} v$. If $H_u = H_{u_2} H_{u_1}$, then

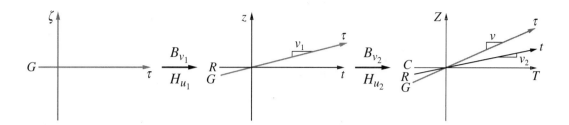

$u = u_1 + u_2$, and thus

$$v = \tanh u = \tanh(u_1 + u_2) = \frac{\tanh u_1 + \tanh u_2}{1 + \tanh u_1 \tanh u_2} = \frac{v_1 + v_2}{1 + v_1 v_2}.$$

For example, in the case of the two galaxies, $v_1 = v_2 = \frac{2}{3}$ and each galaxy sees the other receding from it with velocity

$$v = \frac{\frac{2}{3} + \frac{2}{3}}{1 + \frac{4}{9}} = \frac{\frac{4}{3}}{\frac{13}{9}} = \frac{12}{13} < 1.$$

Since $|v| = |\tanh u| < 1$ for all u, no combination of boosts can ever make the velocity of one observer relative to another exceed the speed of light.

Moving Clocks Run Slow

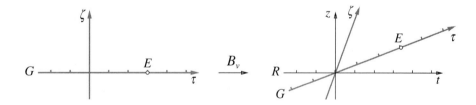

Suppose G is carrying a clock and moving with velocity v relative to R. That clock marks G's **proper time** τ in R's frame. (Here again *proper* means "of one's own," as it does in the term *proper value*, or eigenvalue.) How does R measure G's proper time?

Let $E = (\Delta\tau, 0)$ be the event that happens at time $\Delta\tau$ on G's worldline. We want to know what time R thinks that E occurs. We

Proper time

Calculate the rate of a moving clock

can get the coordinates of E in R's frame by using the velocity boost B_v:

$$\begin{pmatrix} \Delta t \\ * \end{pmatrix} = B_v \begin{pmatrix} \Delta \tau \\ 0 \end{pmatrix} = \frac{1}{\sqrt{1-v^2}} \begin{pmatrix} 1 & v \\ v & 1 \end{pmatrix} \begin{pmatrix} \Delta \tau \\ 0 \end{pmatrix} = \begin{pmatrix} \Delta \tau / \sqrt{1-v^2} \\ * \end{pmatrix},$$

where $*$ stands for a quantity whose value we don't need to know. Thus $\Delta t = \Delta \tau / \sqrt{1-v^2} > \Delta \tau$, or $\Delta \tau = \Delta t \sqrt{1-v^2}$. In other words, R says that G's clock *runs slow* by the factor $\sqrt{1-v^2}$.

For example, when $v = \frac{1}{2}$, the factor is $\sqrt{1-0.25} = \sqrt{0.75} \approx 0.866$, so when R says an hour has passed, G's clock will have advanced slightly less than 52 minutes. If $v = 0.9$, then the factor is $\sqrt{1-0.81} = \sqrt{0.19} \approx 0.436$: A clock moving at nine-tenths the speed of light runs less than half as fast as a stationary clock.

Time dilation is recalibration

Geometrically, time dilation is just a consequence of the calibration of different time axes by hyperbolic rotations, and thus has to do with the preservation of the Minkowski norm. The Minkowski–Pythagorean 3-5-4 triangle provides a particularly simple and lucid example. If a clock is moving with velocity $v = \frac{3}{5}$, then it ticks at only 4/5 the rate of a stationary clock.

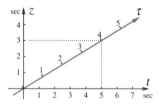

Moving Rulers Contract in Length

Suppose G carries a ruler that is λ seconds in length. To be definite we suppose it lies along the ζ-axis from $\zeta = 0$ to $\zeta = \lambda$. In G the worldline of the ruler is a horizontal band λ seconds wide.

Let G and the ruler move with velocity v relative to R along their common ζ/z-axis. In R's frame the worldline of the ruler is a band with slope v. The z-coordinate of the event E gives the length l of the ruler as R sees it. How is l connected to λ? To make the connection, we shall obtain the ζ-coordinate of E in G's frame in two different ways and compare them. On the one

Obtain E coordinates in G two ways

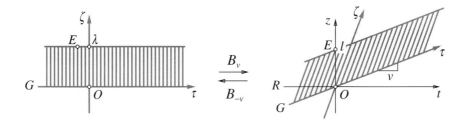

hand, the ζ-coordinate of E must be λ; on the other, we can get the ζ-coordinate by mapping E from R back to G using B_{-v}:

$$\begin{pmatrix} * \\ \lambda \end{pmatrix} = B_{-v} \begin{pmatrix} 0 \\ l \end{pmatrix} = \frac{1}{\sqrt{1-v^2}} \begin{pmatrix} 1 & -v \\ -v & 1 \end{pmatrix} \begin{pmatrix} 0 \\ l \end{pmatrix} = \begin{pmatrix} * \\ l/\sqrt{1-v^2} \end{pmatrix},$$

where $*$ stands for a quantity whose value we don't need to know. From these equations we obtain

$$l = \sqrt{1-v^2}\,\lambda;$$

R considers that the ruler has shrunk by the factor $\sqrt{1-v^2}$.

This is precisely the Fitzgerald contraction. Of course, it had to be, because that is the only result consistent with experience. However, it appears here as a consequence, ultimately, of Einstein's simple postulate that the speed of light is the same for all Galilean observers.

The Fitzgerald contraction

The Doppler Effect

Although the speed of a light signal does not depend on whether the source is moving, its *frequency* does. This is the Doppler effect. If the source is approaching an observer, the frequency increases; visible light is shifted toward the violet. If the source is receding, the frequency decreases; visible light is shifted toward the red. The Doppler effect is part of classical physics, but the size of the shift is different in special relativity.

Light from a moving source changes frequency

To calculate the shift, let us suppose that G is steadily emitting light of frequency ν (this is the Greek letter *nu*). Think of this as an oscillation that reaches peak amplitude once every $1/\nu$ seconds. In G these peaks appear as light cones spaced $\Delta\tau = 1/\nu$ seconds apart along the τ-axis.

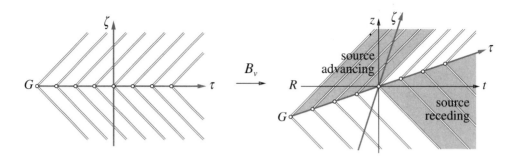

As G moves with velocity v past another observer R, the source is first advancing toward R, then receding. The spacing Δt of the light cones along the t-axis determines the frequency n of the light as R sees it: $n = 1/\Delta t$. The figure suggests that the advancing frequency n_{adv} is greater than ν, while the receding frequency n_{rec} is less than ν. To determine the values, look at the enlargement of R near the origin.

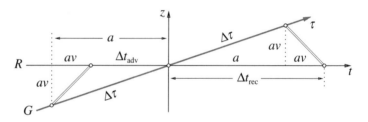

On both sides we have $\Delta\tau^2 = a^2 - a^2v^2$, so $a = \Delta\tau/\sqrt{1-v^2}$. On the advancing side, the time separation between successive advancing peaks is

$$\Delta t_{\text{adv}} = a - av = \Delta\tau\frac{1-v}{\sqrt{1-v^2}}, \quad \text{so} \quad n_{\text{adv}} = \frac{1}{\Delta t_{\text{adv}}} = \nu\sqrt{\frac{1+v}{1-v}} > \nu.$$

On the receding side,

$$\Delta t_{\text{rec}} = a + av = \Delta\tau\frac{1+v}{\sqrt{1-v^2}}, \quad \text{so} \quad n_{\text{rec}} = \frac{1}{\Delta t_{\text{rec}}} = \nu\sqrt{\frac{1-v}{1+v}} < \nu.$$

These calculations are done in R, where the source moves. We get the same result by doing the calculations in G, where the observer moves. See the exercises. It doesn't matter whether we assume that the source or the observer does the moving; it is the relative motion that creates the Doppler effect.

We can even collapse the two formulas to a single one—and abandon the distinction between n_{adv} and n_{rec}—if we take v to be the time rate of change of *distance* between source and observer, rather than time rate of change of the displacement z. Then v is positive when the source and observer are separating and negative when they are approaching one another. To the observer R, the frequency is just

A single formula

$$n = v\sqrt{\frac{1-v}{1+v}}.$$

As in many other areas, the classical theory here approximates the relativistic theory for small velocities. When $|v| \ll 1$, Taylor's theorem gives

The classical approximation

$$\sqrt{\frac{1-v}{1+v}} \approx 1-v,$$

so we have the approximate formula $n \approx v(1-v)$. In classical dynamics, this is the *exact* formula. The difference between the two becomes most striking when $v = -1 = -c$, that is, when the source and observer approach each other at the speed of light. The classical frequency simply doubles, but the relativistic frequency is infinite.

Exercises

1. Suppose galaxies C and G are moving away from the earth with velocities $-v$ and v, respectively. That is, they move with the same speed but in opposite directions. Then G is moving away from C with velocity

$$f(v) = \frac{2v}{1+v^2}.$$

(a) Sketch the graph of $f(v)$ for $0 \leq v \leq 1$. Show that, for v small, $f(v) \approx 2v$, while for $v \approx 1$, $f(v) \approx 1$.

(b) At what speed must C and G be receding from the earth if C considers that G is moving away from C at half the velocity of light?

2. Suppose a subatomic particle that moves at 99% of the speed of light has a lifetime of 10^{-10} seconds in a laboratory frame of reference. That is, after 10^{-10} seconds, it decays into other particles. What is the lifetime of the particle from the particle's own frame of reference?

3. (a) If G moves with velocity v with respect to R, then R says that G's clock runs slow by the factor $\sqrt{1-v^2}$. But then R moves with velocity $-v$ with respect to G, so G will say R's clock runs slow by the same factor $\sqrt{1-(-v)^2} = \sqrt{1-v^2}$. Explain this paradox; that is, explain how the two observers can have symmetric viewpoints. Furthermore, indicate how the figure below both demonstrates and explains this paradox.

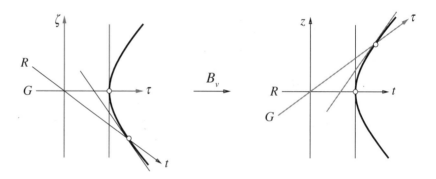

(b) There is a similar paradox about Fitzgerald contraction. If G moves with velocity v relative to R, then R says that G's length contracts by the factor $\sqrt{1-v^2}$ in the direction of motion. However, from G's point of view, R's length contracts by this factor. Draw a diagram that demonstrates and explains this paradox.

4. Determine the first-order Taylor polynomial of $f(v) = \sqrt{\dfrac{1-v}{1+v}}$.

5. Show that the classical Doppler effect, under a Galilean transformation, is $n = v(1-v)$. Here n is frequency recorded by an observer R when an oscillator G of proper frequency v moves with velocity v with respect to R.

6. Use the figure below to calculate the Doppler effect on the moving observer R in the frame G in which a steady signal of frequency ν is emitted. The spacing between light cones is $\Delta \tau = 1/\nu$ seconds on G's worldline.

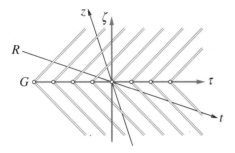

7. Three observers, T, G, and V, are traveling on a high-speed train. Observer T is at the back, G is in the middle, and V is in the front; T and V are each a seconds from G, measured when the train is at rest. They and the train are moving with velocity v with respect to a fourth observer R standing beside the train tracks. (All measurements are made in geometric units.)

At the instant G passes R, they see momentary bright flashes of light from both T and V.

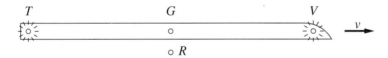

(a) Draw the worldlines of T, G, V, and R and the light cones from T and V in G's frame and again in R's frame.

(b) According to G, were the flashes from T and V simultaneous, or did one happen before the other? If the latter, by how much time does G consider those events to be separated?

(c) According to R, were the flashes from T and V simultaneous, or did one happen before the other? If the latter, by how much time does R consider those events to be separated?

(d) When were the flashes emitted according to G, and when were they emitted according to R?

Further Reading for Chapter 2

The linear algebra needed here can be found in many texts, but the classic by Birkhoff and MacLane [3] has a perspective that is particularly appropriate. Lorentz introduces Lorentz transformations in [19], and Einstein makes them the basis of special relativity in [11]. Minkowski's recasting of Einstein's ideas in geometric terms is described in [23], and a rich and full exposition of Minkowski geometry can be found in the book by Yaglom [30].

CHAPTER 3
Special Relativity – Kinetics

Kinetics is the study of the motion of material objects under the action of forces. Forces cause objects to accelerate, that is, to change their velocity. In spacetime, acceleration makes worldlines curve. In this way the physics of forces is tied to the geometry of spacetime. The starting point is Newton's three laws of motion.

3.1 Newton's Laws of Motion

One of the great scientific problems of the seventeenth century was to understand the motion of the planets and the moon. Newton solved the problem and presented his solution in the *Principia* (*Philosophiæ Naturalis Principia Mathematica*, 1687). Written in the didactic style of a text on Euclidean geometry, it begins with definitions and axioms upon which all the later arguments are based. The axioms are the three laws of motion:

1. "Every body continues in its state of rest, or of uniform motion in a straight line, unless it is compelled to change that state by forces impressed upon it."

2. "The change of motion is proportional to the motive force impressed; and is made in the direction of the straight line in which that force is impressed."

3. "To every action there is always opposed an equal reaction; or, the mutual actions of two bodies upon each other are always equal, and directed to contrary parts."

The first law defines inertia

The first law was actually discovered by Galileo and is called the principle of inertia. The law says that every body has **inertia**, which is the power to resist any change in its velocity. Notice that rest and uniform motion have the same status, as they must according to Galileo's principle of relativity.

Inertial mass

Since a body has inertia, it makes sense to ask how much; that is, how much effort must be expended to change its velocity? Your intuition is helpful here. Imagine a small rowboat tied up beside a 30-foot cabin cruiser at a dock, both of them motionless. By pushing, you can get each boat to move, but the same push will have a much greater effect on the rowboat. The rowboat has less inertia, and this is precisely because it has less mass. The "inertia content" of a body is its mass, sometimes called its **inertial mass** to emphasize the connection between the two.

Momentum and the second law

Newton uses inertial mass in the second law. He defines the term "motion" that appears in the law as mass m times velocity $\mathbf{v}$; this is the vector quantity we today call **momentum**: $\mathbf{p} = m\mathbf{v}$. Furthermore, we take "change in motion" to mean the rate of change of momentum with respect to time: $d\mathbf{p}/dt$. According to the second law, this is proportional to the impressed force $\mathbf{f}$. If we choose units such that the constant of proportionality is 1, the law takes the form

$$\mathbf{f} = \frac{d\mathbf{p}}{dt} = \frac{d(m\mathbf{v})}{dt}.$$

If mass is constant, then $\mathbf{f} = m\mathbf{a}$

If m is constant, then $d(m\mathbf{v})/dt = m\,d\mathbf{v}/dt = m\mathbf{a}$, and the second law takes its more familiar form $\mathbf{f} = m\mathbf{a}$. We shall soon see that the inertia content m must increase with velocity—and thus is not constant; nevertheless, m is essentially constant for small velocities, so in that case we can write the second law in the convenient form $\mathbf{f} = m\mathbf{a}$.

The third law concerns a pair of bodies G_1 and G_2. It says that if G_1 imposes a force $\mathbf{f}_1$ on G_2, then G_2 imposes a force $\mathbf{f}_2$ on G_1, and $\mathbf{f}_2 = -\mathbf{f}_1$. For example, when you push on the rowboat, it pushes back at you with the same strength. Now consider the total momentum $\mathbf{p}_1 + \mathbf{p}_2$ of the system consisting of G_1 and G_2 together. Since

$$\frac{d(\mathbf{p}_1 + \mathbf{p}_2)}{dt} = \frac{d\mathbf{p}_1}{dt} + \frac{d\mathbf{p}_2}{dt} = \mathbf{f}_1 + \mathbf{f}_2 = \mathbf{0},$$

the total momentum does not change over time: According to the third law, *total momentum is conserved*.

The third law says that total momentum is conserved

Difficulties

Newton's laws do not always correspond to reality. For example, an object near the surface of the earth that has no visible forces pushing it accelerates downward: When we let go of things, they fall. A more subtle example is a Foucault pendulum. It swings in a vertical plane, but that plane rotates slowly around a vertical axis, instead of staying fixed. (All rotational motion is accelerated.)

Newton's laws need to be qualified

There are at least two ways to resolve such conflicts between theory and reality. The first is to ascribe an invisible force to each unexplained acceleration. Thus, we say that *gravity* causes things to accelerate downward, and the *Coriolis force* causes the plane of the pendulum to rotate.

The second way is to choose a coordinate system in which the acceleration disappears. For example, imagine a Foucault pendulum at the North Pole. It is easy to see then that with respect to the fixed stars, the earth is doing the rotating, not the plane of the pendulum's motion. In a coordinate frame in which the stars do not move, the pendulum's plane does not move either, and there is no Coriolis force.

Are there coordinate systems that can eliminate gravity in the same way? An earth-orbiting spaceship seems to be gravity-free: Objects just float about when they are released in the cabin. An object initially at rest remains at rest, and one that is set moving with a certain velocity maintains that velocity until it collides

Inertial frames

Special relativity is limited to inertial frames

Inertial mass cannot be constant

Mass increases with velocity

with a wall or other obstruction. The law of inertia holds without qualification.

Thus, whether or not Newton's laws hold depends on the coordinate frame. We single out those in which Newton's laws *do* hold and call them **inertial frames**. Physics is simpler in an inertial frame. The frames we are most familiar with—defined by the walls of an earth-bound laboratory, for example—are *not* inertial.

Henceforth, we assume that the spacetime coordinate systems associated with our observers are inertial. In fact, what makes special relativity *special* is that it is restricted to inertial coordinate frames. So long as our observers are in uniform motion with respect to one another, it is enough to assume that *one* of them has an inertial frame. By Galileo's principle of relativity, they must all formulate physics the same way, and thus must all see Newton's laws as holding in their systems because one of them does.

The second difficulty concerns our intuition that the inertial mass of a body is constant. The velocity limitation If this were so, then a constant force would eventually push the body beyond the speed of light. To see this, suppose a constant force of magnitude k is applied to an object G in the direction of R's positive z-axis. Let v be the (scalar) velocity of G along the z-axis. If the inertial mass m of G were constant, then R could apply Newton's second law to G and write $k = m\,dv/dt$, or $dv/dt = k/m$. Since the acceleration k/m is a constant, velocity is just a linear function of time:

$$v(t) = \frac{k}{m}t + b.$$

Choose T such that $v(T) = 1$, the speed of light. Then applying the force k for T seconds will push G past the speed of light. This contradicts the velocity limitations of special relativity (Section 2.4).

How can we correct this problem? The velocity limitation implies that as G's velocity grows larger and larger, it must get harder and harder to increase that velocity still further. But that effort is a measure of G's inertia content, so G's inertial mass must increase with velocity and, *a fortiori*, be a function of velocity: $m = m(v)$.

In fact, the mass must become infinite as $v \to 1$. If not, there will be an upper bound $m(v) \leq M$ for all v. Then when R invokes the second law in the form $k = d(m(v)v)/dt$, we will still obtain

$$kt + c = m(v)v \leq Mv(t), \qquad \text{implying} \qquad \frac{kt+c}{M} = \frac{k}{M}t + \frac{c}{M} \leq v(t).$$

In other words, $v(t)$ will be bounded below by a linear function (rather than simply be equal to one, as above). If we choose T such that $(k/m)T + c/M = 1$, then we will again have $v \geq 1$ when $t \geq T$. Thus the graph of m as a function of v must look something like this:

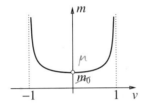

Notice that while R says that G's mass is increasing, G does not. In other words, we must give up the idea that mass is an absolute quantity; we have no way of saying what the mass *really* is. All we can say is what a given observer measures it to be. Mass is relative, just like length or the rate at which a clock ticks. Nonetheless, it is helpful to single out the value $m_0 = m(0)$, which we call G's **rest mass**, or **proper mass**. (This is yet another place where we use *proper* to mean "of one's own.")

Mass is relative

We still have to determine exactly how m depends on v. To this end we take a further look at the third law. The issue for us is Newton's assertion that this is a physical law. By Galilean relativity, if *one* observer sees an interaction in which momentum is conserved, then *all other* observers must agree. If not, then conservation of momentum is not a physical law. Consider the following example in which we assume, for the moment, that mass does *not* depend on velocity.

Is conservation of momentum a physical law?

Let X_1 have mass 1 and travel with G, and let X_2 have mass 2 and travel with R. Suppose they collide at the event O and

Example: an inelastic collision

stick together, forming a new body X of mass 3. (This is called a *completely inelastic* collision.) We assume that total momentum is conserved in G's frame, and we ask whether the same is true in R's.

Suppose, as usual, that G has velocity v with respect to R, and consider the collision in G's frame. In G's frame X_1 is motionless and therefore has zero momentum, while X_2 moves with velocity $-v$ and has momentum $-2v$. The total momentum before the collision is therefore $p = -2v$. After the collision, X moves with some velocity $\bar{v}$, so its momentum is $\bar{p} = 3\bar{v}$. Since G considers that $p = \bar{p}$, it follows that $-2v = 3\bar{v}$, or $\bar{v} = -\frac{2}{3}v$.

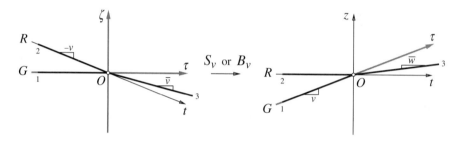

Transform with a Galilean shear

How does R view all this? We answer this question by transforming G's frame to R's. Let us do it first with the Galilean shear $S_v : G \rightarrow R$; to get the velocity of an object in R, just add v to its velocity in G. In R, X_2 is motionless and has zero momentum, while X_1 has velocity v and momentum $1 \cdot v = v$. According to R, the total momentum before the collision is $p = v$. After the collision, X has velocity $\bar{w} = \bar{v} + v = \frac{1}{3}v$ and momentum $\bar{p} = 3\bar{w} = v$. R agrees with G that total momentum is conserved. (Notice that R and G do not agree on the *value* of the conserved momentum. This is only to be expected, because momentum depends on velocity—and is therefore a relative quantity—even before we take special relativity into account.)

Transform with a Lorentz boost

If we use the Lorentz transformation $B_v : G \rightarrow R$, the outcome is different. Before the collision the picture is unchanged; you should check that the total momentum is still $p = v$. After the collision, X has a velocity $\bar{w}$ that we must find using the addition formula for Lorentz boosts:

$$\overline{w} = \frac{\overline{v}+v}{1+\overline{v}v} = \frac{-\frac{2}{3}v+v}{1-\frac{2}{3}v^2} = \frac{\frac{1}{3}v}{1-\frac{2}{3}v^2}.$$

The momentum is

$$\overline{p} = 3\overline{w} = 3\frac{\frac{1}{3}v}{1-\frac{2}{3}v^2} = \frac{v}{1-\frac{2}{3}v^2} \neq v = p.$$

Thus, in R's view, momentum is *not* conserved; R and G now disagree, so conservation of momentum loses its status as a physical law.

What we have learned is that momentum does not transform properly under Lorentz maps—at least when we assume that mass is constant. (For example, we assigned X_2 the mass 2 whether it was at rest or moving with velocity $-v$.) In our search for the function $m(v)$, we will now make it our goal to define $m(v)$ so that momentum *does* transform properly under Lorentz maps. In that way we will preserve conservation of momentum as a physical law.

Goal: Preserve conservation of momentum

Relativistic Mass and Momentum

Lorentz maps act on vectors in spacetime, so if we want momentum to transform properly, we shall first have to express it as a spacetime vector. There is a natural way to do this, starting with velocity in the full $(1+3)$-dimensional spacetime.

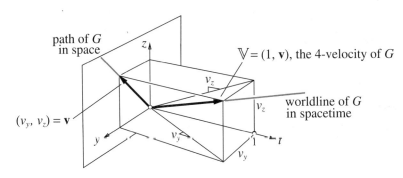

Suppose G's velocity in R's (x, y, z)-coordinates is given by the spatial vector $\mathbf{v} = (v_x, v_y, v_z)$. Then $\mathbb{V} = (1, \mathbf{v})$ is a vector in

4-velocity

spacetime that points in the direction of G's worldline in R. We call $\mathbb{V}$ the **4-velocity of G with respect to** R. If G moves parallel to the z-axis—as we usually assume—then $v_x = v_y = 0$, and the velocity reduces to a scalar $v = v_z$. In this case we write $\mathbb{V} = (1, v)$ (a vector in $\mathbf{R}^2$) but still call $\mathbb{V}$ the 4-velocity of G to emphasize that it is a spacetime vector.

Now suppose that G has mass m according to R, where $m = m(v)$ is the relativistic mass of G when its speed is

$$v = \|\mathbf{v}\| = \sqrt{v_x^2 + v_y^2 + v_z^2}.$$

4-momentum

We define the **4-momentum of G with respect to** R to be the 4-vector $\mathbb{P} = m\mathbb{V} = (m, m\mathbf{v}) = (m, \mathbf{p})$. In a $(1+1)$-dimensional slice of spacetime, the 4-momentum is simply $\mathbb{P} = (m, mv) = (m, p)$.

In G's own frame, the 4-velocity is just $\mathbb{V} = (1, 0)$, and the 4-momentum is $\mathbb{P} = \mu\mathbb{V} = (\mu, 0)$, where μ is G's proper mass $m(0)$. If the 4-momentum we have just defined is to transform properly under all Lorentz transformations, then it must at least transform properly under $B_v : G \to R$. That is, we must have $B_v(\mathbb{P}_G) = \mathbb{P}_R$, where subscripts identify the two 4-momentum vectors:

$$B_v(\mathbb{P}_G) = \frac{1}{\sqrt{1-v^2}} \begin{pmatrix} 1 & v \\ v & 1 \end{pmatrix} \begin{pmatrix} \mu \\ 0 \end{pmatrix} = \frac{\mu}{\sqrt{1-v^2}} \begin{pmatrix} 1 \\ v \end{pmatrix} = m \begin{pmatrix} 1 \\ v \end{pmatrix} = \mathbb{P}_R.$$

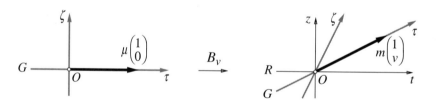

$m(v) = \dfrac{\mu}{\sqrt{1-v^2}}$

Notice that the last equation gives us a formula for **relativistic mass**:

$$m = m(v) = \frac{\mu}{\sqrt{1-v^2}}.$$

Henceforward we use this as the mass function because it has the features we need: $m(0) = \mu$ (the rest mass), $m(v) \to \infty$ as $v \to \pm 1$, and its graph has the right symmetry.

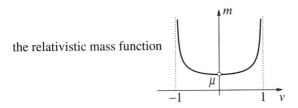
the relativistic mass function

We must now check whether the 4-momentum that incorporates our new mass function transforms properly for *all* pairs of frames. Since we are already using G as the rest frame of the moving body of rest mass μ, we need a new frame in addition to R. Call it C (or "Cap," for *Capital* letters); its coordinates are (T, X, Y, Z). If the velocity of G is v in R's frame and V in C's frame, then the 4-momentum vectors are

$$\mathbb{P}_R = \frac{\mu}{\sqrt{1-v^2}}\begin{pmatrix}1\\v\end{pmatrix} \quad \text{and} \quad \mathbb{P}_C = \frac{\mu}{\sqrt{1-V^2}}\begin{pmatrix}1\\V\end{pmatrix}.$$

The following proposition tells us that 4-momentum does indeed transform properly for all pairs of coordinate frames.

Check *all* pairs of frames

Proposition 3.1 *If $B_w : R \to C$, where w is the velocity of R relative to C, then $B_w(\mathbb{P}_R) = \mathbb{P}_C$.*

PROOF: Here are two proofs. The first links $\mathbb{P}_R$ to $\mathbb{P}_C$ by going through $\mathbb{P}_G$; the second is a brute-force calculation.

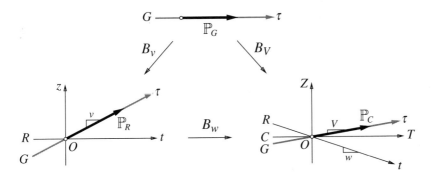

PROOF 1: By definition, each 4-momentum vector is the image of G's own 4-momentum $\mathbb{P}_G = \mu(1, 0)$ under the appropriate Lorentz map:

$$\mathbb{P}_R = B_v(\mathbb{P}_G), \qquad \mathbb{P}_C = B_V(\mathbb{P}_G).$$

In fact, G, R, and C are connected by the commutative diagram shown above, so $B_w B_v = B_V$ and $B_w(\mathbb{P}_R) = B_w(B_v(\mathbb{P}_G)) = B_w B_v(\mathbb{P}_G) = B_V(\mathbb{P}_G) = \mathbb{P}_C$.

PROOF 2: We must evaluate

$$B_w(\mathbb{P}_R) = \frac{1}{\sqrt{1-w^2}} \frac{\mu}{\sqrt{1-v^2}} \begin{pmatrix} 1 & w \\ w & 1 \end{pmatrix} \begin{pmatrix} 1 \\ v \end{pmatrix}$$

$$= \frac{\mu}{\sqrt{1-w^2}\sqrt{1-v^2}} \begin{pmatrix} 1+vw \\ w+v \end{pmatrix} = \frac{\mu(1+vw)}{\sqrt{1-w^2}\sqrt{1-v^2}} \begin{pmatrix} 1 \\ \dfrac{w+v}{1+vw} \end{pmatrix}$$

and show that this vector is equal to $\mathbb{P}_C$. Consider the second component of the vector: Since $B_V = B_w B_v$, the addition formula for velocity boosts gives us

$$V = \frac{w+v}{1+vw}.$$

Now consider the coefficient of the vector; first write it as

$$\frac{1+wv}{\sqrt{1-w^2}\sqrt{1-v^2}} = \frac{1}{\sqrt{\dfrac{(1-w^2)(1-v^2)}{(1+wv)^2}}}.$$

Then

$$\frac{(1-w^2)(1-v^2)}{(1+wv)^2} = \frac{1-w^2-v^2+w^2v^2}{(1+wv)^2}$$

$$= \frac{1+2wv+w^2v^2-w^2-2wv-v^2}{(1+wv)^2}$$

$$= \frac{(1+wv)^2-(w+v)^2}{(1+wv)^2}$$

$$= 1 - \frac{(w+v)^2}{(1+wv)^2} = 1 - V^2.$$

It follows that

$$B_w(\mathbb{P}_R) = \frac{\mu(1+vw)}{\sqrt{1-w^2}\sqrt{1-v^2}} \begin{pmatrix} 1 \\ \frac{w+v}{1+vw} \end{pmatrix} = \frac{\mu}{\sqrt{1-V^2}} \begin{pmatrix} 1 \\ V \end{pmatrix} = \mathbb{P}_C.$$

END OF PROOF

Covariance

In Chapter 1 we introduced the viewpoint that coordinates like (t, x, y, z) and (T, X, Y, Z) are the "names" that different observers give to events, and that the maps

$$B_w : R \to C \quad \text{and} \quad B_w^{-1} : C \to R$$

are pairs of "dictionaries" that allow us to translate one "name" to the other. Thus, as soon we know R's name for an event E, we can find C's name for the same event by using the dictionary B_w.

B_w is a dictionary to translate coordinates

Proposition 3.1 is really about names and dictionaries, too. What is says is that if we know R's "name" for the 4-momentum of a moving body G, then we can use it to find C's name; moreover, the same "translation dictionary" B_w that we used for coordinates does the job here, too. In other words, the coordinates of a 4-momentum vector vary from one observer to another in exactly the same way that the coordinates of an event do; we say that 4-momentum is **covariant**.

4-momentum is covariant

Since $\mathbb{P} = m(v)(1, \mathbf{v})$, it follows that the components $m(v)$ and $\mathbf{v}$ are covariant; since $\mathbb{V} = m(v)^{-1}\mathbb{P}$, 4-velocity must also be covariant. You can check that time, length, and speed are likewise covariant. That is, if we know the value of any of these in R's frame, we can deduce the value in C's frame by using the map $B_w : R \to C$ that transforms coordinates. It is impossible to formulate meaningful physical laws without using covariant quantities. This is called the **principle of covariance**. Because we are working now only with inertial frames, this is called *special* covariance. Later on, we will look at Einstein's general theory of relativity, in which *all* coordinate frames are put on an equal footing.

Principle of covariance

In Chapter 1 we saw that part of the crisis that led to Einstein's theory of special relativity was the observation that Maxwell's equations fail to be covariant under Galilean shears but *are* covariant under Lorentz transformations. Since the validity of Maxwell's laws was not in doubt, the principle of special covariance meant that Galilean shears could not be the correct maps for connecting the spacetime frames of different observers.

Conservation of 4-Momentum

Example 1: an elastic collision

Now that we know that 4-momentum is covariant, it is easy to confirm that conservation of momentum regains its status as a physical law. Take two bodies G_1 and G_2 whose rest masses are μ_1 and μ_2, respectively, and have them collide elastically at the event O. (In an *elastic* collision the bodies bounce off each other with no loss of energy.) We shall assume that R says that total 4-momentum is conserved and test to see whether C agrees.

Consider first what happens in R's frame. Suppose the 4-momenta of G_1 and G_2 before the collision are $\mathbb{P}_1$ and $\mathbb{P}_2$, respectively, while afterwards they are $\overline{\mathbb{P}}_1$ and $\overline{\mathbb{P}}_2$. According to R, total momentum is conserved through the collision: $\mathbb{P}_1 + \mathbb{P}_2 = \overline{\mathbb{P}}_1 + \overline{\mathbb{P}}_2$. Does C agree?

Suppose C is connected to R by the Lorentz map $B_w : R \to C$. Then, since 4-momentum is covariant, each 4-momentum vector in C is the image of the corresponding one in R under the map B_w.

	BEFORE	AFTER
G_1:	$\mathbb{Q}_1 = B_w(\mathbb{P}_1)$	$\overline{\mathbb{Q}}_1 = B_w(\overline{\mathbb{P}}_1)$
G_2:	$\mathbb{Q}_2 = B_w(\mathbb{P}_2)$	$\overline{\mathbb{Q}}_2 = B_w(\overline{\mathbb{P}}_2)$

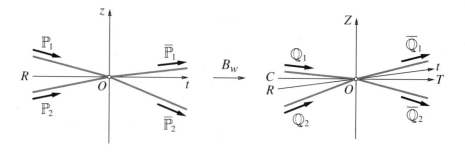

Proposition 3.2 $\mathbb{Q}_1 + \mathbb{Q}_2 = \overline{\mathbb{Q}}_1 + \overline{\mathbb{Q}}_2$.

PROOF: This is a straightforward consequence of the fact that B_w is linear and that $\mathbb{P}_1 + \mathbb{P}_2 = \overline{\mathbb{P}}_1 + \overline{\mathbb{P}}_2$:

$$\mathbb{Q}_1 + \mathbb{Q}_2 = B_w(\mathbb{P}_1) + B_w(\mathbb{P}_2) = B_w(\mathbb{P}_1 + \mathbb{P}_2) = B_w(\overline{\mathbb{P}}_1 + \overline{\mathbb{P}}_2)$$
$$= B_w(\overline{\mathbb{P}}_1) + B_w(\overline{\mathbb{P}}_2) = \overline{\mathbb{Q}}_1 + \overline{\mathbb{Q}}_2. \qquad \text{END OF PROOF}$$

Thus, as soon as one Galilean observer determines that 4-momentum is conserved in a particular collision, all other observers agree. We have therefore reestablished Newton's third law in special relativity as the *conservation of 4-momentum*.

Here is another example; it is a purely *inelastic* collision involving two identical objects traveling with equal but opposite velocities. The objects are two 10-ton trucks traveling toward each other at about 60 miles per hour. After the collision they stick together; by symmetry, they are also at rest.

Example 2: an inelastic collision

$$v_1 = +60 \text{ mph} \qquad\qquad v_2 = -60 \text{ mph}$$
$$G_1 \longrightarrow \qquad\qquad \longleftarrow G_2$$
$$\mu_1 = 10 \text{ tons} \qquad\qquad \mu_2 = 10 \text{ tons}$$

The rest masses are $\mu_1 = \mu_2 = 10$ tons $\approx 10^4$ kg $= 10^7$ gm. The speed is

$$60 \text{ mph} \approx 88 \text{ feet/sec} \approx 30 \text{ m/sec} \approx 10^{-7} \times \underbrace{3 \times 10^8 \text{ m/sec}}_{\text{speed of light}},$$

so $v_1 = -v_2 = 10^{-7}$ in geometrized units. By Taylor's theorem we have

$$\frac{1}{\sqrt{1-v_1^2}} = \frac{1}{\sqrt{1-v_2^2}} \approx 1 + \tfrac{1}{2}v_1^2 = 1 + \tfrac{1}{2}10^{-14}.$$

Before the collision the 4-momenta of the two trucks are

$$\mathbb{P}_1 = \left(\frac{\mu_1}{\sqrt{1-v_1^2}}, \frac{\mu_1 v_1}{\sqrt{1-v_1^2}} \right) \approx \left(10^7 + \tfrac{1}{2}10^{-7}, 1 \right) \text{ gm},$$

$$\mathbb{P}_2 = \left(\frac{\mu_2}{\sqrt{1-v_2^2}}, \frac{\mu_1 v_2}{\sqrt{1-v_2^2}} \right) \approx \left(10^7 + \tfrac{1}{2}10^{-7}, -1 \right) \text{ gm.}$$

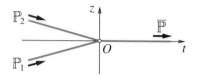

If $\overline{\mathbb{P}}$ is the 4-momentum of the two trucks stuck together after the collision, then, by the conservation of 4-momentum,

$$\overline{\mathbb{P}} = \mathbb{P}_1 + \mathbb{P}_2 \approx \left(2 \times 10^7 + 10^{-7}, 0 \right) \text{ gm.}$$

The second component tells us that the velocity after the collision is zero, so the first component, which is the mass, must be the *rest mass* of the two trucks together. But

combined rest mass = sum of individual rest masses + 10^{-7} gm.

The extra 10^{-7} grams means that rest mass is *created* in the collision, not conserved. Where has that extra mass come from?

Energy is converted to mass

The answer is that the kinetic energy of the moving trucks was converted to mass when the trucks came to rest. To see why this is so, consider the relativistic mass of one of the moving trucks. By Taylor's theorem we can write it in the form

$$\frac{\mu}{\sqrt{1-v^2}} = \mu \left(1 + \tfrac{1}{2}v^2 + O(v^4) \right) \approx \mu + \tfrac{1}{2}\mu v^2 = \underbrace{10^7}_{\text{rest mass}} + \underbrace{\tfrac{1}{2} \times 10^7 \times 10^{-14}}_{\text{kinetic energy}}.$$

Thus, the main contributions to relativistic mass when v is small are rest mass and the classical kinetic energy based on rest mass. The extra mass that shows up in the collided trucks is the sum of the two equal kinetic energies

$$\tfrac{1}{2}\mu v^2 = \tfrac{1}{2} \times 10^{-7} \text{ gm};$$

this energy has been converted directly into mass. Nothing like this is contemplated in Newtonian physics; it is a consequence of the fact that relativistic mass increases with velocity.

Now imagine what happens when we reverse the direction of time in the last example. At the start there is a single object at rest. Then, at the event O, the object bursts into two equal fragments that speed away from each other. The rest masses of the fragments will add up to less than the original mass; the difference is the kinetic energy of the fragments. In this process, called **fission**, some mass is converted directly into energy. If the velocity of the fragments is large—as it is, for example, in the fission of uranium or plutonium nuclei—then the amount of energy released can be substantial. Because of the interconvertibility of matter and energy, we sometimes use the single term **matter–energy** to refer to either.

Fission: mass is converted into energy

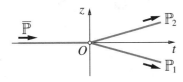

Conventional Units

The geometric units that we have been using can mask important information. For example, the famous equation $E = mc^2$ collapses to the rather cryptic $E = m$ in geometric units. But we can still recover the conventional equation by noticing that $E = m$ is out of balance dimensionally: Energy has the dimensions of mass × velocity². By attaching a factor of c^2 to the mass term on the right we restore the balance and get the correct equation in conventional units.

Converting equations to conventional units

This method will always convert an equation properly, but it does not solve all our problems. For example, when we switch to conventional units, the last three components of 4-velocity $\mathbb{V} = (1, \mathbf{v})$ acquire the appropriate dimensions of meters per second, but the first remains dimensionless. This confuses the physics and compounds the mathematical difficulties we face, especially when we take up general relativity in the later chapters.

But here, too, we have a simple remedy: Just give an event $E = (t, x, y, z)$ new coordinates (x_0, x_1, x_2, x_3) that have the same dimensions. By using the method of attaching an appropriate

Dimensionally homogeneous coordinates

factor of c, we can do this two different ways. One is to divide the three spatial coordinates by c; then all components of the 4-vector of position will be measured in seconds. A better way, though, is to multiply the time coordinate by c; then 4-velocity and 4-momentum have the right physical units:

$$\mathbb{X} = (x_0, x_1, x_2, x_3) = (ct, x, y, z) = (ct, \mathbf{x}) \quad \text{meters},$$

$$\mathbb{V} = \frac{\Delta \mathbb{X}}{\Delta t} = \left(\frac{\Delta x_0}{\Delta t}, \frac{\Delta x_1}{\Delta t}, \frac{\Delta x_2}{\Delta t}, \frac{\Delta x_3}{\Delta t}\right) = (c, \mathbf{v}) \quad \text{m/sec},$$

$$\mathbb{P} = m\mathbb{V} = (mc, m\mathbf{v}) = (mc, \mathbf{p}) \quad \text{kg-m/sec}.$$

We shall call (ct, x, y, z) the **dimensionally homogeneous coordinates** for the frame R and (t, x, y, z) its **traditional coordinates**.

Dimensional homogeneity has some inevitable consequences. When we use dimensionally homogeneous coordinates in R, R's own 4-velocity is $\mathbb{V} = (c, 0, 0, 0)$ rather than $(1, 0, 0, 0)$. Furthermore, the first component of the 4-momentum of an observer G is no longer simply the relativistic mass of G but is the mass multiplied by c.

The Minkowski norm

The Minkowski norm in dimensionally homogeneous coordinates is determined, as it was in traditional coordinates, by the equation of the light cone,

$$x_0^2 - x_1^2 - x_2^2 - x_3^2 = c^2 t^2 - x^2 - y^2 - z^2 = 0.$$

Hence $Q(\mathbb{X}) = x_0^2 - x_1^2 - x_2^2 - x_3^2$ (and $Q(\mathbb{X})$ has the dimensions of meters^2), so the Minkowski inner product matrix is the familiar

$$J_{1,3} = \begin{pmatrix} 1 & 0 & 0 & 0 \\ 0 & -1 & 0 & 0 \\ 0 & 0 & -1 & 0 \\ 0 & 0 & 0 & -1 \end{pmatrix},$$

and its entries are all dimensionless. With this inner product the 4-speed also has the right units:

$$\|\mathbb{V}\| = \sqrt{\mathbb{V} \cdot \mathbb{V}} = \sqrt{c^2 - v^2} = c\sqrt{1 - (v/c)^2} \quad \text{m/sec}.$$

Hyperbolic rotations and velocity boosts

Hyperbolic rotations are those linear transformations M that leave the Minkowski inner product invariant: $M^t J_{1,3} M = J_{1,3}$.

Since $J_{1,3}$ is unchanged when we switch from traditional to dimensionally homogeneous coordinates, hyperbolic rotation by u hyperbolic radians in a $(1+1)$-dimensional spacetime (where we use $J_{1,1}$) is likewise unchanged:

$$H_u = \begin{pmatrix} \cosh u & \sinh u \\ \sinh u & \cosh u \end{pmatrix}.$$

Suppose $H_u : G \to R$, where G and R have dimensionally homogeneous coordinates. Let us derive the connection between the hyperbolic angle u and G's velocity v relative to R. That velocity is

$$v = \frac{\Delta z}{\Delta t} = \frac{\Delta x_3}{\Delta x_0} \frac{\Delta x_0}{\Delta t} = c \frac{\Delta x_3}{\Delta x_0}.$$

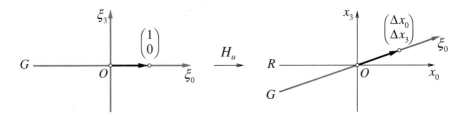

The fraction $\Delta x_3 / \Delta x_0$ is the slope of G's worldline in R. That worldline contains the event

$$\begin{pmatrix} \Delta x_0 \\ \Delta x_3 \end{pmatrix} = \begin{pmatrix} \cosh u & \sinh u \\ \sinh u & \cosh u \end{pmatrix} \begin{pmatrix} 1 \\ 0 \end{pmatrix} = \begin{pmatrix} \cosh u \\ \sinh u \end{pmatrix},$$

so we can take $\Delta x_3 / \Delta x_0 = \sinh u / \cosh u = \tanh u$. Hence $v = c \tanh u$. Incidentally, since $v = \tanh u$ in geometric units and $\tanh u$ is dimensionless, our method for converting expressions already dictates that we attach a factor of c to $\tanh u$ to get v in conventional units. From here it is possible to see that the velocity boost corresponding to H_u is

$$B_v = \frac{1}{\sqrt{1 - (v/c)^2}} \begin{pmatrix} 1 & v/c \\ v/c & 1 \end{pmatrix}.$$

The time dilation factor $\Delta t / \Delta \tau$ that relates the proper times of R and G emerges when we apply the velocity boost B_v to the

Time dilation

event $(\xi_0, \xi_3) = (c\Delta\tau, 0)$:

$$\begin{pmatrix} x_0 \\ * \end{pmatrix} = \begin{pmatrix} c\Delta t \\ * \end{pmatrix} = \frac{1}{\sqrt{1-(v/c)^2}} \begin{pmatrix} 1 & v/c \\ v/c & 1 \end{pmatrix} \cdot \begin{pmatrix} c\Delta\tau \\ 0 \end{pmatrix} = \frac{c\Delta\tau}{\sqrt{1-(v/c)^2}} \begin{pmatrix} 1 \\ * \end{pmatrix},$$

$$\text{time dilation factor}: \quad \frac{\Delta t}{\Delta\tau} = \frac{1}{\sqrt{1-(v/c)^2}}.$$

Proper 4-velocity

Because of the connection between t and τ provided by time dilation, we can regard G's position vector $\mathbb{X}$ in R as a function of G's proper time τ. This then gives us a second covariant way to describe G's velocity, as the rate of change of $\mathbb{X}$ with respect to τ:

$$\mathbb{U} = \frac{\Delta \mathbb{X}}{\Delta \tau} = \frac{\Delta \mathbb{X}}{\Delta t} \frac{\Delta t}{\Delta \tau} = \frac{1}{\sqrt{1-(v/c)^2}} \mathbb{V} = \frac{1}{\sqrt{1-(v/c)^2}} (c, \mathbf{v}).$$

We call this the **proper 4-velocity of G in R**. Note that the norm of proper 4-velocity is always c:

$$\|\mathbb{U}\| = \frac{\|\mathbb{V}\|}{\sqrt{1-(v/c)^2}} = \frac{\sqrt{c^2 - v^2}}{\sqrt{1-(v/c)^2}} = \frac{c\sqrt{1-(v/c)^2}}{\sqrt{1-(v/c)^2}} = c.$$

In traditional coordinates $\mathbb{U}$ is therefore a *unit* 4-vector.

If μ is G's proper mass, then its relativistic mass is $m = \mu/\sqrt{1-(v/c)^2}$ in conventional units. This leads to a particularly simple and elegant formula for G's 4-momentum:

$$\mathbb{P} = m\mathbb{V} = \frac{\mu}{\sqrt{1-(v/c)^2}} \mathbb{V} = \mu \mathbb{U}.$$

By expressing $\mathbb{P}$ in terms of two manifestly covariant objects — G's proper mass and proper 4-velocity — we see once again that 4-momentum is covariant.

Summary data

Here is a summary of the data we have about an observer G whose proper mass is μ and proper time is τ when viewed in R's $(1+3)$-dimensional spacetime with dimensionally homogeneous coordinates:

$$\text{4-position}: \quad \mathbb{X} = (ct, x, y, z) = (ct, \mathbf{x}),$$

$$\text{4-velocity}: \quad \mathbb{V} = \frac{\Delta \mathbb{X}}{\Delta t} = \left(c, \frac{\Delta x}{\Delta t}, \frac{\Delta y}{\Delta t}, \frac{\Delta z}{\Delta t}\right) = (c, \mathbf{v}),$$

§3.1 Newton's Laws of Motion

$$\text{4-speed}: \quad \|\mathbb{V}\| = \sqrt{c^2 - v^2} = c\sqrt{1 - (v/c)^2},$$

$$\text{time dilation}: \quad \frac{\Delta t}{\Delta \tau} = \frac{1}{\sqrt{1 - (v/c)^2}},$$

$$\text{proper 4-velocity}: \quad \mathbb{U} = \frac{\Delta \mathbb{X}}{\Delta \tau} = \frac{\Delta t}{\Delta \tau}\frac{\Delta \mathbb{X}}{\Delta t} = \frac{1}{\sqrt{1 - (v/c)^2}}\mathbb{V},$$

$$\text{4-momentum}: \quad \mathbb{P} = m\mathbb{V} = \frac{\mu}{\sqrt{1 - (v/c)^2}}\mathbb{V} = \mu\mathbb{U}.$$

Sometimes, the change from geometric to conventional units does more than alter the formulas; it can help clarify the physics. For example, in our discussion of the colliding trucks, we looked at the relativistic mass of a truck and noted that part of the mass was kinetic energy. However, since energy has dimensions mass × velocity^2, mass and energy are dimensionally unequal in conventional units (though they are equal in geometric). It is therefore incorrect to say "part of the mass was kinetic energy." But suppose we convert the original expression for mass into an energy in the usual way, by multiplying by the velocity-squared factor c^2:

Mass-to-energy conversion in conventional units

$$E = \frac{\mu c^2}{\sqrt{1-(v/c)^2}} = \mu c^2 \left(1 + \frac{1}{2}\frac{v^2}{c^2} + \frac{3}{8}\frac{v^4}{c^4} + \cdots\right)$$

$$= \underbrace{\mu c^2}_{\text{rest energy}} + \underbrace{\frac{1}{2}\mu v^2 + \frac{3}{8}\frac{v^4}{c^2} + \cdots}_{\text{kinetic energy}}.$$

Now all the terms are energies, so the equation makes sense dimensionally. Since μ is the rest mass, we call μc^2 the **rest energy**. Since the second term is the *classical* kinetic energy, and since all terms from the second onward are "kinetic" (because they involve v and thus concern motion), we refer to them collectively as the **relativistic kinetic energy**. Thus, in conventional terms we say that the combined rest energy of the two trucks is more than the sum of their individual rest energies; the excess is the kinetic energy they had when they were moving. If we call the entire expression E the **total energy**, then we see that total energy is conserved in the collision.

Rest energy and kinetic energy

The energy–momentum vector

Since relativistic mass is $m = \mu/\sqrt{1 - (v/c)^2}$, total energy is $E = mc^2$ in conventional units. Therefore, we can write 4-momentum in dimensionally homogeneous coordinates as

$$\mathbb{P} = (mc, \mathbf{p}) = \left(\frac{E}{c}, \mathbf{p}\right).$$

For this reason $\mathbb{P}$ is often called the **energy–momentum vector**.

Exercises

1. Suppose $\overline{p} = \dfrac{v}{1 - \frac{2}{3}v^2}$; show that $\overline{p} = v(1 + \frac{2}{3}v^2 + \cdots) \approx v$ when v is small.

2. (a) Show that if $v = c \tanh u$, then $\cosh u = 1/\sqrt{1 - (v/c)^2}$.
   (b) Determine $\sinh u$ in terms of v and show that the velocity boost B_v has the following form in dimensionally homogeneous coordinates:
   $$B_v = \frac{1}{\sqrt{1 - (v/c)^2}} \begin{pmatrix} 1 & v/c \\ v/c & 1 \end{pmatrix}.$$

3. In geometric units, successive boosts by the velocities v and w is a single boost of velocity $V = (v + w)/(1 + vw)$. Find the formula for V in conventional units.

4. Let dimensionally homogeneous coordinates for the event $E = (t, x, y, z)$ be defined using the alternative approach where
   $$\mathbb{X} = (t, x/c, y/c, z/c) \text{ seconds.}$$
   Determine the 4-velocity $\mathbb{V} = \Delta \mathbb{X}/\Delta t$, the Minkowski norm, the 4-speed, the proper 4-velocity, and the 4-momentum. How do these compare to the same quantities constructed with the dimensionally homogeneous coordinates $\mathbb{X} = (ct, x, y, z)$?

It is a relatively common practice to continue to use traditional coordinates (t, x, y, z) after converting from geometric to conventional units. This means that the components of 4-vectors have different dimensions; for example,

$$\mathbb{X} = (t \text{ sec}, x \text{ m}, y \text{ m}, z \text{ m}).$$

§3.1 Newton's Laws of Motion

In the remaining exercises use conventional units with traditional coordinates.

5. (a) Following the rules for converting any expression from geometric to conventional units, obtain the following formula for hyperbolic rotation $H_u : G \to R$ in a $(1+1)$-dimensional spacetime:

$$H_u = \begin{pmatrix} \cosh u & \frac{1}{c}\sinh u \\ c \sinh u & \cosh u \end{pmatrix}, \quad \begin{pmatrix} t \\ z \end{pmatrix} = \begin{pmatrix} \cosh u & \frac{1}{c}\sinh u \\ c \sinh u & \cosh u \end{pmatrix} \begin{pmatrix} \tau \\ \zeta \end{pmatrix}.$$

   (b) Determine the relation between u and G's velocity v relative to R.

   (c) Obtain the formula for the velocity boost B_v that corresponds to H_u.

6. (a) Define the Minkowski quadratic form $Q(\mathbb{X})$ so that it has the dimensions seconds^2.

   (b) Write the Minkowski inner product matrix $J_{1,3}$ that corresponds to this choice of $Q(\mathbb{X})$. What are the dimensions of the components of $J_{1,3}$? In particular, are the dimensions all the same or do they depend on the component considered?

   (c) Derive the formula for H_u in Exercise 5 directly from the relation

$$H_u^t J_{1,1} H_u = J_{1,1}, \qquad J_{1,1} = \begin{pmatrix} 1 & 0 \\ 0 & -\frac{1}{c^2} \end{pmatrix}.$$

7. (a) Using the conventional form of $B_v : (\tau, \zeta) \to (t, z)$ from Exercise 5, calculate the velocity boost B_v when $v = 30$ meters per second (and $c = 3 \times 10^8$ m/sec).

   (b) Calculate the Galilean shear $S_v : (\tau, \zeta) \to (t, z)$ in conventional units when $v = 30$ m/sec. Compare the matrices B_v and S_v and show that $B_v \to S_v$ if $c \to \infty$.

3.2 Curves and Curvature

Parametrized Curves and Arc Length

Bodies that move under the influence of forces experience acceleration—that is, changes in velocity. Since velocity is, in the simplest circumstances, the slope of a worldline, this changing slope means that the worldline will be curved. Thus, to deal with forces and accelerations, we must use the differential geometry of curves that is developed in multivariable calculus.

A parametrized curve

Let us first consider curves in the ordinary Euclidean plane. We start with a parametrized path, which is a smooth map of a closed interval to the plane:

$$\mathbf{x} : [a, b] \to \mathbf{R}^2, \qquad \mathbf{x} : q \mapsto \mathbf{x}(q) = (x(q), y(q)).$$

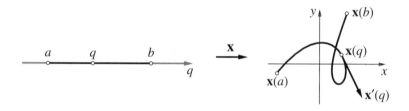

As q moves from a to b, the point $\mathbf{x}(q)$ traces out a path in the plane. We can think of q as providing a "coordinate" that allows us to label points along the path. For this reason q is called a **parameter**; the word comes from *para-* ("beside") and *meter* ("measure").

The tangent vector

The parametrization is *smooth*, which means that we can calculate derivatives of $\mathbf{x}(q)$ of all orders. The first derivative $d\mathbf{x}/dq = \mathbf{x}'(q) = (x'(q), y'(q))$ is a vector tangent to the path at the point $\mathbf{x}(q)$, at least when $\mathbf{x}'(q) \neq \mathbf{0}$. If we think of q as a *time* variable, then $d\mathbf{x}/dq$ gives the rate of change of position with respect to time and is thus a velocity, called the **tangent velocity vector** of the curve.

The tangent vector gives us valuable information. Consider the image under **x** of a small segment $[q_0, q_0 + \Delta q]$ of the interval $[a, b]$. According to Taylor's theorem,

$$\mathbf{x}(q_0 + \Delta q) \approx \mathbf{x}(q_0) + \mathbf{x}'(q_0)\Delta q.$$

If $\mathbf{x}'(q_0) \neq \mathbf{0}$, then the point $\mathbf{x}(q_0 + \Delta q)$ will be very near the continuation of $\mathbf{x}(q_0)$ a distance of $\|\mathbf{x}'(q_0)\|\Delta q$ along the tangent vector. In particular, the map **x** will be one-to-one in a sufficiently small neighborhood of q_0.

By contrast, if $\mathbf{x}'(q_0) = \mathbf{0}$, then **x** may fail to be one-to-one on any neighborhood of q_0. Here is an example to illustrate. Let $\mathbf{x}(q) = (q^2, q^2)$ for $-\varepsilon \leq q \leq \varepsilon$, where ε is any positive number. Notice that $\mathbf{x}'(q) = (2q, 2q)$, so $\mathbf{x}'(0) = \mathbf{0}$. In fact, **x** behaves in a singular way when $q = 0$: The interval $[-\varepsilon, \varepsilon]$ is folded double there, and is then mapped two-to-one onto the straight-line segment in $\mathbf{R}^2$ from $(0, 0)$ to $(\varepsilon^2, \varepsilon^2)$.

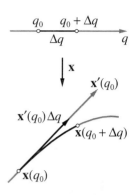

Singular parametrizations

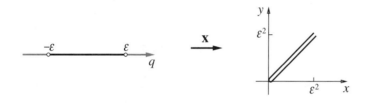

To ensure that our parametrizations $\mathbf{x}(q)$ are locally one-to-one (that is, one-to-one on a sufficiently small neighborhood of each point), we henceforth require that $\|\mathbf{x}'(q)\| > 0$ for all q in $[a, b]$. We say then that **x** is **nonsingular**. This is usually not a severe limitation, because a given path has infinitely many different parametrizations; they differ from one another in the speed with which the parameter moves along the curve. For example, any function $f(q)$ that maps $[a, b]$ onto $[0, 1]$ can be used to parametrize the straight line segment from $(0, 0)$ to $(1, 1)$ in the plane: Just let $\mathbf{x}(q) = (f(q), f(q))$. If $f'(q) \neq 0$ on $[a, b]$, then $\mathbf{x}(q)$ satisfies our new criterion. A more interesting example is the unit circle. Besides the familiar $\mathbf{x}(q) = (\cos(\alpha q), \sin(\alpha q))$, in which the parameter moves around the circle with constant speed $\|\mathbf{x}'\| = \alpha$,

Require **x** to be nonsingular ($\|\mathbf{x}'\| > 0$)

there is

$$\mathbf{X}(q) = \left(\frac{2q}{1+q^2}, \frac{1-q^2}{1+q^2}\right), \quad -\infty < q < \infty.$$

This maps the entire real line in a one-to-one fashion to the circle minus the point $(x, y) = (0, -1)$. The parameter moves clockwise nonuniformly around the circle and has its greatest speed $\|\mathbf{X}'\|$ when $q = 0$.

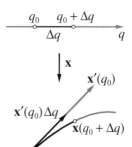

Let us return to the image of the interval $[q_0, q_0 + \Delta q]$ by the map $\mathbf{x}$. Since $\mathbf{x}'(q_0) \neq \mathbf{0}$, the equation

$$\mathbf{x}(q_0 + \Delta q) \approx \mathbf{x}(q_0) + \mathbf{x}'(q_0)\Delta q$$

tells us that the *straight-line* distance $\|\Delta \mathbf{x}\| = \|\mathbf{x}(q_0 + \Delta q) - \mathbf{x}(q_0)\|$ between $\mathbf{x}(q_0)$ and $\mathbf{x}(q_0 + \Delta q)$ is well approximated by the length $\|\mathbf{x}'(q_0)\|\Delta q$ of the vector $\mathbf{x}'(q_0)\Delta q$. Furthermore, if Δq is sufficiently small, this length is a good approximation to the length of the curved arc from $\mathbf{x}(q_0)$ to $\mathbf{x}(q_0 + \Delta q)$ along the curve itself.

$\|\mathbf{x}'\|$ is a local stretch factor

Since the original segment $[q_0, q_0+\Delta q]$ on the q-axis has length Δq, and since its image has approximate length $\|\mathbf{x}'(q_0)\|\Delta q$ on the curve, we can say that the segment is *stretched* by the approximate factor $\|\mathbf{x}'(q_0)\|$ as it is mapped to the curve. (As we did with eigenvalues, we understand "stretch" to be a compression when $\|\mathbf{x}'(q_0)\| < 1$.) Short segments near a different point q_1 are stretched by $\|\mathbf{x}'(q_1)\|$, so the "stretch factor" varies from point to point. We say that $\|\mathbf{x}'(q)\|$ is the *local* stretch factor for the map $\mathbf{x}$ near the point q.

Arc length of an entire curve

To measure the length of the entire curve C, we partition the interval $[a, b]$,

$$a = q_1 < q_2 < \cdots < q_n < q_{n+1} = b,$$

in such a way that each segment $\Delta q_j = q_{j+1} - q_j$ is small enough for

$$\|\mathbf{x}(q_{j+1}) - \mathbf{x}(q_j)\| \approx \|\mathbf{x}'(q_j)\|\Delta q_j$$

to be a good approximation. Then the length of the entire curve is approximately

$$\sum_{j=1}^{n} \|\mathbf{x}'(q_j)\|\Delta q_j.$$

§3.2 Curves and Curvature

This is a Riemann sum; as the partition $a = q_1 < \cdots < q_{n+1} = b$ becomes finer and finer while $n \to \infty$, the sum approaches the integral

$$\int_a^b \|\mathbf{x}'(q)\| \, dq.$$

We summarize these findings in the following definition.

Definition 3.1 *Let* $\mathbf{x} : [a, b] \to \mathbf{R}^2$ *be a parametrization of a curve C. Then the **arc length** of C is*

$$L_\mathbf{x} = \int_a^b \|\mathbf{x}'(q)\| \, dq.$$

Arc length depends on the parametrization chosen, and it is not obvious that the arc lengths computed from two different parametrizations of the same path must agree. In fact, there is a pitfall here. Consider these two parametrizations of the unit circle:

$$\mathbf{x}(q) = (\cos(q), \sin(q)), \qquad \mathbf{X}(Q) = (\cos(2Q), \sin(2Q)),$$

both defined on $[0, 2\pi]$. Then $L_\mathbf{x} = 2\pi$ but $L_\mathbf{X} = 4\pi$. Although both maps are *locally* one-to-one, $\mathbf{X}$ is *globally* two-to-one, so $L_\mathbf{X}$ is twice the true circumference because $\mathbf{X}$ goes twice around the circle. If we avoid this pitfall, then arc length *is* independent of the parametrization, as the following theorem asserts.

Theorem 3.1 *Suppose* $\mathbf{x}(q)$, $a \leq q \leq b$, *and* $\mathbf{X}(Q)$, $A \leq Q \leq B$, *are one-to-one parametrizations of the same curve C. Then the two parametrizations give the same value for the length of C:* $L_\mathbf{x} = L_\mathbf{X}$.

PROOF: To be definite, let us suppose $\mathbf{x}(a) = \mathbf{X}(A)$ (rather than $\mathbf{x}(a) = \mathbf{X}(B)$, which would mean that the parameters traverse C in opposite directions). The crucial step in the proof is to construct the smooth function $Q = \varphi(q)$ that makes the diagram on the following page commutative: $\mathbf{x}(q) = \mathbf{X}(\varphi(q))$. In fact, since $\mathbf{X}$ is one-to-one, it has an inverse $\mathbf{X}^{-1}$ defined on the curve, so we can take $\varphi(q) = \mathbf{X}^{-1}(\mathbf{x}(q))$, or $\mathbf{x}(q) = \mathbf{X}(\varphi(q))$. If φ is smooth, then $\mathbf{x}'(q) = \mathbf{X}'(\varphi(q))\varphi'(q)$ by the chain rule, so

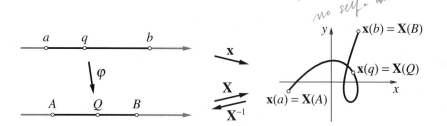

$$\|\mathbf{x}'(q)\|\, dq = \|\mathbf{X}'(\varphi(q))\|\, |\varphi'(q)|\, dq = \|\mathbf{X}'(\varphi(q))\|\, \varphi'(q)\, dq$$
$$= \|\mathbf{X}'(Q)\|\, dQ.$$

The second equality in the sequence holds because φ is an increasing function of q, so $\varphi'(q) \geq 0$ and $|\varphi'(q)| = \varphi'(q)$. Finally, since $\varphi(a) = A$, $\varphi(b) = B$, the formula for the change of variable in an integral gives

$$L_{\mathbf{x}} = \int_a^b \|\mathbf{x}'(q)\|\, dq = \int_A^B \|\mathbf{X}'(Q)\|\, dQ = L_{\mathbf{X}},$$

so the two parametrizations produce the same value for the length of the curve.

The key to finishing the proof is to show that the inverse $\mathbf{X}^{-1}$ is differentiable on C, for then $\varphi(q) = \mathbf{X}^{-1}(\mathbf{x}(q))$ will be differentiable, by the chain rule. The arguments are technical and not essential for the material that follows, so you can skip them without loss of understanding.

Suppose, for the moment, that we already *knew* that $\mathbf{X}^{-1}$ was differentiable and we wanted to know what form the derivative had. If we write $\mathbf{X}(Q) = (X(Q), Y(Q))$, then the fact that $\mathbf{X}^{-1}$ is the inverse of $\mathbf{X}$ implies

$$Q = \mathbf{X}^{-1}(\mathbf{X}(Q)) = \mathbf{X}^{-1}(X(Q), Y(Q)).$$

In particular, $\mathbf{X}^{-1}$ is defined on a subset of $\mathbf{R}^2$, so its derivative is given by the gradient vector

$$\nabla \mathbf{X}^{-1}(X, Y) = \left(\frac{\partial \mathbf{X}^{-1}}{\partial X}, \frac{\partial \mathbf{X}^{-1}}{\partial Y}\right).$$

Furthermore, since $\mathbf{X}^{-1}$ is defined only along the curve, it is reasonable to conjecture that the gradient $\nabla \mathbf{X}^{-1}$ points in the *direction* of the curve. The tangent vector $\mathbf{X}'$ specifies this direction,

§3.2 Curves and Curvature

so $\nabla \mathbf{X}^{-1}$ must be a multiple of $\mathbf{X}'$:

$$\nabla \mathbf{X}^{-1}(\mathbf{X}(Q)) = m(Q)\mathbf{X}'(Q),$$

for some function $m(Q)$. If we now differentiate the equation $Q = \mathbf{X}^{-1}(\mathbf{X}(Q))$ with respect to Q using the multivariable chain rule, we get

$$1 = \frac{dQ}{dQ} = \frac{\partial \mathbf{X}^{-1}}{\partial X}\frac{dX}{dQ} + \frac{\partial \mathbf{X}^{-1}}{\partial Y}\frac{dY}{dQ} = \nabla \mathbf{X}^{-1}(\mathbf{X}(Q)) \cdot \mathbf{X}'(Q).$$

But since $\nabla \mathbf{X}^{-1}(\mathbf{X}(Q)) = m(Q)\mathbf{X}'(Q)$, we now have

$$1 = m(Q)\,\mathbf{X}'(Q) \cdot \mathbf{X}'(Q) = m(Q)\|\mathbf{X}'(Q)\|^2.$$

Since $\|\mathbf{X}'(Q)\| > 0$ for all Q, we can solve for $m(Q)$: $m(Q) = 1/\|\mathbf{X}'(Q)\|^2$. Thus, *if* $\mathbf{X}^{-1}$ is differentiable, its derivative ought to be

$$\nabla \mathbf{X}^{-1}(\mathbf{X}(Q)) = \frac{1}{\|\mathbf{X}'(Q)\|^2}\mathbf{X}'(Q).$$

To prove that $\mathbf{X}^{-1}$ is indeed differentiable, let

$$r(\Delta P) = \mathbf{X}^{-1}(P + \Delta P) - \mathbf{X}^{-1}(P) - \nabla \mathbf{X}^{-1}(P) \cdot \Delta P,$$

where the gradient $\nabla \mathbf{X}^{-1}$ has the form we just deduced. Then, by definition, $\mathbf{X}^{-1}$ is differentiable at P, and its derivative is $\nabla \mathbf{X}^{-1}$, if

$$\frac{r(\Delta P)}{\|\Delta P\|} \to 0 \quad \text{as } \Delta P \to \mathbf{0}.$$

To calculate $r(\Delta P)$ we make use of the fact the $\mathbf{X}$ is a one-to-one map onto the curve; thus there are unique values Q and $Q + \Delta Q$ for which

$$P = \mathbf{X}(Q) \quad \text{and} \quad P + \Delta P = \mathbf{X}(Q + \Delta Q),$$

and $\Delta P \to \mathbf{0}$ if and only if $\Delta Q \to 0$. Thus

$$r(\Delta P) = Q + \Delta Q - Q - \nabla \mathbf{X}^{-1}(\mathbf{X}(Q)) \cdot \Delta P = \Delta Q - \frac{1}{\|\mathbf{X}'(Q)\|^2}\mathbf{X}'(Q) \cdot \Delta P.$$

Since $\mathbf{X}$ is differentiable at Q, Taylor's theorem implies that when we write

$$\Delta P = \mathbf{X}(Q + \Delta Q) - \mathbf{X}(Q) = \mathbf{X}'(Q)\Delta Q + R(\Delta Q),$$

it follows that
$$\frac{R(\Delta Q)}{|\Delta Q|} \to \mathbf{0} \quad \text{as } \Delta Q \to 0.$$
So if we take $\Delta P = \mathbf{X}'(Q)\Delta Q + R(\Delta Q)$, then $r(\Delta P)$ reduces to
$$r(\Delta P) = \Delta Q - \frac{1}{\|\mathbf{X}'(Q)\|^2}\mathbf{X}'(Q) \cdot (\mathbf{X}'(Q)\Delta Q + R(\Delta Q))$$
$$= \Delta Q - \frac{1}{\|\mathbf{X}'(Q)\|^2}(\mathbf{X}'(Q) \cdot \mathbf{X}'(Q)\Delta Q + \mathbf{X}'(Q) \cdot R(\Delta Q))$$
$$= -\frac{\mathbf{X}'(Q) \cdot R(\Delta Q)}{\|\mathbf{X}'(Q)\|^2}.$$
To see what happens when we divide this by $\|\Delta P\|$, first note that
$$\frac{\|\Delta P\|}{|\Delta Q|} = \frac{\|\mathbf{X}'(Q)\Delta Q + R(\Delta Q)\|}{|\Delta Q|} = \left\|\mathbf{X}'(Q)\frac{\Delta Q}{|\Delta Q|} + \frac{R(\Delta)}{|\Delta Q|}\right\|$$
$$= \left\|\pm\mathbf{X}'(Q) + \frac{R(\Delta Q)}{|\Delta Q|}\right\| \to \|\mathbf{X}'(Q)\| \quad \text{as } \Delta Q \to 0.$$
Thus
$$\frac{r(\Delta P)}{\|\Delta P\|} = \frac{r(\Delta P)}{|\Delta Q|}\frac{|\Delta Q|}{\|\Delta P\|} = -\frac{1}{\|\mathbf{X}'(Q)\|^2}\left[\mathbf{X}'(Q) \cdot \frac{R(\Delta Q)}{|\Delta Q|}\right]\frac{|\Delta Q|}{\|\Delta P\|}.$$
Since $\Delta P \to \mathbf{0}$ if and only if $\Delta Q \to 0$,
$$\lim_{\Delta P \to \mathbf{0}}\frac{r(\Delta P)}{\|\Delta P\|} = -\frac{1}{\|\mathbf{X}'(Q)\|^2}\left[\mathbf{X}'(Q) \cdot \left(\lim_{\Delta Q \to 0}\frac{R(\Delta Q)}{|\Delta Q|}\right)\right]\lim_{\Delta Q \to 0}\frac{|\Delta Q|}{\|\Delta P\|}$$
$$= -\frac{\mathbf{X}'(Q) \cdot \mathbf{0}}{\|\mathbf{X}'(Q)\|^2}\frac{1}{\|\mathbf{X}'(Q)\|} = 0.$$
This completes the proof that $\mathbf{X}^{-1}$ is differentiable. We have also proven that the derivative is
$$\nabla \mathbf{X}^{-1}(\mathbf{X}(Q)) = \frac{1}{\|\mathbf{X}'(Q)\|^2}\mathbf{X}'(Q),$$
independently of the conjecture we made earlier about the form that it must take. END OF PROOF

Definition 3.2 *Let* $\mathbf{x} : [a, b] \to \mathbf{R}^2$ *be a parametrization of a curve* C. *Then the **arc-length function** on C is*
$$s(q) = \int_a^q \|\mathbf{x}'(q)\|\, dq.$$

Note that $s(a) = 0$ and $s(b) = L$, the arc length of C. If we write $\mathbf{x}(q) = (x(q), y(q))$, we have another way to express the arc-length function that suggests a more direct connection with the Pythagorean theorem:

$$s(q) = \int_a^q ds = \int_a^q \sqrt{dx^2 + dy^2}.$$

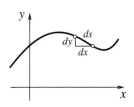

To see how this follows from the definition, first obtain the derivative $s'(q)$. By the fundamental theorem of calculus this is just the integrand of the integral that defines s:

$$s'(q) = \frac{ds}{dq} = \|\mathbf{x}'(q)\| = \left(\left(\frac{dx}{dq}\right)^2 + \left(\frac{dy}{dq}\right)^2\right)^{1/2}.$$

The differential ds is therefore

$$ds = \left(\left(\frac{dx}{dq}\right)^2 + \left(\frac{dy}{dq}\right)^2\right)^{1/2} dq = \sqrt{dx^2 + dy^2},$$

and the integral form follows.

Every curve C has a special parametrization $\mathbf{y}(s)$ that uses arc length s itself as the parameter. What makes $\mathbf{y}$ special is that it is a *unit speed* parametrization: $\|\mathbf{y}'(s)\| \equiv 1$; thus the point $\mathbf{y}(s)$ will be exactly s units from the beginning of the curve.

Arc-length parametrization

We will use the original map $\mathbf{x}$ to construct $\mathbf{y}$; the idea is to alter the speed of the parameter. Since $s'(q) = \|\mathbf{x}'(q)\|$ and since $\|\mathbf{x}'(q)\| > 0$, it follows that $s'(q) > 0$. Thus $s : [a, b] \to [0, L]$ is a smooth monotone increasing function and therefore has a smooth inverse. Let $q = \varphi(s)$ be the inverse of $s = s(q)$. Then

$$\frac{d\varphi}{ds} = \frac{1}{ds/dq} = \frac{1}{\|\mathbf{x}'(q)\|} = \frac{1}{\|\mathbf{x}'(\varphi(s))\|} > 0.$$

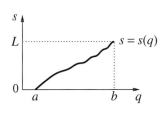
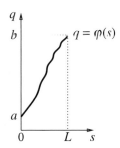

The function $q = \varphi(s)$ gives us the speed-altered parameter that we need. Define $\mathbf{y}(s) = \mathbf{x}(\varphi(s))$ as in the diagram below. Then $\mathbf{y}$ has the same image as $\mathbf{x}$. Moreover, by the chain rule,

$$\mathbf{y}'(s) = \frac{d\mathbf{y}}{ds} = \frac{d\mathbf{x}}{dq}\frac{d\varphi}{ds} = \frac{\mathbf{x}'(q)}{\|\mathbf{x}'(q)\|},$$

so $\|\mathbf{y}'(s)\| = 1$ for all s in $[0, L]$.

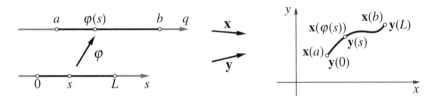

Example: arc length of an ellipse

Let us construct the arc-length parametrization of the ellipse

$$\frac{x^2}{a^2} + \frac{y^2}{b^2} = 1, \qquad 0 < b < a.$$

We can start with $\mathbf{x}(q) = (a \sin q, b \cos q)$. The parameter runs clockwise around the ellipse, and the parameter origin $q = 0$ is on the positive y-axis. The arc-length function is

$$s(q) = \int_0^q \|\mathbf{x}'(q)\| \, dq = \int_0^q \sqrt{a^2 \cos^2 q + b^2 \sin^2 q} \, dq$$

$$= \int_0^q \sqrt{a^2(1 - \sin^2 q) + b^2 \sin^2 q} \, dq = a \int_0^q \sqrt{1 - k^2 \sin^2 q} \, dq,$$

where $k = \sqrt{a^2 - b^2}/a$ and $0 < k < 1$. (Ellipses have different proportions, and k is an index, or *modulus*, of shape. If $k = 0$, then $a = b$, so the ellipse would be a circle. As k increases, the ellipse deviates more and more from a circular shape; for this reason k is called the *eccentricity* of the ellipse.)

Elliptic integrals

The integral $s(q)$ is not one of the elementary function of calculus. However, it appears frequently, along with similar integrals, in a great variety of important problems—for example, in the motion of a pendulum. Such integrals are known collectively as **elliptic integrals**; they get their name from the problem we are considering now. Elliptic integrals and their inverses—which are called *elliptic functions*—have been studied extensively since the eighteenth century. Their values are given in tables and, more recently, by computer algebra systems.

The inverse $q = \varphi(s)$ gives us the arc length parametrization $\mathbf{y}(s) = \mathbf{x}(\varphi(s))$. This is plotted below for $a = 6$, $b = 4$. As you can see, the arc length of a quarter-cycle is just slightly less than 8 units; the total length of a complete cycle is about 31.73 units.

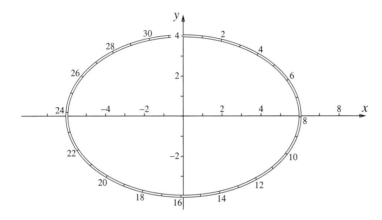

Curvature of a Curve

Intuitively, *curvature* is the rate at which a curve changes direction, that is, the rate at which its tangent vector changes—assuming, of course, that we move along the curve at a steady speed. We can shape these ideas into a precise definition.

Curvature is a rate

Definition 3.3 *Let $\mathbf{y}(s)$ be the arc-length parametrization of the curve C. Then we define:*

$$\textit{unit tangent vector along } C: \quad \mathbf{u}(s) = \frac{d\mathbf{y}}{ds};$$

$$\textit{curvature vector along } C: \quad \mathbf{k}(s) = \frac{d\mathbf{u}}{ds} = \frac{d^2\mathbf{y}}{ds^2};$$

$$\textit{curvature at } s: \quad \kappa(s) = \|\mathbf{k}(s)\|.$$

Theorem 3.2 $\mathbf{k}(s) \perp \mathbf{u}(s)$ *at every point s along C.*

PROOF: Since $\mathbf{u}(s)$ is a unit vector, $1 = \mathbf{u}(s) \cdot \mathbf{u}(s)$ for all s in $[0, L]$. Therefore,

$$0 = \frac{d}{ds}(\mathbf{u}(s) \cdot \mathbf{u}(s)) = \frac{d\mathbf{u}}{ds} \cdot \mathbf{u} + \mathbf{u} \cdot \frac{d\mathbf{u}}{ds} = 2\mathbf{k} \cdot \mathbf{u} \quad \text{for all } s,$$

so $\mathbf{k}(s) \perp \mathbf{u}(s)$, as claimed. END OF PROOF

Example

To get a better idea of what $\mathbf{k}$ tells us about a curve, let us look at an example that we understand well—a circle of radius r:

$$\mathbf{x}(q) = r(\cos(q), \sin(q)), \qquad \mathbf{x}'(q) = r(-\sin(q), \cos(q)).$$

To get a complete circle, we take $0 \leq q \leq 2\pi$. The parameter speed is $\|\mathbf{x}'(q)\| = r$, so the arc length function is

$$s = \int_0^q r \, dq = rq,$$

and the speed correction is $q = \varphi(s) = s/r$. Thus

$$\mathbf{y}(s) = \mathbf{x}(s/r) = r(\cos(s/r), \sin(s/r)), \quad 0 \leq s \leq 2\pi r,$$
$$\mathbf{u}(s) = (-\sin(s/r), \cos(s/r)),$$
$$\mathbf{k}(s) = -\frac{1}{r}(\cos(s/r), \sin(s/r)) = -\frac{1}{r^2}\mathbf{y}(s),$$
$$\kappa(s) = \frac{1}{r}.$$

Notice that the curvature vector $\mathbf{k}$ points toward the center of the circle, and the center itself is at $\mathbf{y}(s) + r^2\mathbf{k}(s)$. Because the curvature κ is the reciprocal of the radius in this example, we make the following definitions for an arbitrary smooth curve C:

§3.2 Curves and Curvature

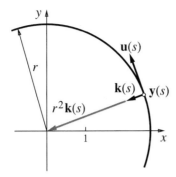

Definition 3.4 *The **radius of curvature** of C at s is $\rho(s) = 1/\kappa(s)$. The **center of curvature** of C at s is the point $\mathbf{c}(s) = \mathbf{y}(s) + \rho^2(s)\mathbf{k}(s)$.*

The center of curvature $\mathbf{c}(s)$ is itself a curve, called the **evolute** of C. However, s is not, in general, the arc-length parameter for $\mathbf{c}(s)$.

A circle has constant curvature, and conversely, any curve of constant curvature is a (portion of a) circle. This is the content of the next theorem.

Theorem 3.3 *The arc-length parametrization of a curve of constant curvature κ is $\mathbf{y}(s) = (\rho \cos(\kappa s + \omega) + a, \rho \sin(\kappa s + \omega) + b)$, where $\rho = 1/\kappa$.*

PROOF: Let $\mathbf{y}(s) = (x(s), y(s))$ be the arc-length parametrization of the curve. Then

$$\|\mathbf{u}(s)\| = 1, \qquad (x')^2 + (y')^2 = 1,$$
$$\|\mathbf{k}(s)\| = \kappa, \qquad (x'')^2 + (y'')^2 = \kappa^2.$$

The first condition implies that the point $\mathbf{u}(s) = (x'(s), y'(s))$ lies on a circle of radius 1, for every s; hence there is a function $\varphi(s)$ for which

$$x'(s) = \cos \varphi(s), \qquad y'(s) = \sin \varphi(s).$$

Then $x''(s) = -\varphi' \sin \varphi$ and $y''(s) = \varphi' \cos \varphi$, so the condition on $\mathbf{k}$ gives us

$$(\varphi')^2 = (x'')^2 + (y'')^2 = \kappa^2.$$

Hence $\varphi'(s) = \kappa$, $\varphi(s) = \kappa s + \omega_0$, where ω_0 is a constant of integration. Thus

(NOTE THAT $\phi'(s) = -k$ IS POSSIBLE, BUT THIS ALSO LEADS TO A CIRCLE OF RADIUS ρ)

$$x' = \cos(\kappa s + \omega_0), \qquad y' = \sin(\kappa s + \omega_0),$$
$$x = \frac{1}{\kappa}\sin(\kappa s + \omega_0) + a, \qquad y = -\frac{1}{\kappa}\cos(\kappa s + \omega_0) + b,$$

where a and b are constants of integration. The change of phase $\omega = \omega_0 - \pi/2$ gives

$$x = \frac{1}{\kappa}\cos(\kappa s + \omega) + a, \qquad y = \frac{1}{\kappa}\sin(\kappa s + \omega) + b.$$

This is a circle of radius $\rho = 1/\kappa$. The constants of integration—which are arbitrary—determine the position of the center and the phase ω of the parameter origin $s = 0$. END OF PROOF

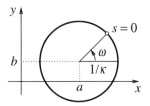

Higher Dimensions

Everything we have done carries over to higher dimensions. A **parametrized curve** in $\mathbf{R}^n$ is a smooth map

$$\mathbf{x} : [a, b] \to \mathbf{R}^n, \qquad \mathbf{x} : q \mapsto \mathbf{x}(q) = (x_1(q), \ldots, x_n(q)).$$

The tangent vector to $\mathbf{x}(q)$ is $\mathbf{x}'(q)$. A given curve C can have different parametrizations $\mathbf{x}$; the arc length of C is

$$s(q) = \int_a^q \|\mathbf{x}'(q)\|\, dq = \int_a^q \sqrt{dx_1^2 + \cdots + dx_n^2}, \qquad L = s(b).$$

When $\|\mathbf{x}'(q)\| > 0$ for all q in $[a, b]$, $s(q)$ is invertible. If we write the inverse as $q = \varphi(s)$, then we can make the following definitions for $0 \le s \le L$:

$$\begin{aligned}
\text{arc-length parametrization of } C: &\quad \mathbf{y}(s) = \mathbf{x}(\varphi(s)), \\
\text{unit tangent vector along } C: &\quad \mathbf{u}(s) = \mathbf{y}'(s), \\
\text{curvature vector along } C: &\quad \mathbf{k}(s) = \mathbf{u}'(s) = \mathbf{y}''(s), \\
\text{curvature at } s: &\quad \kappa(s) = \|\mathbf{k}(s)\|, \\
\text{radius of curvature at } s: &\quad \rho(s) = 1/\kappa(s), \\
\text{center of curvature at } s: &\quad \mathbf{c}(s) = \mathbf{y}(s) + \rho^2(s)\mathbf{k}(s).
\end{aligned}$$

We will still have $\mathbf{k}(s) \perp \mathbf{u}(s)$, but these two vectors can no longer form a basis at $\mathbf{y}(s)$ if $n > 2$.

In Section 5.4 we will take a further look at the curvature of a curve. We will prove there that curvature can be calculated directly from the original parametrization, as

An alternative curvature formula

$$\kappa = \sqrt{\frac{(\mathbf{x}' \cdot \mathbf{x}')(\mathbf{x}'' \cdot \mathbf{x}'') - (\mathbf{x}' \cdot \mathbf{x}'')^2}{(\mathbf{x}' \cdot \mathbf{x}')^3}},$$

without first obtaining the arc-length parametrization.

Exercises

1. For each of the following curves $C: \mathbf{x}(q)$, $a \leq q \leq b$,
   - obtain the formula for arc length s in terms of q: $s = s(q)$;
   - obtain the arc-length parametrization;
   - compute the curvature vector and the radius of curvature function;
   - make a sketch.

   (a) $\mathbf{x}(q) = (3q - 2, 4q + 3)$, $\quad 0 \leq q \leq 5$.
   (b) $\mathbf{x}(q) = (3\cos 2q, 3\sin 2q)$, $\quad 0 \leq q \leq \pi$.
   (c) $\mathbf{x}(q) = (3 + \cos q, 2 + \sin q)$, $\quad 0 \leq q \leq 2\pi$.
   (d) $\mathbf{x}(q) = (\cos^3 q, \sin^3 q)$, $\quad 0 < q < \pi/2$.
   (e) $\mathbf{x}(q) = (\frac{1}{3}q^3, \frac{1}{2}q^2)$, $\quad 0 \leq q \leq 2$.
   (f) $\mathbf{x}(q) = (e^q \cos q, e^q \sin q)$, $\quad 0 \leq q \leq \pi/2$.

2. (a) Sketch the tangent and acceleration vectors, $\mathbf{X}'(q)$ and $\mathbf{X}''(q)$, of the curve

$$\mathbf{X}(q) = \left(\frac{2q}{1+q^2}, \frac{1-q^2}{1+q^2}\right), \quad -\infty < q < \infty,$$

at the points $q = -2, -1, 0, 1, 2$. (Recall that $\mathbf{X}$ is the circle of radius 1 centered at the origin.) Note the relation between the acceleration $\mathbf{X}''$ and the parameter speed at each of these points. Describe how $\mathbf{X}''(q)$ is related to $\mathbf{X}'(q)$ when the parameter point is "speeding up" and when it is "slowing down."

(b) Use the alternative curvature formula to show that the curvature of $\mathbf{X}$ is 1 everywhere.

(c) Determine, directly from the formula, the arc-length function $s(q)$ of $\mathbf{X}$.

(d) Since $\mathbf{X}$ is a circle of radius 1, its total arc length is 2π. Deduce this directly from $s(q)$.

(e) Using the inverse $q(s)$ of the arc-length function, construct the arc-length parametrization $\mathbf{Y}(s) = \mathbf{X}(q(s))$. Indicate clearly the domain $a < s < b$.

(f) Compute the curvature vector $\mathbf{K}(s) = \mathbf{Y}''(s)$ and compare it to the original acceleration vector $\mathbf{X}''$.

3. Describe how the proof of Theorem 3.1 must be modified if $\mathbf{x}(a) = \mathbf{X}(B)$ (so that $\varphi(q)$ is a *decreasing* function).

4. Determine the circumference of each of the following ellipses:

$$\frac{x^2}{16} + \frac{y^2}{9} = 1; \qquad \frac{x^2}{144} + \frac{y^2}{25} = 1.$$

5. (a) Sketch the space curve $\mathbf{x}_a(q) = (\cos q, \sin q, aq)$. Describe its shape in words, and describe what influence the parameter $a \geq 0$ has on the shape. In particular, indicate what happens to the shape when $a \to 0$ and when $a \to \infty$.

(b) Determine the arc-length function $s_a(q)$ for $\mathbf{x}$ and the inverse function $q_a(s)$. Use this to construct the arc-length parametrization $\mathbf{y}_a(s) = \mathbf{x}_a(q_a(s))$ and the curvature vector $\mathbf{k}_a(s)$.

(c) Determine the curvature κ_a two ways: from $\mathbf{k}_a$, and directly from $\mathbf{x}_a$ and its derivatives, using the alternative formula for the curvature.

(d) Your calculation should make it evident that κ_a is constant on a given curve (i.e., for a given a). Sketch κ_a as a function of a for $a \geq 0$ and determine the limiting value of κ_a as $a \to \infty$. Explain the connection between the shape of $\mathbf{x}_a$ and its curvature κ_a.

6. (a) The curve $\mathbf{x}_a(q) = (e^{aq} \cos q, e^{aq} \sin q)$ is called an *equiangular spiral* because there is a constant angle between the tangent $\mathbf{x}'_a(q)$ and the "radius vector" $\mathbf{x}_a(q)$. Prove this and determine the measure of the angle as a function of the parameter a.

(b) Determine the arc length of $\mathbf{x}_a(q)$ on the infinite interval $-\infty < q \leq 0$, assuming $a > 0$.

7. (a) Show that the curve $\mathbf{x}(q) = (\cos^2 q, \cos q \sin q, \sin q)$ lies on the unit sphere

$$x^2 + y^2 + z^2 = 1.$$

(b) Sketch $\mathbf{x}(q)$ over the interval $-\pi/2 \leq q \leq \pi/2$. At what point does it start; at what point does it end? For which values of the parameter q, if any, does it cross the equator (where $z = 0$)?

(c) In what direction is the curve heading at its starting point and at its ending point? Sketch these direction vectors on the curve.

(d) Determine the length of the curve. (Note: You can do this either as a numerical integration or as an elliptic integral.)

3.3 Accelerated Motion

Worldcurves and Proper Time

We are now ready to consider the motion of an accelerated observer G in an inertial frame R. The worldline of G will, in general, be curved. If G's spatial position at time t is given by the smooth

Worldlines as graphs of motion

3-vector function $\mathbf{x}(t) = (x(t), y(t), z(t))$, then G's worldline is the graph of $\mathbf{x}$ in R's spacetime. This is a special kind of parametrized curve in which the first component is just the parameter:

$$\mathbb{X}(t) = (t, x(t), y(t), z(t)).$$

The velocity 4-vector has the special form

$$\mathbb{V}(t) = \mathbb{X}'(t) = (1, x'(t), y'(t), z'(t)) = (1, \mathbf{v}(t)).$$

In fact, this is a *future* vector, as the following argument shows. The velocity limitation requires $v^2(t) = \|\mathbf{v}(t)\|^2 < 1$ for all t, so $Q(\mathbb{V}(t)) = 1 - v^2(t) > 0$. Thus $\mathbb{V}(t)$ is timelike; it is a future vector because its first component is a positive number.

Worldlines as parametrized curves

But G's worldline may not be given simply as a graph, even if we restrict ourselves to a $(1+1)$-dimensional spacetime. For example, suppose $C : (T, Z)$ is another inertial frame related to R by the hyperbolic rotation $H_u : R \to C$. If G's worldline in R is the graph $\mathbb{X}(t) = (t, z(t))$, then in C it becomes

$$\widehat{\mathbb{X}}(t) = H_u(\mathbb{X}(t)), \qquad \widehat{\mathbb{X}}(t) : \begin{cases} T(t) = t \cosh u + z(t) \sinh u, \\ Z(t) = t \sinh u + z(t) \cosh u. \end{cases}$$

This is not in the simple form $Z = Z(T)$, so $\widehat{\mathbb{X}}$ is not given as a graph. However, it is a parametrized curve; t is the parameter. Furthermore, since $\mathbb{X}'(t)$ is a future vector and H_u preserves future vectors, $\widehat{\mathbb{X}}'(t) = H_u(\mathbb{X}'(t))$ is also a future vector. In other words, C still sees G moving steadily into the future. These observations lead to the following definitions.

Definition 3.5 *A **worldcurve** G is a set of events in spacetime that can be described in any inertial frame R by a smooth parametrized curve*

$$\mathbb{X} : [a, b] \to R, \qquad q \mapsto \mathbb{X}(q) = (t(q), x(q), y(q), z(q)),$$

in which $\mathbb{X}'(q)$ is a future vector, for every q in $[a, b]$.

Since $\mathbb{X}'(q)$ is a future vector, $\mathbb{X}$ is nonsingular. That is, it satisfies the technical requirement $\|\mathbb{X}'\| > 0$ that we imposed on parametrized curves to make them locally one-to-one. In fact, the following proposition shows that worldcurves are globally one-to-one.

§3.3 Accelerated Motion

Proposition 3.3 *If $\mathbb{X}(q) = (t(q), x(q), y(q), z(q))$ parametrizes the worldcurve G, then $t(q)$ is a strictly monotonic increasing function. Consequently, $\mathbb{X}(q)$ is globally one-to-one.*

PROOF: Since $\mathbb{X}'(q) = (t'(q), x'(q), y'(q), z'(q))$ is a future timelike vector,
$$t'(q) > \sqrt{(x'(q))^2 + (y'(q))^2 + (z'(q))^2} \geq 0.$$
Thus the slope of the graph of $t(q)$ is positive everywhere, so $t(q)$ is strictly monotonic increasing. This means that if $q_1 < q_2$, then $t(q_1) < t(q_2)$. In particular, $\mathbb{X}(q_1) \neq \mathbb{X}(q_2)$, so different parameter values can never be mapped to the same point by $\mathbb{X}$. END OF PROOF

At the heart of special relativity is the observation that observers in motion relative to one another measure time differently. When G moves uniformly in R's frame, elapsed proper time for G is given by the Minkowski–Pythagorean theorem (Theorem 2.9) along G's straight worldcurve:

$$\Delta\tau = \sqrt{\Delta t^2 - (\Delta x^2 + \Delta y^2 + \Delta z^2)}.$$

Proper time

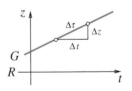

Now that we permit G to move *nonuniformly* in R's frame, how should R measure elapsed proper time along G's worldcurve?

Proper time in relativity corresponds to arc length in Euclidean geometry, and arc length along a curve is the integral of the differential $ds = \sqrt{dx^2 + dy^2 + dz^2}$. This suggests that we turn to the differential form of $\Delta\tau$:

$$d\tau = \sqrt{dt^2 - (dx^2 + dy^2 + dz^2)}.$$

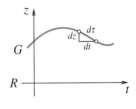

Then, if E_1 and E_2 are two events on the worldcurve G, and R parametrizes G as $\mathbb{X}(q)$, with $\mathbb{X}(q_i) = E_i$, $i = 1, 2$, the proper-time interval between E_1 and E_2 is

$$\Delta\tau = \int_{E_1}^{E_2} d\tau = \int_{q_1}^{q_2} \sqrt{dt^2 - (dx^2 + dy^2 + dz^2)}$$

$$= \int_{q_1}^{q_2} \sqrt{\left(\frac{dt}{dq}\right)^2 - \left(\left(\frac{dx}{dq}\right)^2 + \left(\frac{dy}{dq}\right)^2 + \left(\frac{dz}{dq}\right)^2\right)}\, dq$$

$$= \int_{q_1}^{q_2} \|\mathbb{X}'(q)\|\, dq.$$

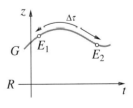

As the following theorem shows, elapsed proper time defined this way is physically meaningful; that is, it has the same value in all inertial frames.

Theorem 3.4 *Let R and C be two inertial frames, and suppose G has the parametrization $\mathbb{X}(q)$ in R and $\widehat{\mathbb{X}}(Q)$ in C. Suppose also that $E_i = \mathbb{X}(q_i) = \widehat{\mathbb{X}}(Q_i)$, $i = 1, 2$. Then*

$$\int_{q_1}^{q_2} \|\mathbb{X}'(q)\| \, dq = \Delta\tau = \int_{Q_1}^{Q_2} \|\widehat{\mathbb{X}}'(Q)\| \, dQ.$$

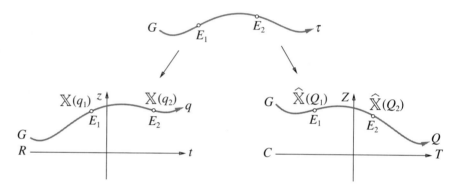

PROOF: This is essentially the same as Theorem 3.1. Its proof is left to the exercises.

Definition 3.6 *If $\mathbb{X}(q) = (t(q), x(q), y(q), z(q))$, $a \leq q \leq b$, is a worldcurve G in R, then the **proper-time parameter** on G is the function*

$$\tau(q) = \int_a^q \|\mathbb{X}'(q)\| \, dq = \int_a^q \sqrt{t'(q)^2 - (x'(q)^2 + y'(q)^2 + z'(q)^2)} \, dq.$$

If we use this function, then the elapsed time between two events $\mathbb{X}(q_1)$ and $\mathbb{X}(q_2)$ on G's worldcurve is just $\Delta\tau = \tau(q_2) - \tau(q_1)$.

Theorem 3.5 *In an inertial frame R, any worldcurve G can be parametrized by proper time τ. In the proper-time parametrization $\mathbb{G}(\tau)$, velocity is a unit 4-vector: $\|\mathbb{G}'(\tau)\| \equiv 1$.*

PROOF: This is entirely analogous to the theorem about the arc-length parametrization of a Euclidean curve. Suppose we obtain the proper time on G from the parametrization $\mathbb{X}(q)$:

$$\tau(q) = \int_a^q \|\mathbb{X}'(q)\|\, dq.$$

Since $d\tau/dq = \|\mathbb{X}'(q)\| > 0$, the function $\tau = \tau(q)$ is invertible. Let $q = \varphi(\tau)$ be the inverse, and set $\mathbb{G}(\tau) = \mathbb{X}(\varphi(\tau))$. This is the proper-time parametrization, and the following calculation shows that $\|\mathbb{G}'(\tau)\| \equiv 1$:

$$\frac{d\mathbb{G}}{d\tau} = \frac{d\mathbb{X}}{dq}\frac{d\varphi}{d\tau} = \frac{d\mathbb{X}}{dq}\frac{1}{d\tau/dq} = \frac{\mathbb{X}'}{\|\mathbb{X}'\|}; \qquad \left\|\frac{d\mathbb{G}}{d\tau}\right\| = \frac{\|\mathbb{X}'\|}{\|\mathbb{X}'\|} \equiv 1.$$

END OF PROOF

We construct the proper-time parametrization when G oscillates sinusoidally on the z-axis. In our first attempt, the computations are simple but involve a worldcurve on which G attains the speed of light momentarily in each oscillation. The second attempt removes this defect but at the cost of complicating the computations.

Examples: proper time for oscillatory motion

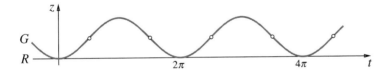

Let G's worldcurve be the graph of $z = 1 - \cos t$: $\mathbb{X}(t) = (t, 1 - \cos t)$. Then $\mathbb{X}'(t) = (1, \sin t)$ and $\|\mathbb{X}'(t)\| = \sqrt{1 - \sin^2 t} = |\cos t|$, so G's speed is less than 1 except when $t = (2n+1)\pi/2$, n an integer (at the points marked). A complete cycle consists of four portions, each congruent to the segment $0 \leq t \leq \pi/2$. If we restrict ourselves to this interval, then $\|\mathbb{X}'(t)\| = \cos t$, and the proper-time function is

Example 1

$$\tau(t) = \int_0^t \|\mathbb{X}'(t)\|\, dt = \int_0^t \cos t\, dt = \sin t, \qquad 0 \leq t \leq \pi/2.$$

The inverse function is $t = \varphi(\tau) = \arcsin \tau$, $0 \leq \tau \leq 1$, so the proper-time parametrization is

$$\mathbb{G}(\tau) = \mathbb{X}(\arcsin \tau) = (\arcsin \tau, 1 - \cos(\arcsin \tau))$$
$$= (\arcsin \tau, 1 - \sqrt{1 - \tau^2}).$$

Note how proper time varies along $\mathbb{G}(\tau)$ in the figure below. We have placed ticks every 0.1 seconds along both G's worldcurve and the t-axis (which is R's worldline). For comparison, we also show how proper time is marked along worldline segments at three fixed velocities. We have also taken the pattern that appears for $0 \leq \tau \leq 1$ on $\mathbb{G}$ and reflected it symmetrically to the interval $1 \leq \tau \leq 2$. Notice that proper-time intervals for the two observers are nearly the same when G's speed is small, but they become very different when G's speed is large.

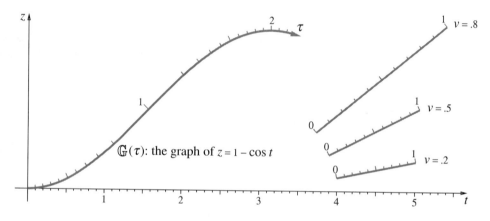

In the larger scale picture below, we see that a full cycle takes 4 seconds according to G but $2\pi \approx 6\frac{1}{4}$ seconds according to R. Thus, on average, G's clock runs at only about two-thirds the rate of R's.

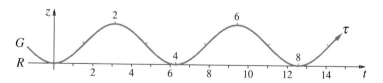

Example 2   To reduce G's top speed, we can just "scale back" the entire motion by a factor $k < 1$. That is, let G move along the curve $z = k(1 - \cos t)$; then G accelerates only to k times the speed of

light before decelerating. The initial parametrization is $\mathbb{X}(t) = (t, k(1 - \cos t))$, and $\|\mathbb{X}'(t)\| = \sqrt{1 - k^2 \sin^2 t}$. We can now see how k complicates things: The proper-time function

$$\tau(t) = \int_0^t \sqrt{1 - k^2 \sin^2 t}\, dt$$

becomes an elliptic integral—exactly the same sort we encountered in calculating the Euclidean arc length of an ellipse. If $t = \varphi(\tau)$ denotes its inverse, then the proper-time parametrization takes the form

$$\mathbb{G}(\tau) = (\varphi(\tau), k(1 - \cos \varphi(\tau))).$$

In the following plot of $\mathbb{G}(\tau)$ we have taken $k = 0.8$. You can see that there is still considerable time expansion at the steepest part of the worldcurve, where $v \approx 0.8$, but not as much as in the first example.

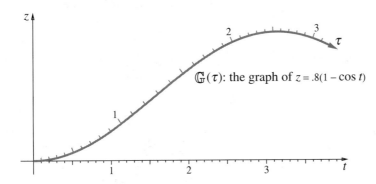

A complete cycle now takes slightly more than 5 seconds according to G. According to R, it still takes 2π seconds. On average, G's clock runs at about five-sixths the rate of R's.

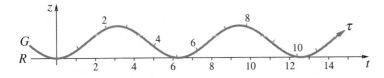

Imagine G and R in the second example are twins, and that the time unit is a year rather than a second. Imagine also that G

The moving twin ages less

The paradox

leaves R on a spaceship that accelerates to 80% of the speed of light before turning around and returning. During one complete journey, R ages about $\Delta t = 6\frac{1}{4}$ years. But G ages only about $\Delta \tau = 5$ years. The twins are no longer the same age! This is a real physical difference: Even the atoms in G's body have gone through fewer oscillations than those in R's.

G ages less because G moves. However, we know that motion is relative, so why don't we reverse the roles of G and R? If we view everything in G's frame, then R will age less. This is a paradox: From apparently equally valid assumptions, we reach contradictory conclusions. The paradox vanishes once we realize that the apparent symmetry between G and R is an illusion: G accelerates and R does not. Therefore, R has an inertial frame but G does not. In special relativity we still restrict our analysis to inertial frames, so we cannot interchange R and G; there is no symmetry.

The following theorem shows that the twin paradox is not limited to our example. Elapsed proper time along *any* worldcurve connecting two events is less than along the straight worldline connecting those events. Incidentally, the twin paradox is just the continuous version of the reverse triangle inequality.

Theorem 3.6 *("Twin paradox") Suppose the separation between two events O and E is timelike. Then, among all worldcurves joining O and E, the* straight *worldline has the longest elapsed proper time.*

PROOF: Because the separation between O and E is timelike, we can choose an inertial frame $R : (t, x, y, z)$ in which these events are on the t-axis at $O : (0, 0, 0, 0)$ and $E : (\bar{t}, 0, 0, 0)$. This is the straight worldline, and t is the proper-time parameter. The elapsed proper time between O and E is just $\bar{t}$.

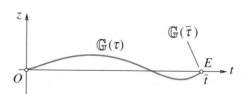

Suppose G is another worldcurve connecting O and E, and suppose the elapsed proper time $\Delta\tau$ between O and E along G is $\bar{\tau}$. Let $\mathbb{G}(\tau) = (t(\tau), x(\tau), y(\tau), z(\tau))$ be the proper-time parametrization of G in R's frame; then

$$1 = \|\mathbb{G}'(\tau)\|^2 = \left(\frac{dt}{d\tau}\right)^2 - \left(\frac{dx}{d\tau}\right)^2 - \left(\frac{dy}{d\tau}\right)^2 - \left(\frac{dz}{d\tau}\right)^2 \leq \left(\frac{dt}{d\tau}\right)^2,$$

and the inequality is *strict* at any point where G's velocity is nonzero. Thus

$$\bar{\tau} = \int_0^{\bar{\tau}} d\tau < \int_0^{\bar{\tau}} \frac{dt}{d\tau} d\tau = \int_0^{\bar{t}} dt = \bar{t};$$

the key step is the change of variable $\tau \mapsto t$ in the integration and the changes $0 \mapsto 0$ and $\bar{\tau} \mapsto \bar{t}$ to the limits of integration.

END OF PROOF

Newton's Second Law

Suppose G is an observer with proper mass μ and proper time τ. In any inertial frame $R : (t, x, y, z)$ we have the following data associated with G:

$$\begin{aligned}
\text{worldcurve of } G: &\quad \mathbb{G}(\tau), \\
\text{proper 4-velocity of } G: &\quad \mathbb{U}(\tau) = \mathbb{G}'(\tau), \\
\text{4-momentum of } G: &\quad \mathbb{P}(\tau) = \mu \mathbb{U}(\tau), \\
\text{4-acceleration of } G: &\quad \mathbb{A}(\tau) = \mathbb{U}'(\tau) = \mathbb{G}''(\tau).
\end{aligned}$$

Keep in mind the similarities between worldcurves in spacetime and ordinary curves in Euclidean space. Proper 4-velocity $\mathbb{U} = \mathbb{G}'$ corresponds to the unit-speed tangent vector $\mathbf{u} = \mathbf{y}'$ and 4-acceleration $\mathbb{A} = \mathbb{G}''$ to the curvature vector $\mathbf{k} = \mathbf{y}''$. We can also introduce the scalar acceleration $\alpha(\tau) = \|\mathbb{A}(\tau)\|$ to correspond to the curvature function $\kappa(s)$.

If we write $\mathbb{G}(\tau) = (t(\tau), \mathbf{x}(\tau))$, then

$$\mathbb{U}(\tau) = \left(\frac{dt}{d\tau}, \frac{d\mathbf{x}}{d\tau}\right) = \left(\frac{dt}{d\tau}, \frac{d\mathbf{x}}{dt}\frac{dt}{d\tau}\right) = \frac{dt}{d\tau}(1, \mathbf{v}),$$

$$1 \equiv \|\mathbb{U}(\tau)\|^2 = \left(\frac{dt}{d\tau}\right)^2 (1 - \|\mathbf{v}\|^2) = \left(\frac{dt}{d\tau}\right)^2 (1 - v^2).$$

Chapter 3 Special Relativity — Kinetics

Therefore,

$$\frac{dt}{d\tau} = \frac{1}{\sqrt{1-v^2}} \quad \text{and} \quad \mathbb{U}(\tau) = \frac{1}{\sqrt{1-v^2}}(1, \mathbf{v}).$$

Local time dilation

The derivative $dt/d\tau$ tells us the rate of change of local time t with respect to proper time τ and is thus the smooth analogue of the time dilation factor $\Delta t/\Delta \tau$. We call it the **local time dilation factor** to emphasize that it varies from point to point along G's worldcurve.

4-momentum

The 4-momentum vector $\mathbb{P}(\tau) = \mu \mathbb{U}(\tau)$ has a norm that is constant and equal to G's rest mass: $\|\mathbb{P}(\tau)\| \equiv \mu$. Moreover, its components are mass and 3-momentum in their covariant, relativistic form:

$$\mathbb{P}(\tau) = \frac{\mu}{\sqrt{1-v^2}}(1, \mathbf{v}) = \left(\frac{\mu}{\sqrt{1-v^2}}, \frac{\mu \mathbf{v}}{\sqrt{1-v^2}}\right) = (m, m\mathbf{v}) = (m, \mathbf{p}).$$

The various 4-vectors we have just defined are essentially the same as their linear counterparts summarized near the end of Section 3.1. We get new information, however, when we differentiate $\mathbb{P}(\tau)$ with respect to τ:

$$\frac{d\mathbb{P}}{d\tau} = \frac{dt}{d\tau}\frac{d\mathbb{P}}{dt} = \frac{dt}{d\tau}\left(\frac{dm}{dt}, \frac{d\mathbf{p}}{dt}\right) = \frac{1}{\sqrt{1-v^2}}\left(\frac{dm}{dt}, \frac{d\mathbf{p}}{dt}\right).$$

According to Newton's second law of motion, the classical 3-force vector acting on G is $\mathbf{f} = d\mathbf{p}/dt$. We write

$$\frac{d\mathbb{P}}{d\tau} = \frac{1}{\sqrt{1-v^2}}\left(\frac{dm}{dt}, \mathbf{f}\right) = \left(\frac{1}{\sqrt{1-v^2}}\frac{dm}{dt}, \frac{\mathbf{f}}{\sqrt{1-v^2}}\right) = \mathbb{F}(\tau)$$

$\mathbb{F} = \mu \mathbb{A}$

to define the 4-vector $\mathbb{F}$. Since the 3-vector $\mathbf{f}/\sqrt{1-v^2}$ that forms the last three components of $\mathbb{F}$ is the covariant form of the 3-force on G, it is reasonable to call $\mathbb{F}$ the **4-force** acting on G. Then (by definition!)

$$\mathbb{F} = \frac{d\mathbb{P}}{d\tau} = \mu \mathbb{A},$$

the relativistic 4-vector form of *Newton's second law*. We have recovered the classical statement: Force equals mass times acceleration.

Theorem 3.7 $\mathbb{A}(\tau) \perp \mathbb{U}(\tau)$ and $\mathbb{F}(\tau) \perp \mathbb{U}(\tau)$ at all points $\mathbb{G}(\tau)$ on G's worldcurve.

PROOF: Since $\mathbb{A}$ and $\mathbb{U}$ are spacetime vectors, orthogonality is in the sense of Minkowski geometry. Differentiate $1 \equiv \mathbb{U}(\tau) \cdot \mathbb{U}(\tau)$ by the product rule to get

$$0 \equiv 2\frac{d\mathbb{U}}{d\tau} \cdot \mathbb{U} = 2\mathbb{A} \cdot \mathbb{U},$$

so $\mathbb{A} \perp \mathbb{U}$. Since $\mathbb{F} = \mu\mathbb{A}$, $\mathbb{F} \perp \mathbb{U}$ as well. END OF PROOF

This corresponds to the result $\mathbf{k}(s) \perp \mathbf{u}(s)$ for ordinary curves. Since the curvature function $\kappa(s)$ in geometry corresponds to the scalar acceleration function $\alpha(\tau)$ here, we can expect that the size of α will be reflected in the curvature of G's worldcurve.

To see what this means concretely, let us plot the acceleration 4-vector of an observer G undergoing the oscillatory motion $z = k(1 - \cos t)$. The proper-time parametrization of G's worldcurve is

Example: oscillatory motion

$$\mathbb{G}(\tau) = (\varphi(\tau), k(1 - \cos\varphi(\tau))),$$

where $t = \varphi(\tau)$ is the inverse of the elliptic integral

$$\tau = \int_0^t \sqrt{1 - k^2 \sin^2 t}\, dt.$$

Then $\mathbb{A}(\tau) = \left(\varphi'', k(\varphi'' \sin\varphi + (\varphi')^2 \cos\varphi)\right)$, and some further calculations (see the exercises) show that

$$\mathbb{A}(\tau) = \frac{k\cos\varphi}{(1 - k^2 \sin^2 \varphi)^2}(k\sin\varphi, 1) = \frac{k\cos t}{(1 - k^2 \sin^2 t)^2}(k\sin t, 1).$$

The two forms are equivalent because $t = \varphi(\tau)$. However, if we use the second form, then we can plot the vectors directly on the curve $z = k(1 - \cos t)$. We do that in the figure below, which shows what it means for vectors to be everywhere orthogonal to the curve in the sense of the Minkowski inner product. Compare this with the figure in Section 2.3 that shows several pairs of Minkowski-orthogonal vectors.

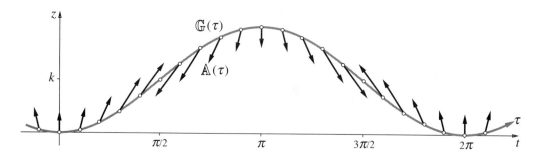

At first glance, it looks as if the acceleration 4-vector $\mathbb{A}$ may be discontinuous at $t = \pi/2$. However, a detailed view shows that this is not the case.

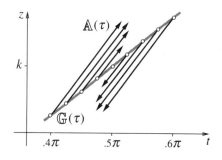

Moreover, the Minkowski norm $\|\mathbb{A}\| = k|\cos t|/(1 - k^2 \sin^2 t)^{3/2}$ is continuous everywhere and equals 0 when $t = \pi/2$. Note that the norm has this form because $\mathbb{A}$ is a spacelike 4-vector.

Example: constant scalar acceleration

In the example we have just considered, the scalar acceleration $\alpha(\tau) = \|\mathbb{A}(\tau)\|$ varies periodically over time. Suppose instead that G is subjected to *constant* scalar acceleration in R's frame. How will G move then?

Recall that scalar acceleration in Minkowski geometry corresponds to curvature in Euclidean geometry. We have seen that a Euclidean plane curve of constant curvature is a portion of a circle, so it is natural to think that a worldcurve of constant acceleration will be a Minkowskian "circle"—that is, a hyperbola. This is established in the following theorem, whose proof carries over directly from the Euclidean analogue.

Theorem 3.8 *The proper-time parametrization of a worldcurve of constant nonzero acceleration α in a $(1 + 1)$-dimensional spacetime*

is
$$\mathbb{G}(\tau) = \left(\frac{1}{\alpha}\sinh(\alpha\tau + \tau_0) + t_0, \pm\frac{1}{\alpha}\cosh(\alpha\tau + \tau_0) + z_0\right).$$

PROOF: If $\mathbb{G}(\tau) = (t(\tau), z(\tau))$ is the proper-time parametrization of the worldcurve we seek, then

$$\|\mathbb{U}(\tau)\| = 1, \quad (t')^2 - (z')^2 = 1,$$
$$\|\mathbb{A}(\tau)\| = \alpha, \quad (z'')^2 - (t'')^2 = \alpha^2.$$

The first condition implies that the point $\mathbb{U}(\tau)$ lies on the unit hyperbola in the future set $\mathcal{F}$, for every τ; hence there is a function $f(\tau)$ for which

$$t'(\tau) = \cosh f(\tau), \quad z'(\tau) = \sinh f(\tau).$$

Then $t''(\tau) = f' \sinh f$ and $z'' = f' \cosh f$, so the condition on $\mathbb{A}$ implies

$$(f')^2 = (z'')^2 - (t'')^2 = \alpha^2.$$

Hence $f' = \pm\alpha$, $f(\tau) = \pm\alpha(\tau + \tau_1)$, where τ_1 is a constant of integration. Thus

$$t' = \cosh(\alpha(\tau + \tau_1)), \quad z' = \pm\sinh(\alpha(\tau + \tau_1)),$$
$$t = \frac{1}{\alpha}\sinh(\alpha(\tau + \tau_1)) + t_0 \quad z = \pm\frac{1}{\alpha}\cosh(\alpha(\tau + \tau_1)) + z_0$$
$$= \frac{1}{\alpha}\sinh(\alpha\tau + \tau_0) + t_0, \quad = \pm\frac{1}{\alpha}\cosh(\alpha\tau + \tau_0) + z_0,$$

where $\tau_0 = \alpha\tau_1$, t_0 and z_0 are constants of integration, and we have used the fact that $\sinh(\pm u) = \pm\sinh(u)$ and $\cosh(\pm u) = \cosh(u)$.

END OF PROOF

The constant-acceleration worldcurves are the upper and lower branches of hyperbolas centered at an arbitrary point $(t, z) = (t_0, z_0)$ in R's frame. They can be written as the graphs

Acceleration is linked to curvature

$$z = z_0 \pm \frac{1}{\alpha}\sqrt{1 + \alpha^2(t - t_0)^2}.$$

As $t \to \infty$, the velocity $z' \to \pm 1$. Plotted below are the worldcurves of two objects with different accelerations $|\alpha_1| < |\alpha_2|$; the larger acceleration occurs on the hyperbola that is more sharply curved.

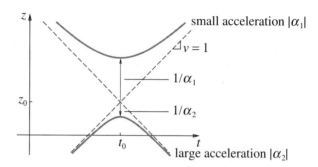

Hyperbolas versus parabolas

In Newtonian mechanics, the worldcurve of an object that experiences a constant (nonrelativistic) acceleration is a parabola, not a hyperbola. But parabolas violate the velocity limitation of special relativity because their slopes increase without bound. By contrast, the slope of a hyperbola cannot exceed the slope of its asymptotes. Nevertheless, there is a nice connection between the two, provided by Taylor's theorem; for $t - t_0$ small, the equation of the hyperbola is approximated by the equation of a parabola:

$$z = z_0 \pm \frac{1}{\alpha}\sqrt{1 + \alpha^2(t - t_0)^2}$$

$$\approx z_0 \pm \frac{1}{\alpha}\left(1 + \frac{\alpha^2}{2}(t - t_0)^2\right) = z_0 \pm \frac{1}{\alpha} \pm \frac{\alpha}{2}(t - t_0)^2.$$

The last equation is the familiar one for motion of an object under constant acceleration in Newtonian physics—for example, under the force of gravity near the surface of the earth.

Mass and Energy

Let us return to the equation $\mathbb{F} \cdot \mathbb{U} = 0$ and look at it in terms of the components of $\mathbb{F}$ and $\mathbb{U}$. These are circumstances where dimensionally homogeneous coordinates (ct, x, y, z) will bring added clarity, so we write

§3.3 Accelerated Motion

$$\mathbb{F} \cdot \mathbb{U} = \begin{pmatrix} \frac{c\, dm/dt}{\sqrt{1-(v/c)^2}} \\ \frac{\mathbf{f}}{\sqrt{1-(v/c)^2}} \end{pmatrix} \cdot \begin{pmatrix} \frac{c}{\sqrt{1-(v/c)^2}} \\ \frac{\mathbf{v}}{\sqrt{1-(v/c)^2}} \end{pmatrix} = \frac{c^2 \frac{dm}{dt} - \mathbf{f} \cdot \mathbf{v}}{1-(v/c)^2}.$$

Here $\mathbf{f} \cdot \mathbf{v}$ is the ordinary Euclidean inner product of 3-vectors. The equation $\mathbb{F} \cdot \mathbb{U} = 0$ thus becomes the scalar equation

$$c^2 \frac{dm}{dt} = \mathbf{f} \cdot \mathbf{v}.$$

Proposition 3.4 *Let $K(t) = \tfrac{1}{2}\mu v^2(t)$ be the classical kinetic energy of G in R's frame. Then*

$$\frac{dK}{dt} = \mathbf{f} \cdot \mathbf{v},$$

where $\mathbf{f}$ is the classical 3-force acting on G.

PROOF: Since $v^2 = \mathbf{v} \cdot \mathbf{v}$, we can write $K(t) = \frac{\mu}{2} \mathbf{v} \cdot \mathbf{v}$. Therefore,

$$\frac{dK}{dt} = \frac{\mu}{2}(\mathbf{v} \cdot \mathbf{v})' = \frac{\mu}{2}(2\mathbf{v}' \cdot \mathbf{v}) = \mu \mathbf{v}' \cdot \mathbf{v} = \mu \mathbf{a} \cdot \mathbf{v} = \mathbf{f} \cdot \mathbf{v}.$$

END OF PROOF

The scalar form of $\mathbb{F} \cdot \mathbb{U} = 0$ can thus be written

$$c^2 \frac{dm}{dt} = \frac{dK}{dt};$$

dimensionally, both sides of the equation are rates of change of energy with respect to time. We can integrate this equation immediately:

$$c^2 m = K + \text{const}.$$

To determine the constant of integration, let $v = 0$. Then $m = \mu$ and $K = 0$, so const $= \mu c^2$, the rest energy of G. We can summarize the previous discussion in the following definition and corollary.

Definition 3.7 *The **total energy** of G in R's frame is the sum of its kinetic and rest energies: $E = K + \mu c^2$.*

E = mc²

Corollary 3.1 $E = mc^2$, where m is the relativistic mass of G in R's frame.

We also have an expression for the relativistic kinetic energy K, THAT AGREES WITH THE ONE WE USED IN SECTION 3.1 FOR NON-ACCELERATED MOTION:

$$K = mc^2 - \mu c^2 = \mu c^2 \left(\frac{1}{\sqrt{1-(v/c)^2}} - 1 \right)$$

$$= \mu c^2 \left(1 + \frac{v^2}{2c^2} + \frac{3v^4}{8c^4} + \frac{5v^6}{16c^6} + \cdots - 1 \right)$$

$$= \underbrace{\frac{1}{2}\mu v^2}_{\text{classical}} + \mu \frac{3v^4}{8c^2} + \mu \frac{5v^6}{16c^4} + \cdots .$$

When $v \ll c$, the largest term in K is the first, and this is the classical kinetic energy.

Energy–momentum vector

As in Section 3.1, the relation $E = mc^2$ allows us to write $\mathbb{P} = (mc, \mathbf{p})$ in the alternative form $\mathbb{P} = (E/c, \mathbf{p})$ that justifies our calling $\mathbb{P}$ the *energy-momentum vector*. The following theorem makes use of this relation between energy and momentum.

Theorem 3.9 $E = c\sqrt{p^2 + \mu^2 c^2}$, where $p = \|\mathbf{p}\|$, the Euclidean length.

PROOF: Just calculate $\|\mathbb{P}\|^2$ using dimensionally homogeneous coordinates (ct, x, y, z) and solve for E:

$$\mu^2 c^2 = \|\mathbb{P}\|^2 = \left(\frac{E}{c}\right)^2 - \|\mathbf{p}\|^2 = \frac{E^2}{c^2} - p^2. \qquad \text{END OF PROOF}$$

It is especially useful to express the first component of $\mathbb{P}$ as an energy rather than a mass, because objects that move at the speed of light have rest mass 0; see the exercises. However, they *do* have energy.

The energy of light

The energy of light was determined by the photoelectric effect (which was first explained by Einstein in 1905, the year of special relativity). Light shining on a metal surface causes electrons to be ejected with a certain velocity, and thus a certain energy. The only possible source of this energy is the light photons that strike the metal.

It was noticed that the energy of the electrons did *not* depend on the intensity of the light, but only on its frequency (i.e., color). The electrons ejected by a bright light have no more energy than those ejected by a dim light of the same frequency; they are simply more numerous. After accounting for the fact that electrons had to expend some of the energy imparted to them by light in becoming "unstuck" from the metal, it was determined that the energy of the electrons—and thus the energy of the photons—was simply proportional to the frequency of the light. If v denotes the frequency of a light photon, in cycles per second, then its energy is $E = hv$, where the proportionality constant $h = 6.625 \times 10^{-34}$ kg m^2/sec is called *Planck's constant*.

$E = hv$

Photons also have momentum. From Theorem 3.9 and the fact that $\mu = 0$ it follows that $p = E/c = hv/c$. This is often written in terms of the *wavelength* λ of the light photon. For any wave motion, the wavelength λ, in meters per wave, times the frequency v, in waves per second, gives the velocity of the wave, in meters per second. For light waves this is $\lambda v = c$, so $v/c = 1/\lambda$. In terms of its wavelength, the momentum of a light photon is therefore $p = h/\lambda$.

The momentum of light

Exercises

1. (a) Describe the shape of the worldcurve $\mathbb{X}(t) = (t, r\cos\omega t, r\sin\omega t)$ in terms of the parameters r and ω.

   (b) Calculate the proper time function $\tau(t)$ for $\mathbb{X}$. Compare proper time τ to coordinate time t.

2. Show that any worldcurve $\mathbb{X}(q) = (t(q), x(q), y(q), z(q))$ can be given a new parametrization as a graph: $\widehat{\mathbb{X}}(t) = (t, \widehat{x}(t), \widehat{y}(t), \widehat{z}(t))$.

3. Prove Theorem 3.4 assuming the results of Theorem 3.1. That is, show that different parametrizations of the same worldcurve yield the same measure of proper time along the worldcurve.

4. Suppose $t = \varphi(\tau)$ is the inverse of the elliptic integral

$$\tau(t) = \int_0^t \sqrt{1 - k^2 \sin^2 t}\, dt.$$

Show that

$$\varphi' = \frac{1}{\sqrt{1 - k^2 \sin^2 \varphi}} \quad \text{and} \quad \varphi'' = \frac{k^2 \sin \varphi \cos \varphi}{\left(1 - k^2 \sin^2 \varphi\right)^2}.$$

Hence show that $\mathbb{A}(\tau) = \left(\varphi'', k(\varphi'' \sin \varphi + (\varphi')^2 \cos \varphi)\right)$ reduces to

$$\mathbb{A}(\tau) = \frac{k \cos \varphi}{(1 - k^2 \sin^2 \varphi)^2} \left(k \sin \varphi, 1\right).$$

5. The curve $\mathbb{X}(q) = (t(q), z(q))$, $a \leq q \leq b$ is said to be **spacelike** if $\mathbb{X}'(q)$ is a spacelike vector for all q in $[a, b]$. We define the **Minkowski length** of $\mathbb{X}$ as

$$l = \int_a^b \|\mathbb{X}'(q)\| dq = \int_a^b \sqrt{\left(\frac{dz}{dq}\right)^2 - \left(\frac{dt}{dq}\right)^2}\, dq.$$

(a) Construct a parametrization of the branch of the hyperbola $t^2 - z^2 = 1$ that lies in the half plane $t > 0$ and use it to prove that the hyperbola is a spacelike curve.

(b) Show that the Minkowski length of the arc of the hyperbola from $(1, 0)$ to $(\cosh q, \sinh q)$ is exactly q.

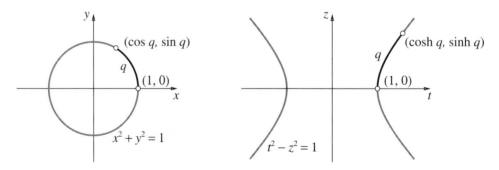

Minkowski length allows us to construct a definition of the hyperbolic functions that runs parallel to the usual definition of the circular functions. The circular functions give the coordinates of the point $(\cos q, \sin q)$ that lies at the distance q from the point $(1, 0)$ on the unit circle $x^2 + y^2 = 1$. See the diagram above. By the previous exercise, we can now say that the hyperbolic functions give the coordinates of the point $(\cosh q, \sinh q)$ that lies at the *Minkowski* distance q from the point $(1, 0)$ on the unit hyperbola $t^2 - z^2 = 1$.

6. Show that $pc = Ev/c$, where $p = \|\mathbf{p}\|$, $v = \|\mathbf{v}\|$.
7. Photons move at the speed of light, $v = c$, with finite energy and momentum. Show that the rest mass μ of such bodies must be 0. (Suggestion: Consider $\|\mathbb{P}\|^2$.)

Further Reading for Chapter 3

Newtonian mechanics is part of the core of physics and is discussed in many texts; *The Feynman Lectures* [13] and *Spacetime Physics* [28] have the advantage of working in the context of special relativity. Einstein introduces relativistic mass in a very brief and readable paper [9].

Proposition 3.4 *If* $\widetilde{K}(t) = \frac{1}{2}\mu v^2(t)$ *is the classical kinetic energy of G in R's frame and* $\widetilde{\mathbf{f}} = \mu \mathbf{a}$ *is the classical 3-force acting on G, then*

$$\frac{d\widetilde{K}}{dt} = \widetilde{\mathbf{f}} \cdot \mathbf{v}.$$

PROOF: Since $v^2 = \mathbf{v} \cdot \mathbf{v}$, we can write $\widetilde{K}(t) = \frac{\mu}{2} \mathbf{v} \cdot \mathbf{v}$. Therefore,

$$\frac{d\widetilde{K}}{dt} = \frac{\mu}{2}(\mathbf{v} \cdot \mathbf{v})' = \frac{\mu}{2}(2\mathbf{v}' \cdot \mathbf{v}) = \mu \mathbf{v}' \cdot \mathbf{v} = \mu \mathbf{a} \cdot \mathbf{v} = \widetilde{\mathbf{f}} \cdot \mathbf{v}. \quad \text{END OF PROOF}$$

Let us therefore interpret $\mathbf{f} \cdot \mathbf{v}$ (which involves the *relativistic* 3-force $\mathbf{f}$) as the time rate of change of *relativistic* kinetic energy K of G in R's frame. Thus $\mathbb{F} \cdot \mathbb{U} = 0$ becomes

$$c^2 \frac{dm}{dt} = \frac{dK}{dt}, \quad \text{implying} \quad c^2 m = K + \text{const}.$$

If we further require that $K = 0$ when $v = 0$, as in the classical case, then we can determine the constant of integration in the last equation: Since $m = \mu$ and $K = 0$ when $v = 0$, it follows that const $= \mu c^2$, the rest energy of G. We can summarize the previous discussion in the following definition and corollary.

CHAPTER 4

Arbitrary Frames

How can the inherent subjectivity of individual observations lead to conclusions that are objectively real, that is, to conclusions that become physical laws? Our answer is the principle of relativity: Any physical law must be formulated the same way by all observers. So far, though, "all observers" has been limited to Galilean observers in inertial frames. Why, Einstein asked, should there be such a limitation? In particular, can it be justified objectively, on physical grounds? In the decade following 1905 he found that gravity could not be described using only the inertial frames of special relativity; he became convinced that "there is nothing for it but to regard all imaginable systems of coordinates, on principle, as equally suitable for the description of nature" ([10], page 117). This is the principle of *general* relativity. Using it as a foundation, Einstein was able to create a new and revolutionary theory of gravity.

    In this chapter we first consider two common types of noninertial frames and explore their properties. Next, we survey Newtonian gravity briefly and then see why it fails to be compatible with special relativity. The remaining chapters then show how Einstein builds a new theory of gravity.

Admit accelerating observers

4.1 Uniform Rotation

A rotating frame is noninertial

We begin, as always, with two observers R and G whose frames are in relative motion. This time, though, the motion is rotational rather than linear. We assume that R and G remain at the same place with their x- and ξ-axes coinciding for all time. We assume also that R's frame is inertial but that G's (η, ζ)-plane rotates at a steady angular velocity around the ξ-axis. Now consider an object that moves freely (that is, obeying Galileo's law of inertia) in the spatial plane defined by $x = \xi = 0$. In R's $(1+2)$-dimensional spacetime with coordinates (t, y, z), the worldcurve of this object will be a straight line because R is inertial. However, in G's $(1+2)$-dimensional spacetime with coordinates (τ, η, ζ), the worldcurve will be a spiral—not a straight line. Thus G's frame is noninertial.

The spacetime map is nonlinear

To compare the two frames we must determine the map $M : G \to R$ that assigns to the Greek coordinates (τ, η, ζ) of an event E the Roman coordinates (t, y, z) of the same event. Since the observer G is stationary in relation to R, there is no time dilation, so we can write $t = \tau$. To determine y and z, we shall assume that the η-axis coincides with the y-axis when $\tau = t = 0$ and the (η, ζ)-plane rotates at an angular velocity of ω radians per second. Then, after τ seconds, the position of the (η, ζ)-plane in the (y, z)-plane is given by the rotation $R_{\omega\tau}$:

$$\begin{pmatrix} y \\ z \end{pmatrix} = R_{\omega\tau} \begin{pmatrix} \eta \\ \zeta \end{pmatrix} = \begin{pmatrix} \cos\omega\tau & -\sin\omega\tau \\ \sin\omega\tau & \cos\omega\tau \end{pmatrix} \begin{pmatrix} \eta \\ \zeta \end{pmatrix}.$$

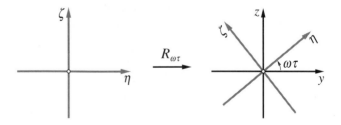

Therefore, $M : G \to R$ is the nonlinear map

$$M : \begin{cases} t = \tau, \\ y = \eta \cos \omega\tau - \zeta \sin \omega\tau, \\ z = \eta \sin \omega\tau + \zeta \cos \omega\tau. \end{cases}$$

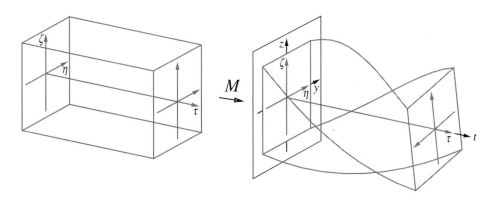

Since M is nonlinear, it is certainly not a Lorentz map. However, the "wire-frame" model of G makes the geometric action clear: M twists the (τ, η, ζ)-space around the τ-axis in mapping it to the (t, y, z)-space R.

Now suppose C is an observer who lies at a distance $r = \sqrt{\eta^2 + \zeta^2}$ from G and is fixed with respect to G's frame. In G, C's worldcurve is a straight line parallel to the τ-axis. In R, however, C has a nonzero speed and has a spiral worldcurve. To be definite, we place C at the point $(\eta, \zeta) = (0, r)$; then C's worldcurve in R can be parametrized as

Speed depends on radius

$$\mathbb{X}(\tau) = (t, y, z) = (\tau, -r \sin \omega\tau, r \cos \omega\tau).$$

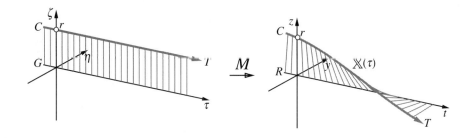

The linear velocity of C in R's frame is then

$$\mathbf{v} = \left(\frac{dy}{dt}, \frac{dz}{dt}\right) = (-r\omega \cos \omega t, -r\omega \sin \omega t)$$

(recall $\tau = t$), so speed is proportional to the radius: $v = \|\mathbf{v}\| = r\omega$.

The noninertial frame has a bounded domain

The velocity limitation of special relativity requires that $r\omega < 1$, implying $r < 1/\omega$. Thus *G's noninertial frame has a spatial boundary*: It can describe only those events $E = (\tau, \eta, \zeta)$ that lie *inside* the circular cylinder $r = \sqrt{\eta^2 + \zeta^2} = 1/\omega$ whose axis is G's worldline. An inertial frame has no boundary; it extends to infinity in all directions. This is the first of several important differences between inertial and noninertial frames.

The next difference will appear when we calculate C's proper time T. We must, as always, carry out the calculation in an inertial frame; in this case, we use R. Since

$$\mathbb{X}'(\tau) = (1, -r\omega \cos \omega \tau, -r\omega \sin \omega \tau) = (1, \mathbf{v}),$$

the proper-time function is

$$T(\tau) = \int_0^\tau \|\mathbb{X}'(\tau)\| \, d\tau = \int_0^\tau \sqrt{1 - r^2\omega^2} \, d\tau = \sqrt{1 - r^2\omega^2} \, \tau.$$

Proper time depends on radius

In particular, τ is *not* the proper time on C, though it is proportional to the true proper time T. The proportionality constant $\sqrt{1 - r^2\omega^2}$ is always less than 1, so C's clock runs more slowly than G's. We know that moving clocks run slow; this shows that rotational motion is no exception. Furthermore, since speed increases with r, so does time dilation. This is illustrated in the figure on the left below, which shows the worldlines of several observers at successively greater distances from G. Precisely when a fixed proper time T_0 occurs on the worldline that lies a distance r from G is given by the function

$$\tau = \frac{T_0}{\sqrt{1 - r^2\omega^2}},$$

whose graph is shown on the right. Note that $\tau \to \infty$ as $r \to 1/\omega$.

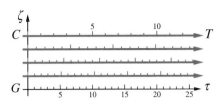
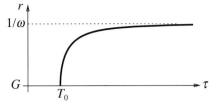

The Meaning of Coordinates

We are faced with an unprecedented development: Clocks that are stationary in G's frame are generating a multiplicity of time variables T that compete with G's own time variable τ. To understand this development, we need to pause and make explicit some of our assumptions about coordinates *in inertial frames*. For simplicity we work with $(1+1)$-dimensional spacetime.

So suppose G is a Galilean observer whose inertial frame has coordinates (τ, ζ). These coordinates have always served two purposes. First, they provide a system of *labels*, or names, by which we can distinguish one event from another. For example, when we regard a Lorentz transformation $H_u : G \to R$ as a *dictionary* that translates from one system of names to another, we are thinking of coordinates in this sense. Indeed, if we carry out the transformation and set

$$\begin{pmatrix} t \\ z \end{pmatrix} = \begin{pmatrix} \cosh u & \sinh u \\ \sinh u & \cosh u \end{pmatrix} \begin{pmatrix} \tau \\ \zeta \end{pmatrix},$$

then the coordinates (t, z) would serve G equally well as a means of distinguishing between events.

Coordinates as labels

But we don't use coordinates simply as labels. The second purpose they serve is to provide a system of *measurements*. That is, if E_1 is the event with coordinates (τ_1, ζ_1), then G considers that E_1 actually happened τ_1 seconds after the event $O = (0,0)$. Similarly, ζ_1 is the distance from G to E_1. Here τ_1 and ζ_1 are more than labels; they report the result of measurements that G makes. The coordinates (τ, ζ) are, in this sense, *natural*; G will usually prefer them to other coordinates—like (t, z)—that are merely labels in G. (Of course, back in R, (t, z) are natural coordinates that measure time and distance for R.)

Coordinates as measurements

Chapter 4 Arbitrary Frames

Timing by radar

There are at least two ways for G to measure the time of the event E_1. The first works like radar: G sends out a succession of light signals and waits for one to be reflected back from E_1. Suppose that signal is emitted at the event $E' = (\tau', 0)$ and returns at the event $E'' = (\tau'', 0)$, where G's own clock determines the values of τ' and τ''. Then $\tau_1 = (\tau' + \tau'')/2$. The important thing here is that the times of *all* events are related back to the times of events that happen on G's worldline.

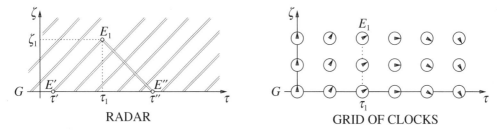

RADAR   GRID OF CLOCKS

Timing by a grid of clocks

The second way is to put a string of clocks, all identical to G's own, at fixed locations along the ζ-axis, and synchronize them. Since the clocks are stationary in relation to G, synchronization will present no theoretical difficulties. When E_1 occurs, just check the time on the clock located at the place $\zeta = \zeta_1$ where E_1 occurs; that will be the time coordinate τ_1. For this to work, all clocks in the grid must continue to stay synchronized with G's own clock. That is, all the clocks on a vertical line $\tau = $ constant must agree. We know that this happens in an inertial frame.

Measuring distances

Similarly, there are two ways for G to measure distances. One is radar. If the event $E_1 = (\tau_1, \zeta_1)$ is detected by the signal emitted at the event $E' = (\tau', 0)$ and returned at $E'' = (\tau'', 0)$, then $\zeta_1 = (\tau'' - \tau')/2$ (if $\zeta_1 > 0$; otherwise, $\zeta_1 = -(\tau'' - \tau')/2$). The second way is to install a grid of rulers along the ζ-axis. Since the rulers are stationary in G, they do not contract, and the grid remains unchanged over time. To determine ζ_1, just note the mark on the ruler at the place where E_1 occurs.

Radar and clocks disagree in a noninertial frame

It is clear now what happens when we move back to the rotating, noninertial frame G: the two ways of measuring time give different results! On the one hand, the grid of clocks gives us the T variables of the various observers C who occupy different places along the ζ-axis. On the other, radar gives us the τ variable, because it relates the time of an event anywhere in spacetime back to events along the τ-axis.

The observers C collectively define a set of coordinates (T, Z) that are related to G's original coordinates by the map $F : C \to G : (T, Z) \to (\tau, \zeta)$,

$$F : \begin{cases} \tau = \dfrac{T}{\sqrt{1 - Z^2 \omega^2}}, \\ \zeta = Z. \end{cases}$$

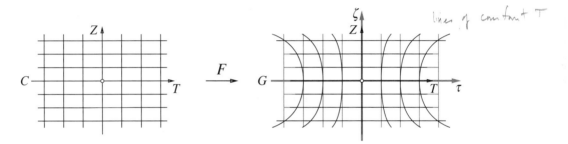

This is a nonlinear map, but notice that it is the identity on the T-axis itself. Here the two times T and τ agree. Away from the T-axis the image "flares out"; as $|Z|$ increases, T time runs more and more slowly in relation to τ time.

Which is right, C or G? Since both are valid extensions of correct procedures for marking coordinates in an inertial frame, we have no grounds for preferring one to the other. And in fact, we do not have to choose between them. Our experience with ordinary maps of the world will show us how to reconcile the two frames. There is a nice analogy between the two ways we measure time here and the two ways we measure east–west position on a map: by longitude and by mileage along parallels of latitude.

Both times are valid

Latitude and Longitude

Every map is equipped with a *scale* that tells you how a given length on the map translates to a distance on the ground. But traditional maps that cover nearly the whole world are noticeably

Maps and scales

A grid of scales

distorted toward the poles—Greenland and Antarctica look much too large. What has happened to the scale?

The distortions occur because these projections show meridians of longitude as parallel lines, implying that a degree of longitude represents a fixed distance on the ground. But this is just not so: At 60° north latitude, one degree of longitude covers an east–west distance of about 35 miles; at the equator, the distance is twice that. Thus a single scale is inadequate; these projections need a multiplicity of scales—one for each latitude. In fact, you can sometimes find them on such a map; five or six scales are stacked into a curved grid, which is then tucked away in an empty spot somewhere. In the map below (a Miller cylindrical projection), you can see the grid in the South Atlantic, just above an oversized Antarctica. The six scales cover 1000 miles of east–west distance, and the grid flares out at higher latitudes because you have to traverse more of the map—cross more longitude meridians—to cover those 1000 miles.

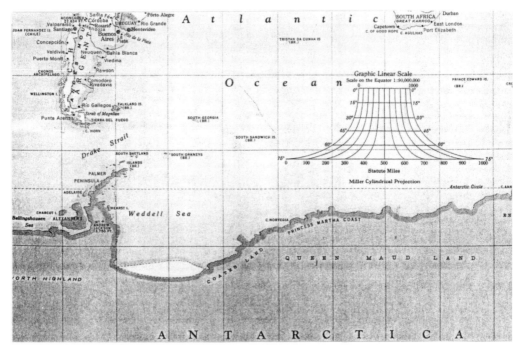

§4.1 Uniform Rotation

Calculate the grid

To understand the grid better, let us construct it ourselves. Along the parallel at latitude φ, an east–west distance of d miles spans a longitudinal arc θ. We want to determine θ as a function of d and φ. For ease of calculation we measure θ and φ in radians, and we suppose the earth to be a sphere whose radius is R miles. The equator ($\varphi = 0$) is therefore a circle of radius R, so a longitudinal arc θ along the equator covers a distance of $d = \theta R$ miles (by definition of radian measure!). Thus $\theta = d/R$ when $\varphi = 0$.

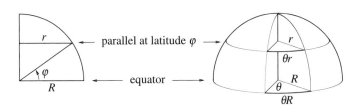

Now consider the circle that is parallel to the equator at latitude $\varphi \neq 0$; it has radius $r = R \cos \varphi$. The longitudinal arc θ along this circle covers a distance of $d = \theta r = \theta R \cos \varphi$ miles. Therefore,

$$\theta = \theta(d, \varphi) = \frac{d}{R \cos \varphi} = \frac{d}{R} \sec \varphi.$$

Notice that θ is defined only for $|\varphi| < \pi/2$, and $\theta \to \infty$ as $\varphi \to \pm \pi/2$. The grid itself is the image of the map $F^* : (d, \varphi) \to (\theta, \varphi)$ given by

$$F^* : \begin{cases} \theta = \dfrac{d}{R} \sec \varphi, \\ \varphi = \varphi. \end{cases}$$

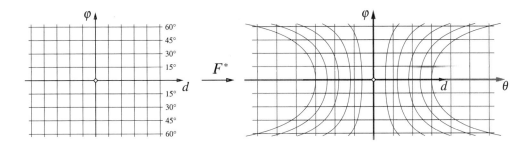

152 Chapter 4 Arbitrary Frames

Is spacetime curved?

Even though the functions defining F^* and F are not the same, they look remarkably similar. Could this similarity spring from a common cause? The reason for the F^* grid is intuitively clear: It is impossible to make a flat map of the curved earth without distortion, that is, without using different scale factors in different places. The F^* grid tells us, ultimately, that the earth is curved. Does the F grid tell us that spacetime is curved? What, indeed, can this mean? We move on now to gather more evidence.

Circumferential Distances

The ratio of circumference to diameter

Return to the (τ, η, ζ) frame of G that we assume to be rotating in relation to the inertial (t, y, z) frame of R. To measure distance, let us install a grid of rulers in the (η, ζ)-plane, arranging them on radial lines and concentric circles like polar coordinates. We shall assume that the rulers lying on a given circle are much shorter than the radius, so that they can follow the circle closely and measure its length accurately. Now use this grid to measure the circumference and the diameter of a circle, and then calculate the ratio.

Rulers contract along the circumference

We look at all this from the inertial frame R. All the rulers are moving, so they contract *in the direction of motion*. This has no effect on the radial rulers, but it makes the circumferential rulers shorter because they are aligned in the direction of motion. Specifically, on the circle of radius r, the circumferential rulers are moving with speed $v = r\omega$, so they contract in length by the factor $\sqrt{1-v^2} = \sqrt{1-r^2\omega^2} < 1$.

The (η, ζ)-plane is non-Euclidean

Consider the circle of radius $r > 0$ centered at the origin. There is no ambiguity here; R and G agree on radial lengths. From the point of view of R's own (fixed) rulers, the circumference has length $2\pi r$. But R considers the circumferential rulers in G's frame to be shorter by the factor $\sqrt{1-r^2\omega^2}$. Therefore, with *these* rulers we obtain the ratio

$$p = \frac{\text{circumference}}{\text{diameter}} = \frac{2\pi r/\sqrt{1-r^2\omega^2}}{2r} = \frac{\pi}{\sqrt{1-r^2\omega^2}} > \pi.$$

In other words, if we just count the rulers, we find there are $p > \pi$ times as many around the circumference as across the diameter. Initially, this is R's count, but it is clear that G's count must be the same. Since these are the rulers that G uses for measuring (because they are at rest in G's frame), there are distance measurements that G makes in the (η, ζ)-plane that do not obey the laws of Euclidean geometry.

Once again, a map of the earth can help us understand what is happening in G. Consider a map of the north polar region: What is the circumference C of the latitude circle that lies at a distance d miles from the North Pole? On the flat map, the distance will appear to be $2\pi d$ miles, but the actual distance on the earth is less. To calculate C, note that the latitude circle lies in a plane, and the center of the circle lies where that plane intersects the polar axis. Let the radius measured from this center be r miles. If the radial arc of length d miles (from the pole to the circle) subtends an angle of ψ radians from the center of the earth and the radius of the earth is R miles, then $d = R\psi$, $r = R\sin\psi$, and therefore

Polar maps are non-Euclidean, too

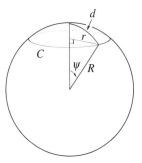

$$C = 2\pi r = 2\pi R \sin\psi.$$

The ratio we want is

$$\frac{C}{2d} = \frac{2\pi R \sin\psi}{2R\psi} = \frac{\sin\psi}{\psi}\pi.$$

Since $\sin\psi/\psi < 1$ when $\psi > 0$, this ratio is always *less than* π. Thus, distance measurements in the polar map do not obey the laws of Euclidean geometry.

Here the reason is obvious: Euclidean geometry applies to *flat* planes, and the earth is curved. The evidence therefore suggests that the (η, ζ)-plane is curved, too—but perhaps in a different way: While the ratio of circumference to diameter is less than π on a sphere, it is greater than π on the rotating plane.

Exercises

1. (a) Consider the function $\tau = T_0/\sqrt{1 - r^2\omega^2}$ that gives the proper time of an observer at a distance r from the center

of a coordinate frame rotating with angular velocity ω with respect to an inertial frame. Find the inverse $r = r(\tau)$ that expresses r in terms of τ.

(b) Confirm that $r(\tau)$ is defined only for $\tau \geq T_0$ and that $r \to 1/\omega$ as $\tau \to \infty$.

(c) Determine $r'(\tau)$ and show that $r' \to \infty$ as $\tau \to T_0$, while $r' \to 0$ as $\tau \to \infty$. Plot the graph $r = r(\tau)$ and confirm that it agrees with the graph in the text.

2. (a) Confirm that the function

$$F : \begin{cases} \tau = \dfrac{T}{\sqrt{1 - Z^2\omega^2}}, \\ \zeta = Z \end{cases}$$

maps a rectangular grid in the (T, Z)-plane in the way shown in the text.

(b) Carry out a similar analysis to confirm that the map

$$F : \begin{cases} \theta = \dfrac{d}{R} \sec\varphi, \\ \varphi = \varphi \end{cases}$$

that scales longitude at different latitudes has the form indicated for it in the text.

3. Show that for small r, we can write

$$\frac{\pi}{\sqrt{1 - r^2\omega^2}} = \pi + \frac{\pi\omega^2}{2} r^2 + O(r^4).$$

Hyperbolic and elliptic geometries

This is the ratio of the circumference to the diameter of a circle of radius r in a rotating coordinate frame. Since the ratio exceeds π, the non-Euclidean geometry that prevails in the rotating frame is said to be *hyperbolic*. By contrast, the non-Euclidean geometry of the sphere is said to be *elliptic* because the ratio is less than π.

4. Areas on a sphere give a further indication that the geometry on a sphere is elliptic. Consider the spherical cap bounded by the circle that lies at distance d from the north pole on a sphere

of radius R. It can be shown (see the exercises in Section 5.2) that the area of the cap is equal to

$$A = 2\pi R^2 \left(1 - \cos \frac{d}{R}\right).$$

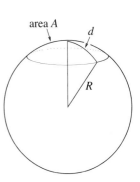

Use Taylor's theorem to show that

$$\frac{A}{\pi d^2} = 1 - \frac{d^2}{12R^2} + O(d^4),$$

and hence conclude that, for small d, the area of the cap is less than it would be in Euclidean geometry.

5. While the geometries of the sphere and of the rotating frame are both non-Euclidean, they are nonetheless "infinitesimally Euclidean" in the following sense.

   (a) Show that if we take a circle of radius r *as measured on the surface of a sphere*, then the ratios

   $$\frac{\text{circumference}}{2\pi r} \quad \text{and} \quad \frac{\text{area}}{\pi r^2}$$

   both approach 1 as $r \to 0$. Thus a sufficiently small portion of a sphere is indistinguishable from a flat Euclidean plane.

   (b) Show that a sufficiently small portion of a rotating coordinate frame is likewise indistinguishable from a flat Euclidean plane.

4.2 Linear Acceleration

In this section we explore some of the properties of a coordinate frame that undergoes constant linear acceleration with respect to an inertial frame.

Let $R : (t, z)$ be the inertial frame of a Galilean observer and suppose that a second observer G moves with a constant acceleration α with respect to R in the direction of the positive z-axis. We saw in Section 3.3 that the worldcurve of G is the graph of the function

Equation of linear acceleration

$$z = z_0 + \frac{1}{\alpha}\sqrt{1 + \alpha^2(t - t_0)^2}.$$

This is a hyperbola whose asymptotes are the lines $z - z_0 = \pm(t - t_0)$; G has velocity 0 with respect to R at the event $(t, z) = (t_0, z_0 + 1/\alpha)$.

The Radar Frame

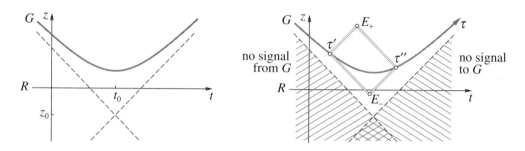

Coordinates by radar

What can we say about G's noninertial frame, and how is it connected with R's frame? We shall use radar to determine the (τ, ζ)-coordinates of an event in G's frame. According to the discussion in the previous section, if G sends out a light signal at proper time τ' that is reflected from the event E and received back by G at the proper time τ'', then G's coordinates for E are

$$E_\pm = \left(\frac{\tau' + \tau''}{2}, \pm\frac{\tau'' - \tau'}{2} \right).$$

Two possibilities arise, as we indicate here, when we ignore the direction from which the reflected signal returns: E_+ lies on the positive side of the ζ-axis and E_- on the negative. The ζ-coordinates $\pm(\tau'' - \tau')/2$ are the negatives of each other. If we take into account the direction of the reflected signal, E is not ambiguous.

The limits of G's coordinate frame

There are some events that G will never detect by radar. No signal from G can reach the region below the asymptote that runs from upper left to lower right, no matter how far back in time we push G's worldcurve. Furthermore, no signal reflected from a point in the lower right region will ever reach G, no matter how far into the future G's worldcurve extends. Therefore G can assign coordinates only to those events in the unshaded region: G's coordinate frame covers only a portion of spacetime. Exactly the same was true for the rotating noninertial frame we studied in the previous section; the details were different, however. We are seeing further evidence that noninertial frames are necessarily limited in scope.

To understand how G describes events, we want to connect G's frame to R's as we always have, by an appropriate map $M : G \to R$. To simplify our analysis, we move the point (t_0, z_0) to the origin. Then G's worldcurve in R has following proper-time parametrization:

Find G's grid in R

$$\mathbb{G}(\tau) = \left(\frac{1}{\alpha} \sinh \alpha \tau, \frac{1}{\alpha} \cosh \alpha \tau \right).$$

Theorem 4.1 *The map $M : G \to R$ is given by the equations*

$$M : \begin{cases} t = \dfrac{e^{\alpha \zeta}}{\alpha} \sinh \alpha \tau, \\ z = \dfrac{e^{\alpha \zeta}}{\alpha} \cosh \alpha \tau. \end{cases}$$

PROOF: Suppose the event E has coordinates (τ^*, ζ^*) in G and (t^*, z^*) in R. We must determine how t^* and z^* depend on τ^* and ζ^*. There are two steps: First, we use radar to connect E to the emission and reception events E' and E'' along G's worldline; second, we use the parametrization $\mathbb{G}$ to connect the Greek and Roman coordinates along G's worldcurve.

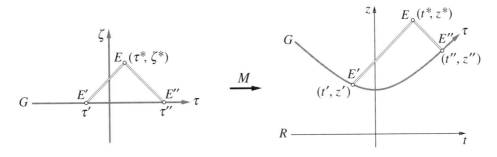

In G we first connect the coordinates of E to the emission and reception times τ' and τ'':

Step 1: connect E to E' and E''

$$\tau^* = \frac{\tau' + \tau''}{2}, \qquad \zeta^* = \frac{\tau'' - \tau'}{2}.$$

Note that for the sake of definiteness we have taken E above G's worldline, so $\zeta^* > 0$. In step 2 we will need τ' and τ'' in terms of τ^* and ζ^*:

$$\tau' = \tau^* - \zeta^*, \qquad \tau'' = \tau^* + \zeta^*.$$

In R there is a corresponding connection between the coordinates of E and those of the emission and reception events. Since E' and E lie on a line of slope 1, while E and E'' lie on a line of slope -1, we have

$$\frac{z^* - z'}{t^* - t'} = 1, \qquad \frac{z^* - z''}{t^* - t''} = -1.$$

Now solve these equation for t^* and z^*:

$$t^* = \frac{t'' + t' + z'' - z'}{2}, \qquad z^* = \frac{t'' - t' + z'' + z'}{2}.$$

Step 2: connect Greek and Roman coordinates along G's worldcurve

On G's worldcurve we have the following connection between the Greek and Roman coordinates of E' and E'':

$$t' = \frac{1}{\alpha} \sinh \alpha \tau', \qquad t'' = \frac{1}{\alpha} \sinh \alpha \tau'',$$

$$z' = \frac{1}{\alpha} \cosh \alpha \tau', \qquad z'' = \frac{1}{\alpha} \cosh \alpha \tau''.$$

Therefore,

$$t^* = \frac{t'' + t' + z'' - z'}{2} = \frac{\sinh \alpha \tau'' + \sinh \alpha \tau' + \cosh \alpha \tau'' - \cosh \alpha \tau'}{2\alpha}$$

$$= \frac{e^{\alpha \tau''} - e^{-\alpha \tau'}}{2\alpha} = \frac{e^{\alpha(\tau^* + \zeta^*)} - e^{-\alpha(\tau^* - \zeta^*)}}{2\alpha} = \frac{e^{\alpha \tau^*} e^{\alpha \zeta^*} - e^{-\alpha \tau^*} e^{\alpha \zeta^*}}{2\alpha}$$

$$= \frac{e^{\alpha \zeta^*}}{\alpha} \sinh \alpha \tau^*.$$

In going from the first line to the second we use the fact that

$$\sinh A + \cosh A = e^A \quad \text{and} \quad \sinh A - \cosh A = -e^{-A}.$$

An entirely similar argument shows that

$$z^* = \frac{e^{\alpha \zeta^*}}{\alpha} \cosh \alpha \tau^*.$$

This establishes the formulas for $M: G \to R$. **END OF PROOF**

The image of M

The map M is nonlinear, but its image has a form that is easy to visualize. The vertical grid lines $\tau = k$ are mapped to rays through the origin:

$$t = K_1 e^{\alpha \zeta}, \qquad z = K_2 e^{\alpha \zeta},$$

where $K_1 = \sinh \alpha k / \alpha$ and $K_2 = \cosh \alpha k / \alpha$. Notice that $K_2 > 0$, so $z > 0$, and the rays all lie in the upper half plane. In fact, since $z = K_2 t / K_1$ and $|K_2/K_1| = |\coth \alpha k| \geq 1$, the rays all lie inside the region bounded by the lines $z = \pm t$.

The horizontal grid lines $\zeta = k$ are mapped to hyperbolas:

$$t = K \sinh \alpha \tau, \qquad z = K \cosh \alpha \tau,$$

where $K = e^{\alpha k}/\alpha$. Since $z^2 - t^2 = K^2$, these are concentric hyperbolas that have the same asymptotes $z = \pm t$. Kinematically, these are the worldcurves of observers who experience a fixed acceleration $1/K = \alpha e^{-\alpha k}$ from the point of view of R and, consequently, any inertial observer.

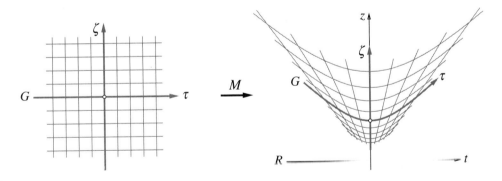

However, the amount of acceleration depends on the parameter k; the observer with the larger value of k experiences the smaller acceleration. Consequently, a fixed value of ζ does *not*

Worldcurves at fixed distances from G

represent a fixed distance from G from R's point of view. This is essentially Fitzgerald contraction; the amount of contraction increases over time because the observers collectively speed up in R's frame.

Where *are* the worldcurves that experience zero acceleration in relation to G? In R, they appear as the hyperbolas that are parallel to G's hyperbola; they are the family given by the equation

$$\alpha^2(z-z_0)^2 - \alpha^2 t^2 = 1$$

that is parametrized by the constant z_0. To find the equation of the family in G, we just use the values of t and z given by the map M. Thus,

$$\alpha^2 \left(\frac{e^{\alpha \zeta}}{\alpha} \cosh \alpha \tau - z_0 \right)^2 - \alpha^2 \left(\frac{e^{\alpha \zeta}}{\alpha} \sinh \alpha \tau \right)^2 = 1.$$

This reduces to

$$e^{2\alpha \zeta} \left(\cosh^2 \alpha \tau - \sinh^2 \alpha \tau \right) - 2\alpha z_0 e^{\alpha \zeta} \cosh \alpha \tau + \alpha^2 z_0^2 - 1 = 0.$$

Now let $u = e^{\alpha \zeta}$ and use the fact that $\cosh^2 \alpha \tau - \sinh^2 \alpha \tau = 1$ to get the ordinary quadratic equation

$$u^2 - 2(\alpha z_0 \cosh \alpha \tau) u + \alpha^2 z_0^2 - 1 = 0.$$

Because the worldcurves are the *upper* branches of hyperbolas in R, we choose the *positive* square root in the quadratic formula in G:

$$u = \frac{2\alpha z_0 \cosh \alpha \tau + \sqrt{4\alpha^2 z_0^2 \cosh^2 \alpha \tau - 4\alpha^2 z_0^2 + 4}}{2}$$

$$= \alpha z_0 \cosh \alpha \tau + \sqrt{\alpha^2 z_0^2 (\cosh^2 \alpha \tau - 1) + 1}$$

$$= \alpha z_0 \cosh \alpha \tau + \sqrt{\alpha^2 z_0^2 \sinh^2 \alpha \tau + 1}.$$

We now have the equation of the family of worldcurves in G:

$$\zeta = \frac{1}{\alpha} \ln \left(\alpha z_0 \cosh \alpha \tau + \sqrt{\alpha^2 z_0^2 \sinh^2 \alpha \tau + 1} \right).$$

These are shown below as the dark overlay.

§4.2 Linear Acceleration

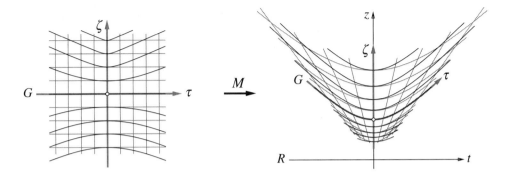

Now suppose we have objects at rest in R's frame; their worldcurves are the horizontal lines $z = b$. Since G accelerates past them, in G's frame they should appear to be accelerating downward. To show this, we determine their worldcurves in G. They will be defined by the equation

Bodies motionless in R accelerate downward in G

$$\frac{e^{\alpha\zeta}}{\alpha} \cosh \alpha\tau = b,$$

which implies $e^{\alpha\zeta} = \alpha b \operatorname{sech} \alpha\tau$ and thus

$$\zeta = \frac{1}{\alpha} \ln(\alpha b \operatorname{sech} \alpha\tau) = \frac{1}{\alpha} \ln(\operatorname{sech} \alpha\tau) + \frac{\ln(\alpha b)}{\alpha}.$$

The second term, $\ln(\alpha b)/\alpha$, is a constant, so these curves are all vertical translates of the particular curve $\zeta = \ln(\operatorname{sech} \alpha\tau)/\alpha$. Since

$$\frac{d\zeta}{d\tau} = -\tanh \alpha\tau \to \mp 1 \quad \text{as } \tau \to \pm\infty,$$

the curves have slopes that approach -1 as $\tau \to +\infty$ and $+1$ as $\tau \to -\infty$. For τ small, however, Taylor's theorem shows that

$$\zeta \approx -\frac{\alpha}{2}\tau^2 + \frac{\ln(\alpha b)}{\alpha};$$

therefore, in G these objects look like bodies falling vertically downward with the same acceleration $-\alpha$. They are at rest when $\tau = 0$, which is when G is at rest relative to R.

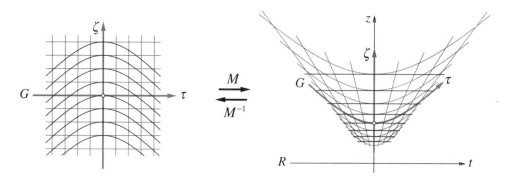

The inverse map M^{-1}

These curves in G come, in effect, from the inverse map M^{-1} that is defined on the region V where $z > |t|$. The exercises ask you to verify that $M^{-1} : V \to G$ is given there by the formulas

$$M^{-1} : \begin{cases} \tau = \dfrac{1}{\alpha} \tanh^{-1}\left(\dfrac{t}{z}\right), \\[2pt] \zeta = \dfrac{1}{2\alpha} \ln\left(\alpha^2 (z^2 - t^2)\right). \end{cases}$$

Light cones under M and M^{-1}

Even though M and M^{-1} are nonlinear, they preserve the *form* of the light cones. In particular, M maps straight lines of slope ± 1 to straight lines of slope ± 1. For suppose $\zeta = \pm \tau + k$; then

$$t = \frac{e^{\pm \alpha \tau} e^{\alpha k}}{\alpha} \sinh \alpha \tau,$$

$$z = \frac{e^{\pm \alpha \tau} e^{\alpha k}}{\alpha} \cosh \alpha \tau,$$

so

$$z \mp t = \frac{e^{\pm \alpha \tau} e^{\alpha k}}{\alpha} (\cosh \alpha \tau \mp \sinh \alpha \tau) = \frac{e^{\pm \alpha \tau} e^{\alpha k} e^{\mp \alpha \tau}}{\alpha} = \frac{e^{\alpha k}}{\alpha}.$$

In other words, $z = \pm t + \dfrac{e^{\alpha k}}{\alpha}$, a straight line of slope ± 1 in the (t, z)-plane.

A Frame of Rulers and Clocks

Coordinates by rulers and clocks

Suppose we use a grid of rulers and clocks instead of radar to specify the coordinates of G's linearly accelerating frame. When

§4.2 Linear Acceleration

we made this change in the rotating frame, we saw that the "clock" time at each place $\zeta \neq 0$ is different from its "radar" time—which always relates back to G's own time on the τ-axis. Exactly the same thing happens here; Einstein demonstrated this using the Doppler effect.

To carry out this demonstration, place clocks at certain fixed distances along the ζ-axis. To the clock at the location $\zeta = h$ we associate the observer C; let T be the proper time for C, as kept by this clock. Quick calculations (cf. Theorem 3.8) show that the worldcurves of G and C are the graphs of

$$z_G = \frac{1}{\alpha}\sqrt{\alpha^2 t^2 + 1} \quad \text{and} \quad z_C = \frac{e^{\alpha h}}{\alpha}\sqrt{\frac{\alpha^2}{e^{2\alpha h}}t^2 + 1},$$

respectively. These are *concentric*, rather than *parallel*, hyperbolas in R. In the form they are written we can see that the accelerations of G and C are $1/\alpha$ and $e^{\alpha h}/\alpha$, respectively. When $t = 0$, their separation on the z-axis is

$$z_C - z_G = \frac{e^{\alpha h} - 1}{\alpha} = h + O(h^2).$$

The worldcurves are concentric hyperbolas

In other words, their separation in R agrees with their separation in G, at least to first order in h.

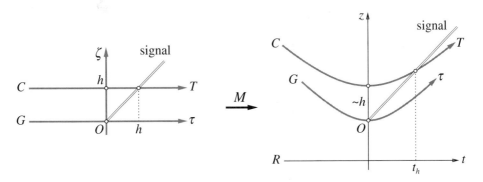

So suppose G emits a light signal of frequency ν at the event O (which is the origin in G's frame but not R's). In the inertial frame R, G's velocity is 0, so there is no shift at time of emission. At the time of arrival, however, C has a positive velocity v_h in R's

The Doppler effect

frame, so the frequency will be red-shifted by the Doppler effect. According to C the frequency will be (cf. Section 2.4)

$$N = \nu \sqrt{\frac{1 - v_h}{1 + v_h}}.$$

To determine v_h, note first that the signal has the worldline $\zeta = \tau$ in G, so the arrival time is just $\tau = h$ in G's frame. Let $t = t_h$ be the arrival time in R's frame; in the exercises you are asked to show that

$$t_h = \frac{e^{\alpha h}}{\alpha} \sinh \alpha h, \qquad v_h = \tanh \alpha h, \qquad N = \frac{1}{e^{\alpha h}} \nu.$$

Comparing proper times of C and G

In fact, the time interval between the peaks of oscillation (the *period*) gives us a way to compare the rates of the two clocks, that is, the proper times of C and G. At emission G's clock says that the period is

$$\Delta \tau = \frac{1}{\nu} \text{ sec};$$

at reception, C's clock says that the period is

$$\Delta T = \frac{1}{N} = e^{\alpha h} \Delta \tau \text{ sec}.$$

When we worked in the rotating frame we expressed G's time in terms of C's; to make comparisons easier, we do the same here:

$$\Delta \tau = F(h) \Delta T, \qquad F(h) = e^{-\alpha h}.$$

$F(h)$ tells us how the two time scales compare on the line $\zeta = h$, which is C's worldline: G's clock runs more slowly than C's precisely when $F(h) < 1$, that is, when $h > 0$. The factor F here is completely analogous to the map F in the rotating frame that showed how the time T measured by a grid of identical clocks was related to the time τ measured by G using radar.

Time dilation and compression

As in the rotating frame, the gray vertical lines $\tau = $ const in the figure below represent radar measurements made by G, while the curved lines $T = $ const of the dark overlay represent measurements made by various observers C using identical clocks at different locations along the ζ-axis. As we would expect, the two

§4.2 Linear Acceleration   165

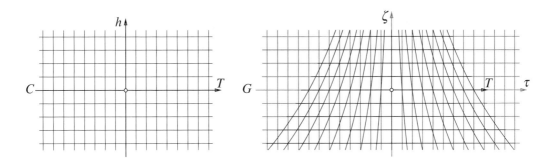

agree on the τ-axis; but they disagree everywhere else. In the rotating frame, the dark grid was symmetric across the τ-axis, but not here. C's clock runs faster than G's when $h > 0$ but slower when $h < 0$. In terms of frequencies, the Doppler effect causes a *red* shift for observers C positioned above G ($h > 0$) but a *violet* shift for those below G.

Once again we find that the radar grid and the rulers-and-clocks grid disagree. We have further evidence that in the noninertial frame of an accelerated observer G, no coordinates simultaneously give measurements of a single ruler and clock—as they naturally do in an inertial frame. A map of the earth suffers the same defect: Measurements on the map cannot all be made proportional to measurements on the surface of the earth. No accurate map of (a substantial portion of) the earth can be made with just a single scale. On the earth we ascribe this defect to curvature—more precisely, to the fact that the earth is curved but the map is not. By analogy, we consider that the same may be true for spacetime: Since measurements within the accelerated frames that we have considered are not proportional to measurements of the corresponding spacetime intervals, perhaps spacetime itself is curved. Our speculations can be summarized this way:

Further evidence for curvature

$$\begin{array}{c}\text{accelerated}\\\text{motions}\end{array} \Longrightarrow \begin{array}{c}\text{noninertial}\\\text{frames}\end{array} \Longrightarrow \begin{array}{c}\text{curved}\\\text{spacetime}\end{array}$$

Exercises

1. Confirm that $\zeta = \frac{1}{\alpha}\ln(\mathrm{sech}\,\alpha\tau) = -\frac{\alpha}{2}\tau^2 + O(\tau^4)$, as claimed in the text.

2. (a) Verify that the following maps are inverses; assume $\alpha > 0$.

$$M : \begin{cases} t = \dfrac{e^{\alpha\zeta}}{\alpha}\sinh\alpha\tau, \\ z = \dfrac{e^{\alpha\zeta}}{\alpha}\cosh\alpha\tau, \end{cases} \qquad M^{-1} : \begin{cases} \tau = \dfrac{1}{\alpha}\tanh^{-1}\left(\dfrac{t}{z}\right), \\ \zeta = \dfrac{1}{2\alpha}\ln\left(\alpha^2(z^2 - t^2)\right). \end{cases}$$

(b) Sketch in the (t, z)-plane the image of the full (τ, ζ)-plane under the map M.

(c) Determine the image of the straight line $z = vt + z_0$ under the map M^{-1}. Indicate how the parameters v and z_0 influence the image.

(d) Consider the grid in R shown on the right in the figure below. It is formed by the vertical lines $t = t_0$ and the hyperbolas $\alpha^2(z - z_0)^2 - \alpha^2 t^2 = 1$ parallel to the image of the τ-axis. Determine the equations of the curves in the image of this grid under that map M^{-1}. Confirm that the image grid has the form shown on the left below by making an accurate sketch of it.

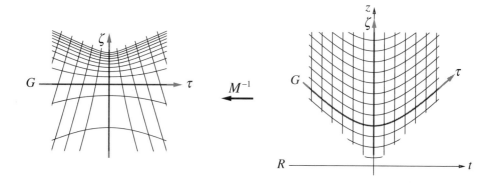

3. (a) Show that in the discussion in the text of the Doppler effect on the observer C, the coordinates of the event represent-

ing the arrival of the signal in R's frame are

$$t_h = \frac{e^{\alpha h}}{\alpha} \sinh \alpha h, \qquad z_h = \frac{e^{\alpha h}}{\alpha} \cosh \alpha h.$$

(b) Show that C's velocity at this event is $v_h = \tanh \alpha h$ and show that this implies

$$F(h) = \sqrt{\frac{1 - v_h}{1 + v_h}} = e^{-\alpha h}.$$

(c) Consider the map $(T, h) \to (\tau, \zeta)$ given by

$$\tau = F(h)T,$$
$$\zeta = h.$$

Verify that the image of the (T, h) grid under this map has the form shown in the text.

4.3 Newtonian Gravity

The Equivalence Principle

In Newtonian mechanics, *mass* appears in two different—and conceptually distinct—contexts. The first is inertia. In Newton's second law of motion, force is proportional to acceleration, and the constant of proportionality is the body's **inertial mass**. The other context is gravity. To help us understand this, let us look first at electricity.

Inertial mass

Any two electrically charged bodies exert an electric force on each other; its magnitude F is proportional to the electric charges q_1 and q_2 of the two bodies and inversely proportional to the square of the distance r between them:

Electrical and gravitational forces

$$F = k \frac{q_1 q_2}{r^2}.$$

But some bodies have no electric charge, so the electric force can be zero. Furthermore, the electric force can attract or repel, depending on the signs of the two charges. By contrast, the gravitational force applies to *all* bodies, and it is always attractive. The formula, Newton's **law of universal gravitation**, is exactly analogous to the formula for the electric force:

$$F = G\frac{m_g M}{r^2}, \qquad G = 6.67 \times 10^{-11} \frac{\text{m}^3}{\text{kg sec}^2}.$$

In this formula the role of the "gravitational" charge of a body is taken over by its **gravitational mass**. There is no reason, on the face of it, why gravitational mass should be the same as inertial mass. To compare the two notions of mass, consider how gravity acts on a body at the surface of the earth. Then r, which is the distance from the body to the center of the earth, is constant; the gravitational mass M of the earth is also constant. It follows that the gravitational force on the body is simply proportional to its gravitational mass m_g:

$$F = Km_g, \qquad K = \frac{GM}{r^2} = \text{constant}.$$

Now, Newton's second law of motion tells us that the same force F is also $m_i g$, where m_i is the body's inertial mass and g is the acceleration due to gravity. Thus

$$Km_g = F = m_i g, \qquad \text{so} \qquad g = K\frac{m_g}{m_i}.$$

If there are two bodies for which the ratio m_g/m_i is different, then g will be, too, and the bodies will have different accelerations as they fall. But this has never been observed; *all bodies experience the same acceleration due to gravity at the earth's surface*. So we take $m_g = m_i = m$.

Bodies that are *freely moving*—that is, subject to no forces other than gravitation—will therefore fall straight down at the same rate. But exactly the same thing can be made to happen, for example, in a spaceship that is far from gravitational influences if its rocket motors drive it with constant acceleration in a fixed direction. Freely moving bodies inside the ship will then all appear to undergo identical acceleration in the opposite direction. Thus, in a coordinate frame that is fixed inside the spaceship, we have created a gravitational field.

Eliminating a gravitational field

We can make a gravitational field go away, too. This is what happens in airplanes that are used as labs to study weightlessness. For 20 or 30 seconds at a time, they are flown ballistically, that is, along the path of a projectile that is shot upward and then drops in the earth's gravity. Inside, freely moving objects fall at the same rate as the plane; therefore, with respect to a coordinate frame fixed inside the plane, they exhibit no acceleration. They float. The earth's gravitational field has been "canceled out." For the same reason, the usual gravitational field is missing from a spacecraft orbiting the earth. It is perpetually dropping in the earth's gravity but has been given a sufficient sideways push to keep it orbiting.

Gravity and acceleration are equivalent

The conclusion is that we cannot readily distinguish between the gravitational field in a coordinate frame K that is fixed on the earth (a *laboratory* frame) and the gravitational field that is created in another coordinate frame K′ undergoing constant linear acceleration. In 1911 ([12], page 100), Einstein turned this conclusion into a physical law—the **equivalence principle**:

> But we arrive at a very satisfactory interpretation of this law of experience, if we assume that the systems K and K′ are physically exactly equivalent, that is, if we assume that we may just as well regard the system K as being in a space free from gravitational fields, if we then regard K as uniformly accelerated. This assumption of exact physical equivalence makes it impossible for us to speak of the absolute acceleration of a system of reference, just as the usual theory of relativity forbids us to talk of the absolute velocity of a system; and it makes the equal falling of all bodies in a gravitational field seem a matter of course.

Local frames

The equivalence principle is based on the assumption that the gravitational acceleration is constant, but this really is not so: It is weaker at higher altitudes, and it points in different directions at different places on the earth. To conceal these differences—and thus make acceleration essentially constant—we must put strict limits on the size of the laboratory frame K. In other words, K and

K′ must be *local* frames that describe only a small portion of space. The equivalence principle really holds only in local frames.

The frames must be limited in time, as well. To see why, consider a spacecraft in a low circular earth orbit. A complete circuit takes about 90 minutes. The physical dimensions of the craft are certainly small enough for the gravitational field to be essentially constant inside it; in fact, the field should be zero in a coordinate frame fixed to the spacecraft. Nevertheless, we can detect an effect of gravity in the following way. Hold two small objects A and B motionless on a line (the z-axis) perpendicular to the plane of the spacecraft's orbit, and then release them.

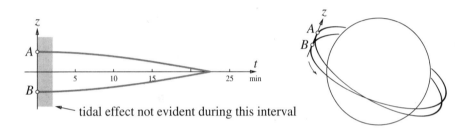

tidal effect not evident during this interval

Tidal effects

They do not remain motionless. Instead, they slowly drift toward each other, and meet after about 22 minutes. This is easy to understand once we recognize that the objects themselves orbit the earth on separate great circles of the same radius. The objects come together because their orbits intersect. They start at points where the circles have their widest separation and travel one-quarter of a full orbit, or about 22 minutes, to the first intersection point. This slow drift is a **tidal effect**; it is due, ultimately, to the fact that the gravitational field is slightly different along the paths of A and B.

Tidal effects give us a way to distinguish between gravity and linear acceleration—but only if we allow enough time for the effects to become apparent. Therefore, if we limit the length of the time axis in our spacetime coordinate frame—to 2 minutes instead of 22 minutes, for example—we will not perceive the tidal effect. In these circumstances, the equivalence principle will continue to hold.

The Gravitational Potential

Suppose we fix a mass M at the point $\mathbf{x}_0 = (x_0, y_0, z_0)$. Then, according to Newton's law of gravitation, this mass exerts a force on a "test particle" of mass m that is placed at any point $\mathbf{x} = (x, y, z)$ in $\mathbf{R}^3$. If we let $\mathbf{r} = \mathbf{x} - \mathbf{x}_0$ be the vector that points from M to m, then the gravitational force that M exerts on m is given by the vector

$$\mathbf{F} = m\mathbf{A} = m\frac{GM}{r^2}\mathbf{u}.$$

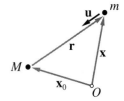

Here $G = 6.67 \times 10^{-11}$ m^3/kg sec^2 is the gravitational constant, $r = \|\mathbf{r}\|$, and $\mathbf{u} = -\mathbf{r}/\|\mathbf{r}\| = -\mathbf{r}/r$ is the unit vector that points from m back to M.

The vectors $\mathbf{F}$ and $\mathbf{A}$ vary from point to point; they are **vector fields**. While $\mathbf{F}$ depends on the test mass m, $\mathbf{A}$ does not; it depends only on the attracting mass M and the position $\mathbf{x}$. We call $\mathbf{A}$ the **gravitational field** determined by M. It has the dimensions of an acceleration.

The gravitational field

Among the simplest vector fields studied in multivariable calculus are the gradient vector fields ∇f. Indeed, we can write $\mathbf{A} = -\nabla \Phi$ where

$$\Phi(x, y, z) = -\frac{GM}{r}.$$

The gravitational potential

(The reason for the minus sign in $\mathbf{A} = -\nabla \Phi$ will become clear later, when we consider the work done by the gravitational field.) To verify that $\mathbf{A} = -\nabla \Phi$, you should first check that

$$\frac{\partial \Phi}{\partial x} = GM\frac{x - x_0}{r^3},$$

and similarly for y and z. Then

$$-\nabla \Phi = -\left(\frac{\partial \Phi}{\partial x}, \frac{\partial \Phi}{\partial y}, \frac{\partial \Phi}{\partial z}\right)$$
$$= -\frac{GM}{r^2}\left(\frac{x - x_0}{r}, \frac{y - y_0}{r}, \frac{z - z_0}{r}\right) = \frac{GM}{r^2}\mathbf{u} = \mathbf{A}.$$

The function $\Phi(x, y, z)$ is called the **gravitational potential** of the field. It has the dimensions of a velocity squared.

Chapter 4 Arbitrary Frames

The gravitational law is linear

A crucial feature of Newton's law of gravitation is *linearity*: The potential created by several point masses is just the sum of the potentials of the individual masses. Since the field is essentially the derivative of the potential, and since differentiation is also linear, the field also depends linearly on the masses. Specifically, consider several masses $M_1, \ldots, M_k$. If M_j is at $\mathbf{x}_j$, then

$$\Phi_j = -\frac{GM_j}{r_j} \quad \text{and} \quad \mathbf{A}_j = \frac{GM_j}{r_j^2}\mathbf{u}_j,$$

where $\mathbf{r}_j = \mathbf{x} - \mathbf{x_j}$, $r_j = \|\mathbf{r}_j\|$, and $\mathbf{u}_j = -\mathbf{r}_j/r_j$. If Φ_{total} and $\mathbf{A}_{\text{total}}$ are the potential and the field of these masses acting together, then

$$\Phi_{\text{total}} = \Phi_1 + \cdots + \Phi_k, \qquad \mathbf{A}_{\text{total}} = \mathbf{A}_1 + \cdots + \mathbf{A}_k.$$

The point masses M_j are called the **sources** of the field. Later we shall see that we can even define the gravitational potential when the sources form a continuous distribution of matter.

The geometric connection between the potential and field

Essentially, the potential is the integral of the field, and for this reason it is not unique: $\Phi(x, y, z) + C$ would serve equally well, for any constant C (since $\nabla C \equiv 0$). It is a standard result of multivariable calculus that the level sets $\Phi(x, y, z) = $ constant are surfaces that are everywhere orthogonal to the gradient field $\mathbf{A} = -\nabla\Phi$. Furthermore, $-\nabla\Phi$ points in the direction in which Φ decreases, and paths that follow this field lead toward the minima of Φ, which therefore determine the positions of the sources of the field.

Example 1: two sources

Here is an example with two point sources whose relative strengths are 5 and 1; the larger mass is at $(0, 0, 0)$, the smaller at $(1, 0, 0)$:

$$\Phi(x, y, z) = -\frac{5}{\sqrt{x^2 + y^2 + z^2}} - \frac{1}{\sqrt{(x-1)^2 + y^2 + z^2}}.$$

The figures below show Φ restricted to the (x, y)-plane, that is, to the 2-dimensional slice $z = 0$ of $\mathbf{R}^3$. On the left is the graph of $\varphi = \Phi(x, y, 0)$; note that the sources are at the bottom of infinitely deep wells: $\Phi(0, 0, 0) = \Phi(1, 0, 0) = -\infty$. The curves in the figure on the right are the level sets $\Phi(x, y, 0) = $ constant. Shown with these curves are the vectors of the gravitational field $\mathbf{A} = -\nabla\Phi$ (although the vectors are not drawn to scale). The full level sets $\Phi(x, y, z) =$ constant in $\mathbf{R}^3$ are the surfaces obtained by rotating the curves $\Phi(x, y, 0) =$ constant around the x-axis.

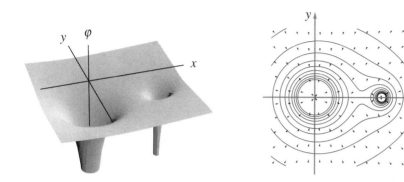

Let $\mathbf{x}(q)$, $a \leq q \leq b$, be a smooth parametrization of a path P in $\mathbf{R}^3$ that avoids the sources. Then the work done by the field $\mathbf{A}$ in moving a test particle of mass m along this path is given by the line integral

Gravitational work

$$W = m \int_P \mathbf{A} \cdot d\mathbf{x}.$$

The basic idea is that work is the product of force $(m\mathbf{A})$ times distance $(d\mathbf{x})$. Since $\mathbf{A} = -\nabla\Phi$, another standard result of multivariable calculus allows us to write

$$W = -m \int_P \nabla\Phi \cdot d\mathbf{x} = -m\Phi\Big|_{\mathbf{x}(a)}^{\mathbf{x}(b)} = -m(\Phi(\mathbf{x}(b)) - \Phi(\mathbf{x}(a))) = -m\Delta\Phi.$$

Here $\Delta\Phi$ is the difference in the values of Φ at the two ends of the path. Hence, the work done is *path-independent*; only the endpoints matter. Furthermore, the field is *conservative*: The work done around a closed loop is zero.

Let $U(x, y, z) = m\Phi(x, y, z)$; since Φ has the dimensions of a velocity squared, U has the dimensions of an *energy*. We call U the **potential energy** of the particle m in the field. In these terms, the work done equals the drop in potential energy: $W = -\Delta U$. The potential energy is constant on a level set $\Phi =$ constant, and it

Potential energy and work

increases as the test particle moves away from the sources. Thus, if a path takes the particle closer to the sources ("downhill" in our two-source example), the potential energy decreases. Therefore, the gravitational field does positive work. If the particle moves away from the sources, the field does *negative* work. What this means is that some other energy source must do an equal amount of positive work against gravity to "lift" the particle out of the gravitational well. The original choice of the sign of Φ was made to yield this relation between work and energy. For a second example consider gravity in a small laboratory frame on earth. The field is constant; if we choose coordinates (x, y, z) in the usual way so the positive z-axis points "up," then

Example 2: a constant field

$$\mathbf{A}(x, y, z) = g(0, 0, -1), \qquad \Phi(x, y, z) = gz,$$

where $g = 9.8$ m/sec^2 is the acceleration due to gravity at the surface of the earth. Note the signs of both $\mathbf{A}$ and Φ. In this field the potential energy of a test particle of mass m is $U(x, y, z) = mgz$; but of course, Φ and U are determined only up to additive constants.

Tides

Tidal forces

We are now in a position to analyze tidal forces. We were able to see tidal effects in the spacecraft because it was "falling" in the earth's gravitational field; the effect was manifested by motions *relative to* that falling frame. This suggests that we look at the field of a single source, make a translation that cancels out the acceleration at one point, and then see what accelerations remain at nearby points.

Place the source M at the point $(k, 0, 0)$ on the x-axis, and suppose units have been chosen such that $GM = 1$. The field is then

$$\mathbf{A}(x, y, z) = -\frac{1}{r^3}(x - k, y, z), \qquad r = \sqrt{(x-k)^2 + y^2 + z^2}.$$

The figure below is the 2-dimensional slice $z = 0$ of a neighborhood of $(0, 0, 0)$. On the left is the original field $\mathbf{A}$; on the right is the tidal field obtained by translating away the acceleration

§4.3 Newtonian Gravity

$\mathbf{A}(0, 0, 0)$ at the origin:

$$\mathbf{T}(x, y, z) = \mathbf{A}(x, y, z) - \mathbf{A}(0, 0, 0).$$

Notice that the tidal acceleration $\mathbf{T}$ is attractive along the y-axis but repulsive along the x-axis. In view of the symmetric relation between y and z in $\mathbf{A}$ and $\mathbf{T}$, the tidal acceleration will be attractive everywhere in the (y, z)-plane.

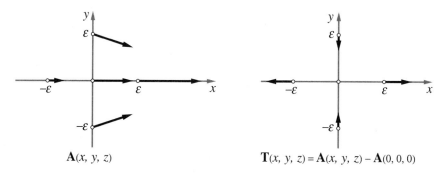

$\mathbf{A}(x, y, z)$

$\mathbf{T}(x, y, z) = \mathbf{A}(x, y, z) - \mathbf{A}(0, 0, 0)$

T along the axes

We can use Taylor's theorem to estimate $\mathbf{T}$ at points ε units from the origin on each of the axes. Along the y-axis we have

$$\mathbf{T}(0, \varepsilon, 0) = \mathbf{A}(0, \varepsilon, 0) - \mathbf{A}(0, 0, 0) \approx \frac{\partial \mathbf{A}}{\partial y}(0, 0, 0)\, \varepsilon,$$

so by the product rule,

$$\frac{\partial \mathbf{A}}{\partial y}(x, y, z) = \frac{3y}{r^5}(x - k, y, z) - \frac{1}{r^3}(0, 1, 0) \quad \text{and} \quad \frac{\partial \mathbf{A}}{\partial y}(0, 0, 0) = -\frac{1}{k^3}(0, 1, 0).$$

Thus

$$\mathbf{T}(0, \varepsilon, 0) \approx -\frac{\varepsilon}{k^3}(0, 1, 0)$$

causes an acceleration *toward* the origin that is proportional to the displacement ε and inversely proportional to the cube of the distance k to the source of the gravitational field.

Along the z-axis the result is the same, but along the x-axis it is different. We have

$$\frac{\partial \mathbf{A}}{\partial x}(x, y, z) = \frac{3(x - k)}{r^5}(x - k, y, z) - \frac{1}{r^3}(1, 0, 0),$$

$$\frac{\partial \mathbf{A}}{\partial x}(0, 0, 0) = \frac{-3k}{k^5}(-k, 0, 0) - \frac{1}{k^3}(1, 0, 0) = \frac{2}{k^3}(1, 0, 0).$$

Therefore,

$$\mathbf{T}(0, 0, \varepsilon) \approx \frac{2\varepsilon}{k^3}(1, 0, 0)$$

causes an acceleration *away from* the origin that is likewise proportional to the displacement and inversely proportional to the cube of the distance to the source. However, the repulsion along the x-axis is twice as strong as the attraction in the (y, z)-plane at the same distance from the origin.

T at an arbitrary point

More generally, to see what **T** looks like at an arbitrary point, suppose (α, β, γ) is a unit vector. Then we can approximate $\mathbf{T}(\varepsilon\alpha, \varepsilon\beta, \varepsilon\gamma)$ using the directional derivative of **A** in the direction (α, β, γ):

$$\mathbf{T}(\varepsilon\alpha, \varepsilon\beta, \varepsilon\gamma) = \mathbf{A}(\varepsilon\alpha, \varepsilon\beta, \varepsilon\gamma) - \mathbf{A}(0, 0, 0)$$
$$\approx \varepsilon \nabla \mathbf{A}(0, 0, 0) \cdot (\alpha, \beta, \gamma) = \varepsilon\alpha \frac{\partial \mathbf{A}}{\partial x} + \varepsilon\beta \frac{\partial \mathbf{A}}{\partial y} + \varepsilon\gamma \frac{\partial \mathbf{A}}{\partial z}.$$

In the slice $z = 0$ we can find **T** for points on the circle

$$(x, y, 0) = (\varepsilon \cos q, \varepsilon \sin q, 0).$$

The result is

$$\mathbf{T}(\varepsilon \cos q, \varepsilon \sin q, 0) \approx \varepsilon \cos q \frac{\partial \mathbf{A}}{\partial x}(0, 0, 0) + \varepsilon \sin q \frac{\partial \mathbf{A}}{\partial y}(0, 0, 0)$$
$$= \frac{2\varepsilon \cos q}{k^3}(1, 0, 0) - \frac{\varepsilon \sin q}{k^3}(0, 1, 0)$$
$$= \frac{\varepsilon}{k^3}(2\cos q, -\sin q, 0),$$

which is itself a vector that lies in the same slice $z = 0$. Here is a sketch of the tidal acceleration field along such a circle. Because of the symmetric way in which y and z appear, you can rotate this figure around the x-axis to see the tidal acceleration on an entire sphere.

Earth tides

This should remind you of pictures of the earth's tides. The moon is off to the right, the gray circle is the theoretical level of the earth's oceans, and the arrows indicate how the water is drawn to its tidal level. As we have just seen, the outward bulge is twice the size of the inward one. (Of course, the tide is enormously

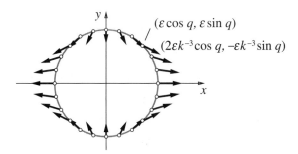

exaggerated.) The sun also causes tides, but the effect is less. The moon dominates because the tides depend on the inverse *cube* of the distance to the gravitational source, not the inverse *square* of gravity itself. The relative masses and distances to the sun and the moon are as follows:

$$M_{\text{sun}} = 2.7 \times 10^7 \, M_{\text{moon}},$$
$$r_{\text{sun}} = 4 \times 10^2 \, r_{\text{moon}}.$$

Therefore, the relative magnitudes of the gravitational and tidal accelerations are

$$\|\mathbf{A}_{\text{sun}}\| = G \frac{M_{\text{sun}}}{r_{\text{sun}}^2} = G \frac{2.7 \times 10^7 \, M_{\text{moon}}}{(4 \times 10^2 \, r_{\text{moon}})^2} = 170 \, \|\mathbf{A}_{\text{moon}}\|,$$

$$\|\mathbf{T}_{\text{sun}}\| = \varepsilon G \frac{M_{\text{sun}}}{r_{\text{sun}}^3} = \varepsilon G \frac{2.7 \times 10^7 \, M_{\text{moon}}}{(4 \times 10^2 \, r_{\text{moon}})^3} = \frac{2}{5} \, \|\mathbf{T}_{\text{moon}}\|.$$

The tidal accelerations **T** depend on the additional parameter ε, which here represents the radius of the earth.

The inverse cube appears in the tides because the tidal acceleration is the difference between nearby gravitational accelerations. By Taylor's theorem, that difference is approximated by the derivative of the gravitational acceleration; since the gravitational acceleration is an inverse square, its derivative will be an inverse cube.

Our example illustrates one of the remarkable properties of the tidal acceleration: It is attractive in some directions but repulsive in others. In fact, the *net* acceleration is exactly zero in a sense that we will now make precise. In our example of a single

The tides "balance out"

gravitating mass at a point on the x-axis, the potential is

$$\Phi(x, y, z) = -\frac{1}{r}, \qquad r = \sqrt{(x-k)^2 + y^2 + z^2},$$

and $\mathbf{A} = -\nabla \Phi$, as usual. We found that the tidal acceleration created by this source at $(0,0,0)$ is attractive along the y- and z-axes but repulsive along the x-axis. We deduced these facts from the value of the derivatives of $\mathbf{A}(0,0,0)$:

$$\frac{\partial \mathbf{A}}{\partial x}(0,0,0) = \left(+\frac{2}{k^3}, 0, 0\right),$$

$$\frac{\partial \mathbf{A}}{\partial y}(0,0,0) = \left(0, -\frac{1}{k^3}, 0\right),$$

$$\frac{\partial \mathbf{A}}{\partial z}(0,0,0) = \left(0, 0, -\frac{1}{k^3}\right).$$

Now, first derivatives of $\mathbf{A}$ are second derivatives of the potential; for example,

$$\frac{\partial \mathbf{A}}{\partial x} = -\left(\frac{\partial^2 \Phi}{\partial x^2}, \frac{\partial^2 \Phi}{\partial x \partial y}, \frac{\partial^2 \Phi}{\partial x \partial z}\right).$$

Consequently, the 3×3 matrix $d^2 \Phi$ of second derivatives of Φ at the origin is therefore

$$d^2 \Phi(0,0,0) = -\begin{pmatrix} +\frac{2}{k^3} & 0 & 0 \\ 0 & -\frac{1}{k^3} & 0 \\ 0 & 0 & -\frac{1}{k^3} \end{pmatrix}.$$

The Laplacian

We can now describe, in terms of the potential Φ, in what sense the tidal accelerations balance out: The trace of the matrix $d^2 \Phi$ is zero. The terms in the trace are the second derivatives that appear in the *Laplacian* of Φ:

$$\nabla^2 \Phi(0,0,0) = \frac{\partial^2 \Phi}{\partial x^2} + \frac{\partial^2 \Phi}{\partial y^2} + \frac{\partial^2 \Phi}{\partial z^2} = \frac{2}{k^3} - \frac{1}{k^3} - \frac{1}{k^3} = 0.$$

$\nabla^2 \Phi(x, y, z) = 0$ in empty space

To this point, all our analysis of this gravitational field has been carried out at the origin. However, you can check that the trace of the matrix of second derivatives is zero at *every* point in $\mathbf{R}^3$ away from the gravitational source:

$$\nabla^2 \Phi(x, y, z) \equiv 0, \qquad (x, y, z) \neq (k, 0, 0).$$

This means that the net tidal acceleration is zero at every point in space away from the gravitational source.

What can we say about tidal accelerations in other gravitational fields? The gravitational potential defined by k masses M_j at the points (x_j, y_j, z_j) has the form

The tides created by several point sources

$$\Phi = -\sum_{j=1}^{k} \frac{GM_j}{r_j}, \qquad r_j = \sqrt{(x - x_j)^2 + (y - y_j)^2 + (z - z_j)^2}.$$

Since

$$\frac{\partial^2}{\partial x^2}\left(\frac{1}{r_j}\right) = -\frac{1}{r_j^3} + \frac{3(x - x_j)^2}{r_j^5}$$

and similarly for the second partial derivatives with respect to y and z, it follows that

$$\nabla^2\left(\frac{1}{r_j}\right) = -\frac{3}{r_j^3} + \frac{3(x - x_j)^2 + 3(y - y_j)^2 + 3(z - z_j)^2}{r_j^5} = 0$$

at every point $(x, y, z) \neq (x_j, y_j, z_j)$. Hence, by linearity of differentiation, $\nabla^2 \Phi = 0$ at all points in empty space. Because $\nabla^2 \Phi = 0$ characterizes the gravitational field in empty space, we call it the **vacuum field equation**.

The Gravitational Field Equations

We can deal with continuous distributions of matter as well as point sources. A continuous distribution is defined by a function $\rho(x, y, z)$ kg/m³ that tells us the density of matter at the point (x, y, z). Suppose that all the matter is concentrated in a bounded region R in space; in other words, $\rho \equiv 0$ outside R. Then we can subdivide R into a collection of small boxes (rectangular parallelepipeds) $P_{i,j,k}$, where $P_{i,j,k}$ is centered at the point (u_i, v_j, w_k) and has sides whose dimensions are Δu_i, Δv_j, and Δw_k meters,

The potential of a continuous distribution

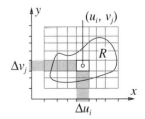

respectively. (For clarity the illustration shows only two dimensions instead of three—rectangles instead of boxes.) The total mass in $P_{i,j,k}$ is approximately

$$M_{i,j,k} \approx \rho(u_i, v_j, w_k)\, \Delta u_i \Delta v_j \Delta w_k,$$

and the gravitational potential induced by the entire mass distribution in R is approximately equal to that of the point masses $M_{i,j,k}$ at (u_i, v_j, w_k):

$$\Phi(x, y, z) \approx -\sum_{k=1}^{N}\sum_{j=1}^{N}\sum_{i=1}^{N} \frac{G\rho(u_i, v_j, w_k)}{\sqrt{(x-u_i)^2 + (y-v_j)^2 + (z-w_k)^2}}\, \Delta u_i \Delta v_j \Delta w_k.$$

The approximation is improved by taking still smaller boxes; in the limit as $N \to \infty$ we have the exact value as an integral:

$$\Phi(x, y, z) = -\iiint_R \frac{G\rho(u, v, w)}{\sqrt{(x-u)^2 + (y-v)^2 + (z-w)^2}}\, du\, dv\, dw.$$

Φ is defined and finite everywhere

If (x, y, z) is not in R, then the denominator in the integrand is never zero, so the integral can be computed and will have a finite value. If (x, y, z) is in R, the integrand appears to be singular at the point $(u, v, w) = (x, y, z)$. However, by changing from Cartesian to spherical coordinates we can remove the singularity, showing that the integral is actually finite. The first step is the translation $T: \mathbf{R}^3 \to \mathbf{R}^3$ defined by

$$\begin{aligned} U &= u - x, & dU &= du; \\ V &= v - y, & dV &= dv; \\ W &= w - z, & dW &= dw. \end{aligned}$$

Let $R^* = T(R)$; then

$$\Phi(x, y, z) = -\iiint_{R^*} \frac{G\rho(x+U, y+V, z+W)}{\sqrt{U^2 + V^2 + W^2}}\, dU\, dV\, dW,$$

and the apparent singularity of the integrand is now at the origin $(U, V, W) = (0, 0, 0)$. Next, introduce spherical coordinates

(r, φ, θ) by the equations

$$U = r \sin \varphi \cos \theta,$$
$$V = r \sin \varphi \sin \theta,$$
$$W = r \cos \varphi,$$
$$dU\, dV\, dW = r^2 \sin \varphi \, dr\, d\varphi\, d\theta.$$

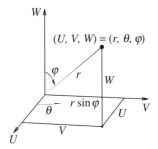

The fourth equation tells us how to convert the volume element from one coordinate system to the other; it is derived in the exercises. In terms of spherical coordinates, $\sqrt{U^2 + V^2 + W^2} = r$, and if we write the cumbersome

$$\rho(x + U, y + V, z + W) = \rho(x + r \sin \varphi \cos \theta, y + r \sin \varphi \sin \theta, z + r \cos \varphi)$$

simply as ρ, the gravitational potential takes the form

$$\Phi(x, y, z) = -\iiint_{R^{**}} \frac{G\rho}{r} r^2 \sin \varphi \, dr\, d\varphi\, d\theta = -\iiint_{R^{**}} G\rho r \sin \varphi \, dr\, d\varphi\, d\theta.$$

Here R^{**} is R^* in spherical coordinates. The integrand is no longer singular at the origin, proving that Φ is defined and finite everywhere.

What can we say about $\nabla^2 \Phi$ when Φ is defined by a *continuous* distribution of matter? In the integral

$\nabla^2 \Phi = 0$ outside R

$$\Phi(x, y, z) = -\iiint_R \frac{G\rho(u, v, w)}{r} du\, dv\, dw,$$

only the term $r = \sqrt{(x-u)^2 + (y-v)^2 + (z-w)^2}$ involves x, y, and z. Therefore, if we carry the Laplace operator

$$\nabla^2 = \frac{\partial^2}{\partial x^2} + \frac{\partial^2}{\partial y^2} + \frac{\partial^2}{\partial z^2}$$

through the integral sign, we get

$$\nabla^2 \Phi(x, y, z) = - \iiint_R G\rho(u, v, w) \, \nabla^2 \left(\frac{1}{r}\right) du \, dv \, dw.$$

We have already seen that $\nabla^2(1/r) = 0$ at any point (x, y, z) where $r \neq 0$. Since $r(x, y, z) = \sqrt{(x-u)^2 + (y-v)^2 + (z-w)^2}$ and since (u, v, w) ranges over the region R, we can be sure that $r \neq 0$ if we take (x, y, z) outside R. Then $\nabla^2(1/r) \equiv 0$ over the entire region R, so $\nabla^2 \Phi(x, y, z) = 0$. Thus, in empty space the average tidal acceleration is zero, whether the gravitational source is continuous or discrete.

Determine $\nabla^2 \Phi$ inside R

But $\Phi(x, y, z)$ is defined for (x, y, z) in R, so it makes sense to determine $\nabla^2 \Phi(x, y, z)$ inside R. To pursue this matter, it is helpful to bring in the gravitational field $\mathbf{A} = -\nabla \Phi$ associated with Φ and interpret tidal acceleration in terms of $\mathbf{A}$. Start with the definition of the differential operator ∇^2 as the inner product of the vector differential operator ∇ with itself, as follows:

$$\nabla \cdot \nabla = \left(\frac{\partial}{\partial x}, \frac{\partial}{\partial y}, \frac{\partial}{\partial z}\right) \cdot \left(\frac{\partial}{\partial x}, \frac{\partial}{\partial y}, \frac{\partial}{\partial z}\right) = \frac{\partial^2}{\partial x^2} + \frac{\partial^2}{\partial y^2} + \frac{\partial^2}{\partial z^2} = \nabla^2.$$

Then $\nabla^2 \Phi = \nabla \cdot \nabla \Phi = -\nabla \cdot \mathbf{A}$. If we write $\mathbf{A} = (A_x, A_y, A_z)$, then

$$-\nabla^2 \Phi = \nabla \cdot \mathbf{A} = \left(\frac{\partial}{\partial x}, \frac{\partial}{\partial y}, \frac{\partial}{\partial z}\right) \cdot (A_x, A_y, A_z) = \frac{\partial A_x}{\partial x} + \frac{\partial A_y}{\partial y} + \frac{\partial A_z}{\partial z}.$$

div $\mathbf{A} = \nabla \cdot \mathbf{A} = 0$ in empty space

For reasons that will emerge below, this particular sum of derivatives of the components of $\mathbf{A}$ is called the **divergence** of $\mathbf{A}$, and is written div $\mathbf{A}$ as well as $\nabla \cdot \mathbf{A}$. The fact that the average tidal acceleration is zero at any point (x, y, z) in empty space can therefore be written in the form

$$\text{div } \mathbf{A}(x, y, z) = \nabla \cdot \mathbf{A}(x, y, z) = 0.$$

The divergence of an incompressible fluid flow

We can get insight into the meaning of divergence by studying fluid flow. For simplicity we take an incompressible fluid and assume that the flow varies from point to point but at a given point does not change over time. Then the fluid has constant density ρ kg/m^3, and its velocity is given by a vector field $\mathbf{V}(x, y, z)$ m/sec at each point (x, y, z) in some region in $\mathbf{R}^3$. The vector field

$$\mathbf{F} = \rho \mathbf{V} \quad \frac{\text{kg/sec}}{\text{m}^2}$$

describes the flow that we are interested in. It tells us the how much matter flows across a surface of unit area in unit time.

Consider the flow $\mathbf{F}$ through a small box centered at (x, y, z) with sides of dimension Δx, Δy, and Δz. What is the net outflow of matter from this box, in kilograms per second? Since the fluid is incompressible, we must see eventually that the net outflow is zero unless fluid is being created or destroyed inside the box—that is, unless the box contains *sources* or *sinks*. Right now, though, we merely want an expression that measures the net outflow, whatever it happens to be.

Flow through a small box

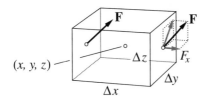

The net outflow is the algebraic sum of the outward flows across the six faces of the box. We assume that the box is so small that the flow vector $\mathbf{F}$ is essentially constant over each face. On each face, we can break down $\mathbf{F}$ into a parallel component and a normal one, as in the figure; only the normal component contributes to the sum. On the right face the normal component is $F_x(x + \Delta x/2, y, z)$. Therefore, the flow from left to right across that face is approximately F_x kg/sec per unit area times the area $\Delta y \Delta z$ of the face:

$$\text{flow} \approx F_x(x + \Delta x/2, y, z)\, \Delta y \Delta z \quad \text{kg/sec}.$$

Flow across a single face

The same argument shows that the flow from left to right on the *left* face is approximately

$$F_x(x - \Delta x/2, y, z)\, \Delta y \Delta z \quad \text{kg/sec}.$$

However, this represents an *inflow*, not an outflow, because the box lies to the right of the face; the outflow is the negative of this quantity. Therefore, the combined outflow across the left and right faces together is approximately

$$[F_x(x + \Delta x/2, y, z) - F_x(x - \Delta x/2, y, z)]\, \Delta y \Delta z.$$

The quantity in brackets approximates the derivative

$$\frac{\partial F_x}{\partial x}(x, y, z)\, \Delta x,$$

so the outflow contribution itself is approximately

$$\frac{\partial F_x}{\partial x}(x, y, z)\, \Delta x \Delta y \Delta z.$$

In a similar way we can approximate the net outflow across the front and back faces and the top and bottom faces, respectively, by

$$\frac{\partial F_y}{\partial y}(x, y, z)\, \Delta x \Delta y \Delta z \quad \text{and} \quad \frac{\partial F_z}{\partial z}(x, y, z)\, \Delta x \Delta y \Delta z.$$

Total outflow is the divergence

Therefore, the total net outflow is approximately

$$\left(\frac{\partial F_x}{\partial x}(x, y, z) + \frac{\partial F_y}{\partial y}(x, y, z) + \frac{\partial F_z}{\partial z}(x, y, z) \right) \Delta x \Delta y \Delta z \quad \text{kg/sec.}$$

Note that this is the sum of derivatives of the components of **F** that we have called the *divergence* of **F**.

We can improve the approximation by making the box smaller; in the limit, the outflow over *any* 3-dimensional region R is given by the integral

$$\iiint_R \frac{\partial F_x}{\partial x} + \frac{\partial F_y}{\partial y} + \frac{\partial F_z}{\partial z}\, dx\, dy\, dz = \iiint_R \operatorname{div} \mathbf{F}\, dx\, dy\, dz.$$

Meaning of divergence

Suppose the scalar function $\operatorname{div} \mathbf{F} = \nabla \cdot \mathbf{F}$ is essentially constant over the region R. Then, according to the integral mean value theorem,

$$\text{flow out of } R = \iiint_R \operatorname{div} \mathbf{F}\, dx\, dy\, dz \approx (\operatorname{div} \mathbf{F})\, \operatorname{vol} R.$$

Therefore, if we let the region R shrink down to a point $\mathbf{u} = (u, v, w)$ (and we write this as $R \downarrow \mathbf{u}$), then

$$\lim_{R \downarrow \mathbf{u}} \frac{\text{flow out of } R}{\text{vol } R} = \text{div } \mathbf{F}(\mathbf{u}) = \nabla \cdot \mathbf{F}(\mathbf{u}).$$

It is in this sense that the function $\nabla \cdot \mathbf{F}(u, v, w) = \text{div } \mathbf{F}(u, v, w)$ is called the *divergence* of $\mathbf{F}$ at the point (u, v, w).

We can now return to the question of determining $\nabla^2 \Phi = -\nabla \cdot \mathbf{A}$ at a point $\mathbf{u} = (u, v, w)$ in the region R where the mass density ρ is positive. We are interested in finding the divergence of $\mathbf{A}$ at $\mathbf{u}$: $\text{div } \mathbf{A}(\mathbf{u})$.

$\nabla^2 \Phi$ inside R

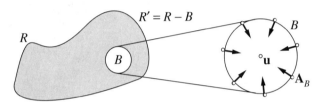

Decompose R into B and $R' = R - B$, where B is the ball of radius b centered at $\mathbf{u}$. Let $\mathbf{A}_{R'}$ be the gravitational field due to the masses in R', and let $\mathbf{A}_B$ be the field due to the masses in B. Together these masses constitute the entire source, so

$$\mathbf{A} = \mathbf{A}_{R'} + \mathbf{A}_B.$$

Since $\mathbf{u}$ is not in R', $\text{div } \mathbf{A}_{R'}(\mathbf{u}) = 0$, as we have already shown. Therefore,

$$\text{div } \mathbf{A}(\mathbf{u}) = \text{div } \mathbf{A}_B(\mathbf{u}),$$

so we now concentrate on $\mathbf{A}_B$.

Make the radius b so small that the value of ρ anywhere in B differs only very little from its value at $\mathbf{u}$. Then the mass inside B is approximately $M_B = \rho(\mathbf{u}) \text{ vol } B$, and since B is small, we can think of this as a point mass concentrated at $\mathbf{u}$. Therefore, to a good approximation $\mathbf{A}_B$ is the field of a point source and has the formula

$$\mathbf{A}_B(\mathbf{x}) = -\frac{GM_B}{r^2} \mathbf{n} = -\frac{G\rho(\mathbf{u}) \text{ vol } B}{r^2} \mathbf{n}, \qquad \mathbf{n} = \frac{\mathbf{x} - \mathbf{u}}{r}, \quad r = \|\mathbf{x} - \mathbf{u}\|.$$

Outflow of A_B from B

If we now interpret A_B as the flow field of a fluid of density ρ, then we can calculate the net outflow of A_B from B two ways. On the surface of B, A_B reduces to

$$A_B(x) = -\frac{G\rho(u)\,\text{vol}\,B}{b^2}\,n.$$

This is a vector whose magnitude is constant and whose direction is everywhere normal to the surface of B. The outflow is therefore the product of that magnitude by the area of the surface. The surface is a sphere of radius b, so its area is $4\pi b^2$ and

$$\text{outflow} = -\frac{G\rho(u)\,\text{vol}\,B}{b^2} \cdot 4\pi b^2 = -4\pi G\rho(u)\,\text{vol}\,B.$$

The second way to calculate the outflow is with the divergence:

$$\text{outflow} = \text{div}\,A_B(u)\,\text{vol}\,B.$$

Comparing these, we obtain

$$\text{div}\,A(u) = \text{div}\,A_B(u) = -4\pi G\rho(u).$$

In fact, this is only an approximation, whose accuracy depends on the size of the ball B. We obtain the exact result by taking the limit as $b \to 0$.

The gravitational field equations

Since $\nabla^2\Phi = -\text{div}\,A$, our final result is $\nabla^2\Phi = 4\pi G\rho$. Since $\nabla^2\Phi = 0$ in empty space and $\rho = 0$ in empty space, the single equation $\nabla^2\Phi = 4\pi G\rho$ holds *everywhere*. The equations

$$A = -\nabla \cdot \Phi,$$

$$\nabla^2\Phi = 4\pi G\rho$$

are called the **gravitational field equations**. The culmination of Newton's theory, they describe exactly how the gravitational field A arises from the sources ρ.

Exercises

1. There are two different ways to describe the gravitational field at the surface of the earth:

$$A = \frac{GM}{r^2}(0, 0, -1) \quad \text{and} \quad A = g(0, 0, -1).$$

Here M is the mass of the earth and r is its radius, while G is the gravitational constant and g is the acceleration due to gravity. Since the two expressions must be equal, $M = gr^2/G$. Using the known values of G and g, and taking the circumference of the earth to be 4×10^7 meters, determine the mass of the earth in kilograms.

2. Let $\Phi(x, y, z) = 1/r$, $r = \sqrt{(x-x_0)^2 + (y-y_0)^2 + (z-z_0)^2}$. Show that
$$\nabla^2 \Phi(x, y, z) = 0 \quad \text{for every } (x, y, z) \neq (x_0, y_0, z_0).$$

3. (a) Sketch the contours of the function $\Phi(x, y, 0)$ when Φ is the potential function of two equal masses M at the points $(\pm 1, 0, 0)$. Take units in which $GM = 1$.

   (b) What is the work done by the field $-\nabla \Phi$ in moving a unit mass from infinity to the origin along the y-axis?

4. What is the work done by the field in Example 1 in this section in moving a unit mass from infinity to the point $(-1, 0, 0)$?

5. Suppose $f(x, y, z)$ is a smooth function, (α, β, γ) is a unit vector (in the Euclidean sense), and $\varepsilon \approx 0$. Explain why
$$f(\varepsilon\alpha, \varepsilon\beta, \varepsilon\gamma) - f(0, 0, 0) \approx \varepsilon \nabla f(0, 0, 0) \cdot (\alpha, \beta, \gamma).$$

6. Suppose (U, V, W) and (r, φ, θ) are related by the equations
$$U = r \sin \varphi \cos \theta, \qquad V = r \sin \varphi \sin \theta, \qquad W = r \cos \varphi.$$

   (a) Show $dU = \sin \varphi \cos \theta \, dr + r \cos \varphi \cos \theta \, d\varphi - r \sin \varphi \sin \theta \, d\theta$, and obtain the corresponding expressions for dV and dW as linear combinations of the differentials dr, $d\varphi$, and $d\theta$.

   (b) To continue, you need to use the algebra of differential forms, in particular the **exterior product**. The exterior product of any number of differentials can be constructed according to the following rules: If dp and dq are arbitrary differentials and f is an arbitrary smooth function, then $dp \, dp = 0$, $dq \, dp = -dp \, dq$, and $(f \, dp + dq) dr = f \, dp \, dr + dq \, dr$. Consult a text on advanced calculus or differential forms for details. Show that $dr \, dq \, dp = -dp \, dq \, dr = -dq \, dr \, dp = -dr \, dp \, dq$. (The exterior product $dp \, dq$ is often written $dp \wedge dq$.)

(c) Show that $dU\,dV = r\sin^2\varphi\,dr\,d\theta + r^2\sin\varphi\cos\varphi\,d\varphi\,d\theta$ and obtain the corresponding expressions for $dV\,dW$ and $dW\,dU$.

(d) Show that $dU\,dV\,dW = r^2\sin\varphi\,dr\,d\varphi\,d\theta$.

(e) Suppose (r,θ) are polar coordinates in the (x,y)-plane; that is, $x = r\cos\theta$ and $y = r\sin\theta$. Obtain dx and dy as linear combinations of dr and $d\theta$ and show that $dx\,dy = r\,dr\,d\theta$.

7. For each of the following 2-dimensional vector fields $\mathbf{A}(x,y)$, make a sketch and calculate the divergence function div $\mathbf{A}$.

(a) $\mathbf{A}(x,y) = (x/5, y/5)$.

(b) $\mathbf{A}(x,y) = ((x-1)/5, (y-2)/5)$.

(c) $\mathbf{A}(x,y) = (y/5, 0)$.

(d) $\mathbf{A}(x,y) = (y/5, x/5)$.

4.4 Gravity in Special Relativity

The Gravitational Red Shift

Gravity causes a Doppler effect

According to the equivalence principle, we must expect the features we have already identified in a linearly accelerating frame to carry over to a frame that is stationary in a gravitational field—at least if the frame covers only a small region in spacetime. One such feature is the Doppler effect.

Thus we take G to be stationary in a constant gravitational field of strength α that points in the direction of the negative ζ-axis. Suppose G emits a light signal of frequency ν. Let a second stationary observer C be positioned at the point $\zeta = h$ on the ζ-axis. What is the frequency of the light signal from G when C receives it?

If G were instead undergoing an upward linear acceleration of constant magnitude α in the direction of the positive ζ-axis, then there is certainly a Doppler effect. In Section 4.2 we used arguments from special relativity to show that the signal C receives will have frequency

$$N = e^{-\alpha h}\nu.$$

By the equivalence principle, the same equation must hold in the new setting, where G and C are stationary in the gravitational field. In other words, gravity causes a Doppler effect. But because we can apply the equivalence principle only in a strictly limited portion of spacetime, it follows that h must be small, and we can replace our formula for N by its linear approximation using Taylor's theorem:

$$N = \nu(1 - \alpha h).$$

Einstein arrived at this point in 1911 ([12], page 105); when he did, he made the following comment (here adapted to our own notation):

Time dilation and compression

> On a superficial consideration, this equation seems to assert an absurdity. If there is a constant transmission of light from G to C, how can any other number of periods per second arrive in C than is emitted in G? But the answer is simple. We cannot regard ν or N simply as frequencies (as the numbers of periods per second) since we have not yet determined the time in the frame G. What ν denotes is the number of periods with reference to the time-unit of the clock τ of G, while N denotes the number of periods per second with reference to the identical clock T of C. Nothing compels us to assume that the clocks ... must be regarded as going at the same rate.

Indeed, we have already seen that identical clocks at different locations $\zeta = h$ in a linearly accelerating frame run at different rates; by the equivalence principle, the same must be true here. In particular, if a time interval lasts ΔT seconds according to C, then G will say it lasts $\Delta \tau = F(h) \Delta T$ seconds, where $F(h) = 1 - \alpha h$. This has the same form as the analogous equation in Section 4.2; we have merely replaced the compensation factor $\Gamma(h)$ by its linearization. The figure below shows us how C's clock, at various levels $\zeta = h$, runs in comparison to G's clock. Since the gravitational field gets stronger as h (or ζ) decreases, what we see is that *clocks slow down in a gravitational field*.

Clocks slow down in a gravitational field

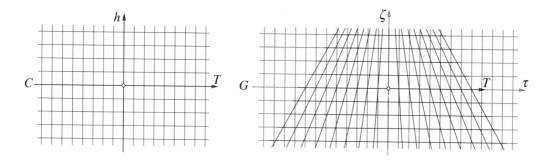

The relative red shift

Suppose we rewrite the equation for the Doppler effect by expressing the difference in frequencies as a fraction of the emission frequency:

$$\frac{N - \nu}{\nu} = -\alpha h.$$

This gives us the *relative* shift, which is independent of the frequency and decreases linearly with h. It says that any light climbing out of the gravitational field has its frequency shifted toward the red in such a way that the relative shift is proportional to the distance climbed and to the strength of the field.

Compare this to what happens to matter: When a mass m climbs out of a gravitational field, it loses potential energy in the amount $\Delta U = m\Delta\Phi$, where Φ is the gravitational potential. Since the potential function of our gravitational field is $\Phi(h) = \alpha h + \text{const}$, we can express the red shift of a photon in terms of potential differences, too:

$$N = \nu(1 - \Delta\Phi), \quad \text{or} \quad \frac{N - \nu}{\nu} = -\Delta\Phi.$$

The red shift is an energy loss

But the energy of a photon is its frequency times Planck's constant, $E = h\nu$, so we can rewrite the red shift as an energy shift: $hN = h\nu - h\nu\Delta\Phi$. If we let $E_{\text{bottom}} = h\nu$ and $E_{\text{top}} = hN$ be the energies of the photon at the start and end of the climb, we can rewrite the last equation as

$$\Delta U = E_{\text{top}} - E_{\text{bottom}} = -E_{\text{bottom}}\Delta\Phi.$$

(For the rest of this paragraph h represents Planck's constant, not the position of C.)

Thus, light photons (which have zero rest mass) and ordinary matter undergo exactly the same loss of potential energy as they rise in a gravitational field.

Incidentally, if we use conventional units instead of geometric, the red shift and energy loss equations become

$$N = \nu\left(1 - \frac{\Delta\Phi}{c^2}\right) \quad \text{and} \quad \Delta U = E_{\text{top}} - E_{\text{bottom}} = -\underbrace{\frac{E_{\text{bottom}}}{c^2}}_{\text{mass}}\underbrace{\Delta\Phi}_{\text{velocity}^2}.$$

The potential function gives us a way to express the red shift in *any* gravitational field, not just a constant one. In 1911 Einstein used this approach to calculate the red shift of sunlight that reaches the earth. The effect is small, though; the gravitational potential difference between the surface of the earth and the surface of the sun is only $\Delta\Phi = 2 \times 10^{-6}$ in geometrical units.

The Bending of Light Rays

The equivalence principle also implies that a gravitational field bends light rays. To see what this means, assume once again that G is stationary in a constant gravitational field of strength α that points in the direction of the negative ζ-axis. Suppose that a photon is emitted perpendicular to the field, in the direction of the positive η-axis. We want to find the track of this photon in space and its worldcurve in spacetime.

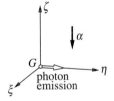

If we replace the photon by a material particle that is launched from the origin with a certain velocity v in the direction of the positive η-axis, we know it will follow a downward parabolic track in the (η, ζ)-plane. When we construct the photon track, it will be instructive to construct this particle track at the same time.

Invoking the equivalence principle, we assume that G is *not* in a gravitational field but is instead undergoing an upward acceleration of constant magnitude α with respect to an inertial frame R. We also assume, as usual, that corresponding spatial axes of R and G have the same orientation. By Theorem 4.1, the two frames are related by the map $M : G \to R$ given by the equations

$$M : \begin{cases} t = \dfrac{e^{\alpha \zeta}}{\alpha} \sinh \alpha \tau, \\ x = \xi, \\ y = \eta, \\ z = \dfrac{e^{\alpha \zeta}}{\alpha} \cosh \alpha \tau. \end{cases}$$

Hereafter we can ignore x and ξ because these axes are perpendicular to the motions; everything will happen in the $(1 + 2)$-dimensional slices of spacetime given by $x = 0$ and $\xi = 0$. Also, remember that the limitation built into the equivalence principle implies that M is valid only in some *small* neighborhood of the origin in G.

Worldlines in the inertial frame R

In the inertial frame R, the photon and the particle follow the same perfectly straight track in the (y, z)-plane — namely, the part of the line $z = 1/\alpha$ where $y > 0$. Their worldlines in the (t, y, z)-spacetime lie in the plane $z = 1/\alpha$; they are straight lines with slopes $c = 1$ and v, respectively, when viewed with respect to the (t, y)-axes. They are space curves that we can parametrize in the following way:

$$(t, y, z) = (q, vq, 1/\alpha).$$

For the photon, $v = c = 1$. The figure below shows two particles moving at different speeds, as well as a photon.

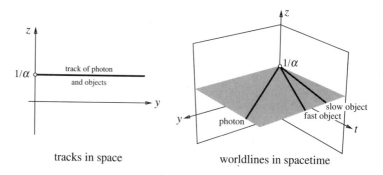

tracks in space   worldlines in spacetime

Worldcurves in the accelerating frame G

The corresponding worldcurves in G are the images of these space curves under the map $M^{-1} : R \to G$ given by

$$M^{-1} : \begin{cases} \tau = \dfrac{1}{\alpha} \tanh^{-1}\left(\dfrac{t}{z}\right), \\ \eta = y, \\ \zeta = \dfrac{1}{2\alpha} \ln\left(\alpha^2(z^2 - t^2)\right). \end{cases}$$

Remember that M^{-1}, like M, is defined only in a limited domain; for M^{-1} it is a small neighborhood of the point $(t, x, y, z) = (0, 0, 0, 1/\alpha)$. Thus, in G the worldcurves have the parametrization

$$(\tau, \eta, \zeta) = \left(\dfrac{1}{\alpha} \tanh^{-1}(\alpha q), vq, \dfrac{1}{2\alpha} \ln\left(1 - \alpha^2 q^2\right)\right).$$

Here is the map M^{-1} showing the image worldcurves in G:

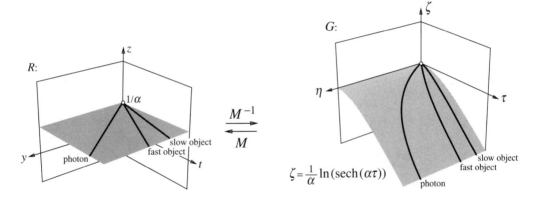

Note that the image worldcurves lie on a surface that is itself the image of the horizontal plane $z = 1/\alpha$ under M^{-1}. In fact, we can show that this surface is the graph of a function $\zeta = \psi(\tau, \eta)$. To determine ψ, note first that the graph of ψ is a cylinder in the η-direction; this means that η is not explicitly involved in ψ. Thus we want to see how the condition $z = 1/\alpha$ determines ζ as a function of τ. Substitute $z = 1/\alpha$ into the equations M^{-1}; this gives us

The image of $z = 1/\alpha$

$$\tau = \dfrac{1}{\alpha} \tanh^{-1}(\alpha t), \qquad \zeta = \dfrac{1}{2\alpha} \ln\left(1 - \alpha^2 t^2\right)).$$

Now solve the first equation for αt and substitute the result into the equation for ζ:

$$\alpha t = \tanh(\alpha\tau), \qquad \zeta = \frac{1}{2\alpha}\ln\left(1 - \tanh^2(\alpha\tau)\right).$$

But $1 - \tanh^2 \alpha\tau = \operatorname{sech}^2 \alpha\tau$, so we can finally write

$$\zeta = \frac{1}{2\alpha}\ln\left(\operatorname{sech}^2(\alpha\tau)\right) = \frac{1}{\alpha}\ln\left(\operatorname{sech}(\alpha\tau)\right) = \psi(\tau,\eta).$$

Tracks in the accelerating frame

The tracks of the photon and the particles are just the projections of the worldcurves in the (τ, η, ζ)-spacetime to the (η, ζ)-plane. In parametric form, the tracks are

$$(\eta, \zeta) = \left(vq, \frac{1}{2\alpha}\ln\left(1 - \alpha^2 q^2\right)\right).$$

Since $q = \eta/v$, we can rewrite them as graphs of functions:

$$\zeta = \frac{1}{2\alpha}\ln\left(1 - \frac{\alpha^2}{v^2}\eta^2\right).$$

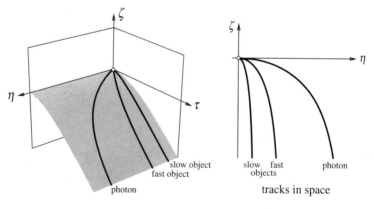

Newtonian formulas for a falling object

By the equivalence principle, the worldcurves we have just found are those that occur when G is in a gravitational field instead of accelerating linearly with respect to an inertial frame—at least if G's frame is small. (Remember, the equivalence principle is *local*.) But then q and η are small, so we can use Taylor's theorem to replace our two descriptions of the worldcurves by simple approximations:

$$(\eta, \zeta) = \left(vq, -\frac{\alpha}{2}q^2\right), \qquad \zeta = -\frac{\alpha}{2}\frac{\eta^2}{v^2}.$$

These are the usual Newtonian formulas for the motion of an object falling in a vertical gravitational field of strength α while moving sideways with velocity v. However, we have discovered something new: A photon moving with velocity $v = c = 1$ is governed by the same formulas as a material object. Light follows a curved path in a gravitational field.

A Basic Incompatibility

The equivalence principle has led us naturally to consider accelerated, noninertial frames in trying to describe gravity. But this does not, in itself, rule out the simpler inertial frames of special relativity. However, we shall now show that any attempt to incorporate gravity directly into special relativity will fail; the two are fundamentally incompatible.

Can we include gravity in an inertial frame?

Suppose that $G : (\tau, \zeta)$ is an inertial frame in which there is a gravitational field. This means that Newton's laws of motion hold in G, and any accelerated motions that cannot be accounted for by obvious tangible forces are explained by the presence of a suitable *gravitational* force. We shall allow the gravitational field to vary from point to point, but require only that it be constant in time at any given point. We also assume that the gravitational potential function Φ takes different values at different points on the ζ-axis.

Let C be a second observer stationary at some point $\zeta \neq 0$; then Φ has different values at C and G ($\Delta\Phi \neq 0$). Now suppose G emits a light signal of period $\Delta\tau$. When C receives the signal, its period will have been changed to

The red shift and proper times

$$\Delta T = \frac{\Delta\tau}{1 - \Delta\Phi} \neq \Delta\tau$$

by the gravitational red shift; see page 190.

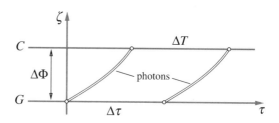

But now consider the parallelogram-like figure formed by the parts of the worldlines of G and C that are bounded by the worldcurves of photons sent at two different times from G to C. The gravitational field causes the sides to be curved; while we do not know the precise form of those curves, we do know that the second has exactly the same shape as the first, because the gravitational field is invariant over time. This implies that the top of the parallelogram must have the same length as the bottom: $\Delta T = \Delta \tau$.

Gravity distorts geometry

This puts the incompatibility in a clear and simple form: Minkowski geometry requires $\Delta T = \Delta \tau$, while gravity requires $\Delta T \neq \Delta \tau$. We cannot put a gravitational field in an inertial frame; distances get distorted. Spacetime with gravity does not obey the laws of Minkowski geometry; we shall come to regard this as a manifestation of the *curvature* of spacetime.

The Meaning of Curvature

Bending and stretching

If you take a portion of a plane and alter it so it takes the shape of a curved surface, two things happen: The plane *bends* up into the third dimension, and it *stretches* to fit the new shape. These are not the same: When you roll a piece of paper into a cylinder, it bends without stretching; when you pull on the edges of a flat rubber sheet, it stretches without bending. It is bending that we usually associate with curvature, but this is unfortunate, for bending requires that we visualize the surface in some still-larger space that contains it. If we try to think of the curvature of spacetime this way, we need that larger space. But none is forthcoming; our own work has certainly not produced it.

Fortunately, in mathematics and in physics we associate curvature with stretching rather than bending. From this point of view, what makes a surface curved is that it cannot be represented by a flat scale model. In other words, if we make a flat chart of the surface—or even just a portion of the surface—the chart will necessarily distort distances, because the laws of plane geometry hold in the chart but not in the surface.

Here is an example. The triangle on the left, below, is a chart that represents a quarter of the northern hemisphere bounded by the North Pole P and two points E_1, E_2 that are 90° apart on the equator. In the chart itself, these three points form an equilateral triangle; in fact, each side represents a distance of 10,000 kilometers. (The meter was originally defined as $1/10,000,000$ of the distance from the equator to the pole!) How long is the altitude A? According to Euclidean plane geometry it should be

$$A = 10000 \times \frac{\sqrt{3}}{2} = 8660 \text{ kilometers.}$$

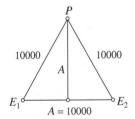
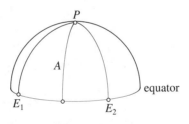

But on the earth itself the altitude runs from the pole to the equator, just like the two sides, so it is actually 10,000 kilometers long, not 8,660. Since the chart represents the earth and distances on the earth, it must give the value of A as 10,000 kilometers, not 8,660. This is how the chart distorts distances. (It also distorts angles: The three 60° angles at the vertices of the triangle actually represent 90° angles on the earth.)

The crucial point here is that we do not need to see the picture on the right—which shows how the surface is embedded in 3-dimensional space—to determine that the surface is curved. It is enough to see just the flat, 2-dimensional, chart on the left, because the metric information that the chart gives (namely, that A equals 10000 rather than 8660) already reveals that the surface is curved, that is, that it does not obey the laws of plane geometry.

In the next two chapters we will study the differential geometry of curved surfaces and develop the tools that will allow us to extract information about curvature directly from flat charts. When we do, it will become evident that there is nothing special about 2-dimensional surfaces or Euclidean distances; the ideas work in any dimension and are readily adapted to distance as defined in Minkowski geometry. With that perspective we will be able to interpret noninertial coordinate frames like those we have created as suitable charts for our 4-dimensional spacetime, and we will be able to connect the distance distortions we find in those charts to the curvature of spacetime.

A New Theory of Gravity

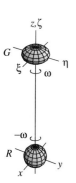

In his 1916 paper that introduces the world to general relativity, Einstein gives two compelling arguments for using arbitrary frames to model gravity. He attributes the first to the physicist Ernst Mach. It begins with a thought experiment: Suppose there are two large fluid masses R and G that are so far from each other and from all other masses that the only appreciable gravitational effects they experience are internal. The internal field will tend to make the fluid a sphere. Suppose, furthermore, that in a frame in which R is motionless, G rotates with constant angular velocity ω around the line that connects the centers of the two bodies. Then, from a frame in which G is motionless, R rotates around the same axis but with constant angular velocity $-\omega$.

So far, the relation between R and G is symmetric. However, the experiment continues by assuming that the bodies have different shapes: R is indeed a sphere, but G has an equatorial bulge. What is the physical cause of this asymmetry? Einstein first brings forward the explanation provided by Newtonian mechanics: Certain coordinate frames—the inertial frames—are singled out as "privileged"; the laws of physics hold only in these frames, and only these frames should be used to study physics. In R's frame,

G spins and bulges as a result; this is consistent with physical law. In G's frame, R spins but does *not* bulge; this is contradictory, so we rule G's frame "out of bounds": R's frame is privileged while G's is not.

But simply declaring that physical laws hold only in certain frames is not an explanation; in particular, it is not an explanation of what *causes* the asymmetry. In fact, no explanation is to be found within the system that consists solely of the two bodies: The symmetry of the situation ensures that. The explanation comes from outside: Only G has a bulge because only G is "really" rotating—that is, rotating with respect to the bulk of distant matter in the universe. It is the observable presence of this other matter that causes the asymmetry. There is no observable justification for asserting that the laws of physics hold only in certain privileged frames.

Privileged frames cannot be justified

On the contrary, Einstein asserts ([10], page 113, and the emphasis is his own): "*The laws of physics must be of such a nature that they apply to systems of reference in any kind of motion.*" This is nothing less than a complete generalization of Galileo's principle: "The laws of physics must be of such a nature that they apply to systems of reference in uniform motion with respect to one another." For this reason the conclusion of Einstein's first argument for arbitrary frames is called **the principle of general relativity**.

The principle of general relativity

Einstein's second argument is based on the principle of equivalence. It says that we can study a local gravitational field by using a coordinate frame undergoing uniform linear acceleration with respect to an inertial frame. But Einstein does more than just *study* gravity; with arbitrary frames he gives it a fundamentally new theoretical foundation. In his own words ([10], page 114), "It will be seen from these reflections that in pursuing the general theory of relativity we shall be led to a theory of gravitation, since we will be able to 'produce' a gravitational field merely by changing the system of coordinates."

Arbitrary frames and gravity

The following chapters are about Einstein's theory of gravitation. It is in many ways remarkably different from Newton's theory. For both Newton and Einstein, it is the gravitational field

Gravity and geometry

that causes objects to move the way they do. For Newton, the field describes a *force*, and the motions we see are the result of forces induced by gravitating masses. But Einstein takes an entirely different view. For Einstein, the gravitational field describes not a force but simply the curvature of spacetime. Gravitating masses induce this curvature; they alter the geometry of spacetime. The worldcurves that objects follow in the presence of gravitating masses are simply the straightest possible paths in the spacetime that has been curved by those masses. Gravity is no longer a force; it is an aspect of geometry. For Einstein, arbitrary noninertial frames are an essential link in the chain that runs from gravity to geometry:

$$\text{gravity} \implies \text{accelerated motions} \implies \text{noninertial frames} \implies \text{curved spacetime}$$

Exercises

1. (a) Explain why the acceleration due to gravity at the surface of the sun is given by the formula

$$\alpha = \frac{GM}{X^2},$$

where G is the gravitational constant, $M = 2 \times 10^{30}$ kilograms is the mass of the sun, and $X = 7 \times 10^8$ meters is its radius. Determine α in m/sec^2 and in geometric units (in which it has the dimensions sec^{-1}).

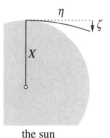

the sun

(b) Consider a photon that grazes the sun, moving along a ray that is initially perpendicular to a radius. Using the formula $\zeta = -\alpha\eta^2/2$ from the text to describe how far the photon drops after it has traveled the horizontal distance η, determine how far the photon drops after it travels the distance $\eta = X$ equal to the radius of the sun. Express the drop in both conventional and geometric units. Note: The formula derived in the text assumes that the gravitational field is constant in both magnitude and direction. Obviously, that is not the case when we take η as large as the radius X of the sun; nevertheless, assume that the formula gives a reasonable approximation for $|\eta| \leq X$.

(c) Suppose θ is the change in *direction* that the photon undergoes when it travels the distance η. Show that

$$\theta \approx \tan\theta = -\alpha\eta \quad \text{radians}.$$

Determine the value of θ when the photon travels the distance $\eta = X$.

(d) In 1911 Einstein calculated that a photon from a distant star will be deflected by about 4×10^{-6} radians if it grazes the sun. He assumed a Newtonian gravitational field that pointed toward the center of the sun and had a magnitude proportional to the inverse square of the distance from the center. Compare your result here with Einstein's; note that you should double your estimate of the change in direction (why?).

2. (a) Why does the Doppler effect have the form $N = \nu\left(1 - \frac{\Delta\Phi}{c^2}\right)$ in conventional units?

(b) Determine the potential difference $\Delta\Phi$ between the earth and the surface of the sun.

(c) Determine the relative red shift of sunlight viewed from the earth.

Further Reading for Chapter 4

Newtonian gravity is discussed in detail in *The Feynman Lectures* [13] and particularly in relation to general relativity by Berry [2]. Boas [4], Section 13.8, has a clear, accessible treatment of the gravitational field equation. Berry and Einstein himself [12], [10] deal with the complications that arise in noninertial frames. Our argument for the incompatibility of gravity and special relativity is adapted from Misner, Thorne, and Wheeler [24], who devote a chapter to the subject.

5 Surfaces and Curvature

CHAPTER

This chapter is about the differential geometry of a 2-dimensional surface in 3-space. It involves measuring angles, lengths of curves, and areas of regions and ultimately finding the curvature at each point on the surface. Then in the following chapter we will see how make all this *intrinsic*, that is, to define a surface as a piece of an ordinary plane endowed with a non-Euclidean metric, and to determine curvature directly from that metric, without reference to any embedding in space.

5.1 The Metric

Our strategy for developing the differential geometry of a curved surface will come from suitable generalizations of what we have already done for curves. Consider, for example, how we define the length of a curve C:

$$\text{length of } C = \int_a^b \|\mathbf{x}'(q)\| \, dq \quad \text{when } C : \mathbf{x}(q),\ a \leq q \leq b.$$

Note that this result depends, first of all, on a parametrization $\mathbf{x}(q)$ of C; second, on a way to define and measure the length of the tangent vector at points along the curve; and finally, on calculating an integral. Thus, the first step here will be to define the parametrization of a surface. From there we will consider the vectors in a tangent plane and establish geometric constructs in that plane. Since it is an ordinary Euclidean plane, the key will be an inner product. Finally, we will develop the tools—like integration—that will allow us to bring geometric features of the tangent planes down to the surface itself.

Parametrizing a Surface

We start with a map $\mathbf{x} : \mathbf{R}^2 \to \mathbf{R}^3 : (q^1, q^2) \to (x, y, z)$ defined by three smooth functions of the parameters q^1 and q^2:

$$\mathbf{x}(q^1, q^2) = \left(x(q^1, q^2), y(q^1, q^2), z(q^1, q^2)\right).$$

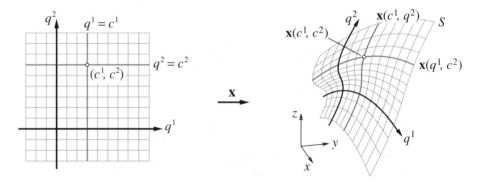

The surface S is the image of $\mathbf{x}$, which we usually visualize as a grid made up of the images of vertical and horizontal lines from the (q^1, q^2)-plane. (The reason for the superscripts will become clear later.) Let us look at these image curves in more detail. The map $\mathbf{x}$, restricted to the horizontal line $q^2 = c^2$, defines a parametrized curve

$$q^1 \mapsto \mathbf{x}(q^1, c^2),$$

with q^1 as parameter. As c^2 varies the images sweep out a family of roughly parallel curves in space. Similarly, if we restrict $\mathbf{x}$ to the vertical lines $q^1 = c^1$,

$$q^2 \mapsto \mathbf{x}(c^1, q^2),$$

we get a second family of parametrized curves with q^2 as parameter. Members of the second family cross the first, though not, in general, at right angles. The two families thus define a *curvilinear* coordinate grid on S in much the same way that latitude and longitude lines define one on the earth. Because q^1 and q^2 are indeed coordinates on S, we often refer to the point $\mathbf{x}(q^1, q^2)$ simply as (q^1, q^2).

q^1 and q^2 define coordinates on S

The coordinate curves have tangent vectors at each point, and these vectors lie in the tangent plane to the surface at that point. Consider first the "horizontal" curve $\mathbf{x}(q^1, c^2)$. It is a function of the single variable q^1, so we can write its tangent vector as $\mathbf{x}_1 = \mathbf{x}'(q^1, c^2)$, where $\mathbf{x}'$ means the derivative with respect to q^1. Similarly, the "vertical" curve is $\mathbf{x}(c^1, q^2)$, so its tangent vector is $\mathbf{x}_2 = \mathbf{x}'(c^1, q^2)$, where $\mathbf{x}'$ now means the derivative with respect to q^2. Since $\mathbf{x}$ is actually a function of *both* q^1 and q^2, it will be clearer if we use partial derivatives. Thus the tangents to S in the two coordinate directions at $(q^1, q^2) = (c^1, c^2)$ are

Tangents to the coordinate curves

$$\mathbf{x}_1 = \frac{\partial \mathbf{x}}{\partial q^1}(c^1, c^2) \quad \text{and} \quad \mathbf{x}_2 = \frac{\partial \mathbf{x}}{\partial q^2}(c^1, c^2).$$

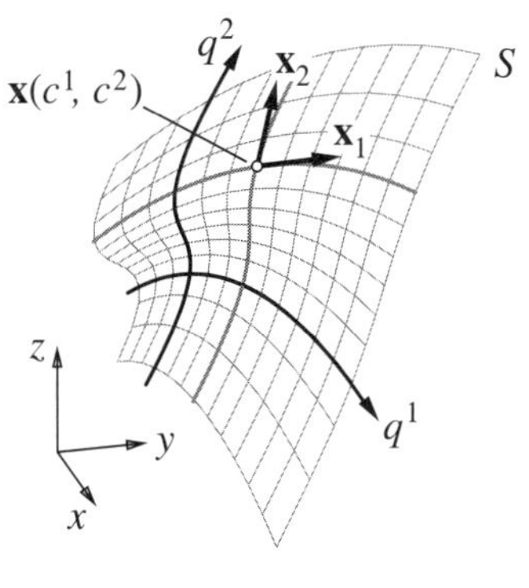

In the figure above, the vectors $\mathbf{x}_1$ and $\mathbf{x}_2$ are linearly independent. But this need not always happen; one of the vectors could collapse to $\mathbf{0}$, or the two could point in the same direction. For example, the "bowtie" map

Singular behavior

$$\mathbf{y}: \begin{cases} x = q^1 + 0.9q^2, \\ y = 0, \\ z = (q^1 - 0.9q^2)(q^1 + 0.9q^2)^2 \end{cases}$$

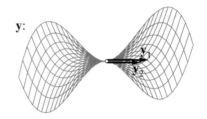

y:

pinches the (q^1, q^2)-plane down to a line at the origin, and you should check that $\mathbf{y}_1$ and $\mathbf{y}_2$ are collinear there: $\mathbf{y}_1 = (1, 0, 0)$, $\mathbf{y}_2 = (0.9, 0, 0)$. Points like this where the tangents in the coordinate directions fail to be linearly independent are said to be **singular**.

Surfaces must be nonsingular

Singular points can have various forms, but they typically are places where the surface is locally not 2-dimensional. To avoid this, we henceforth require that our surface parametrizations be nonsingular everywhere. Since the cross product of two vectors is nonzero precisely when they are linearly independent, we impose the following condition to guarantee that the surface $\mathbf{x}$ has no singular point:

$$\mathbf{x}_1 \times \mathbf{x}_2 = \mathbf{x}_1(q^1, q^2) \times \mathbf{x}_2(q^1, q^2) \neq \mathbf{0}$$

for all (q^1, q^2).

Definition 5.1 *Let $\mathcal{R}$ be a region in $\mathbf{R}^2$; a **surface S parametrized by** $\mathcal{R}$ is a smooth map*

$$\mathbf{x}: \mathcal{R} \to \mathbf{R}^3: (q^1, q^2) \mapsto \big(x(q^1, q^2), y(q^1, q^2), z(q^1, q^2)\big)$$

for which the cross product $\mathbf{x}_1 \times \mathbf{x}_2$ is everywhere nonzero in $\mathcal{R}$.

The unit normals to S

The cross product of two vectors is orthogonal to each of its factors: $(\mathbf{v} \times \mathbf{w}) \cdot \mathbf{v} = (\mathbf{v} \times \mathbf{w}) \cdot \mathbf{w} = 0$. Since $\mathbf{x}_1$ and $\mathbf{x}_2$ span the tangent plane of the surface S parametrized by $\mathbf{x}$, their cross product $\mathbf{x}_1 \times \mathbf{x}_2$ is normal to the tangent plane and thus to the surface. Furthermore, since **x** is nonsingular, we can divide by $\|\mathbf{x}_1 \times \mathbf{x}_2\|$ to get the **unit normal** at each point:

$$\mathbf{n}(q^1, q^2) = \frac{\mathbf{x}_1 \times \mathbf{x}_2}{\|\mathbf{x}_1 \times \mathbf{x}_2\|}.$$

Geometry in the Tangent Plane

Let TS_P denote the set of all vectors in $\mathbf{R}^3$ that are tangent to S at the point $P = (c^1, c^2)$; we call TS_P the **tangent plane to S at P**. The geometric notions we want to establish on TS_P will all follow from the inner product defined there. Since any vectors **v**, **w** in TS_P are vectors in $\mathbf{R}^3$, we can express their inner product in terms of the $\mathbf{R}^3$ coordinates: If $\mathbf{v} = (v_1, v_2, v_3)$ and $\mathbf{w} = (w_1, w_2, w_3)$, then

$$\mathbf{v} \cdot \mathbf{w} = v_1 w_1 + v_2 w_2 + v_3 w_3.$$

The tangent plane at a point

While all this is true, it is not useful. Only some coordinate triples (v_1, v_2, v_3) actually represent vectors in TS_P, and the triples that *do* represent tangent vectors have no connection with the map $\mathbf{x}(q^1, q^2)$ that defines S. Finally, the tangent plane is only 2-dimensional, so its vectors would be better represented by coordinate pairs, rather than triples.

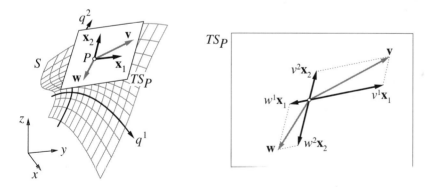

A more useful approach is to use the linearly independent vectors $\mathbf{x}_1$ and $\mathbf{x}_2$ as a basis for TS_P. Then, when we write a tangent vector as a linear combination of $\mathbf{x}_1$ and $\mathbf{x}_2$, we get a *pair*

A basis for TS_P

of coordinates. We shall write these with superscripts, as follows:
$$\mathbf{v} = v^1 \mathbf{x}_1 + v^2 \mathbf{x}_2, \qquad \mathbf{w} = w^1 \mathbf{x}_1 + w^2 \mathbf{x}_2,$$
and we shall also write
$$\mathbf{v} = \begin{pmatrix} v^1 \\ v^2 \end{pmatrix}, \qquad \mathbf{w} = \begin{pmatrix} w^1 \\ w^2 \end{pmatrix}.$$

The coordinates that appear here are unrelated to the coordinates that $\mathbf{v}$ and $\mathbf{w}$ have as vectors in $\mathbf{R}^3$; the superscripts help remind us of the difference (but they have a more important purpose, too, which we shall see presently).

The inner product with components

We can express the inner product of $\mathbf{v}$ and $\mathbf{w}$ directly in terms of these coordinates:
$$\begin{aligned}
\mathbf{v} \cdot \mathbf{w} &= \left(v^1 \mathbf{x}_1 + v^2 \mathbf{x}_2\right) \cdot \left(w^1 \mathbf{x}_1 + w^2 \mathbf{x}_2\right) \\
&= v^1 w^1 (\mathbf{x}_1 \cdot \mathbf{x}_1) + v^1 w^2 (\mathbf{x}_1 \cdot \mathbf{x}_2) + v^2 w^1 (\mathbf{x}_2 \cdot \mathbf{x}_1) + v^2 w^2 (\mathbf{x}_2 \cdot \mathbf{x}_2) \\
&= (v^1, v^2) \begin{pmatrix} \mathbf{x}_1 \cdot \mathbf{x}_1 & \mathbf{x}_1 \cdot \mathbf{x}_2 \\ \mathbf{x}_2 \cdot \mathbf{x}_1 & \mathbf{x}_2 \cdot \mathbf{x}_2 \end{pmatrix} \begin{pmatrix} w^1 \\ w^2 \end{pmatrix} \\
&= \mathbf{v}^t G \mathbf{w}.
\end{aligned}$$

The last step expresses $\mathbf{v} \cdot \mathbf{w}$ as a matrix multiplication of the same sort we found so useful with the Minkowski inner product. In that case the matrix was
$$J_{1,1} = \begin{pmatrix} 1 & 0 \\ 0 & -1 \end{pmatrix};$$
here it is
$$G = \begin{pmatrix} g_{11} & g_{12} \\ g_{21} & g_{22} \end{pmatrix} = \begin{pmatrix} \mathbf{x}_1 \cdot \mathbf{x}_1 & \mathbf{x}_1 \cdot \mathbf{x}_2 \\ \mathbf{x}_2 \cdot \mathbf{x}_1 & \mathbf{x}_2 \cdot \mathbf{x}_2 \end{pmatrix} = \begin{pmatrix} \|\mathbf{x}_1\|^2 & \mathbf{x}_1 \cdot \mathbf{x}_2 \\ \mathbf{x}_2 \cdot \mathbf{x}_1 & \|\mathbf{x}_2\|^2 \end{pmatrix}.$$

The metric

Because G plays the same crucial role in TS_P that $J_{1,1}$ plays in Minkowski geometry, we call G, or g_{ij}, the **metric** on TS_P. Each tangent plane has its own metric G, which therefore is a function of the parameters q^1 and q^2: $G = G(q^1, q^2)$. Also, G is always a *symmetric* matrix ($G^t = G$) because $g_{21} = \mathbf{x}_2 \cdot \mathbf{x}_1 = \mathbf{x}_1 \cdot \mathbf{x}_2 = g_{12}$.

The metric is also called the **metric tensor**, because it is "tensorial"; we shall see what this means when we consider tensors in detail in Chapter 6. Yet another name for the metric is **first**

fundamental form, a term introduced by Gauss in his study of surfaces and now very commonly used.

Proposition 5.1 *The eigenvalues of G are positive.*

PROOF: Since G is symmetric, its eigenvalues λ_1 and λ_2 are real. Since $\det G = \lambda_1 \lambda_2$ and $\operatorname{tr} G = \lambda_1 + \lambda_2$, it is sufficient to show that the trace and determinant are both positive.

But $\operatorname{tr} G = \|\mathbf{x}_1\|^2 + \|\mathbf{x}_2\|^2 > 0$ is immediate. To compute $\det G$, note first that since $\mathbf{x}_1$ and $\mathbf{x}_2$ are linearly independent, the angle θ between them lies strictly between 0 and π, so $\sin\theta > 0$. Therefore,

$$\det G = \|\mathbf{x}_1\|^2 \|\mathbf{x}_2\|^2 - (\mathbf{x}_1 \cdot \mathbf{x}_2)^2 = \|\mathbf{x}_1\|^2 \|\mathbf{x}_2\|^2 - \|\mathbf{x}_1\|^2 \|\mathbf{x}_2\|^2 \cos^2\theta$$
$$= \|\mathbf{x}_1\|^2 \|\mathbf{x}_2\|^2 (1 - \cos^2\theta) = \|\mathbf{x}_1\|^2 \|\mathbf{x}_2\|^2 \sin^2\theta > 0.$$

END OF PROOF

We can express the inner product and the metric using summation notation: *Basic metric quantities*

$$\mathbf{v} \cdot \mathbf{w} = \left(\sum_{i=1}^{2} v^i \mathbf{x}_i\right) \cdot \left(\sum_{j=1}^{2} w^j \mathbf{x}_j\right) = \sum_{i=1}^{2}\sum_{j=1}^{2} v^i w^j (\mathbf{x}_i \cdot \mathbf{x}_j) = \sum_{i=1}^{2}\sum_{j=1}^{2} v^i w^j g_{ij}.$$

In the same way, the length of a tangent vector is

$$\|\mathbf{v}\| = \sqrt{\mathbf{v} \cdot \mathbf{v}} = \sqrt{\sum_{i,j} v^i v^j g_{ij}}.$$

If θ is the angle between two vectors $\mathbf{v}$ and $\mathbf{w}$, then

$$\cos\theta = \frac{\sum_{i,j} v^i w^j g_{ij}}{\sqrt{\sum_{i,j} v^i v^j g_{ij}} \sqrt{\sum_{i,j} w^i w^j g_{ij}}}.$$

These examples begin to suggest why $G = (g_{ij})$ is called the metric, but they also illustrate a tendency for formulas in differential geometry to become visually cluttered. To help relieve the clutter, Einstein proposed that summation signs could usually be removed and replaced by the following   *Einstein summation convention*

Summation convention: *Whenever an index appears twice in an expression, once as a subscript and once as a superscript, sum over it. Summation is understood without a summation sign.*

Thus, for example, we write simply

$$\mathbf{v} = v^i \mathbf{x}_i \quad \text{instead of} \quad \mathbf{v} = \sum_i v^i \mathbf{x}_i,$$

and the metric quantities we have already defined become

$$\mathbf{v} \cdot \mathbf{w} = v^i w^j g_{ij}, \quad \|\mathbf{v}\| = \sqrt{v^i v^j g_{ij}}, \quad \cos\theta = \frac{v^i w^j g_{ij}}{\sqrt{v^i v^j g_{ij}}\sqrt{w^i w^j g_{ij}}}.$$

Oriented areas

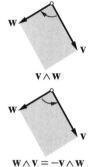

We also use the metric to calculate areas. In fact, we shall define *oriented* regions with areas that can have negative, as well as positive, values. We define $\mathbf{v} \wedge \mathbf{w}$ to be the oriented parallelogram spanned by the vectors $\mathbf{v}$ and $\mathbf{w}$, *in that order*; then $\mathbf{w} \wedge \mathbf{v} = -\mathbf{v} \wedge \mathbf{w}$ is that parallelogram with the opposite orientation. Furthermore, it will always be true that

$$\text{area } \mathbf{w} \wedge \mathbf{v} = -\text{area } \mathbf{v} \wedge \mathbf{w}.$$

Finally, we require that the parallelogram $U^* = \mathbf{x}_1 \wedge \mathbf{x}_2$ have a positive area; its value will be determined in the following proposition. For convenience in future work we define $g = \det G$.

Proposition 5.2 area $U^* = $ area $\mathbf{x}_1 \wedge \mathbf{x}_2 = \|\mathbf{x}_1 \times \mathbf{x}_2\| = \sqrt{g}$.

PROOF: By definition, $\|\mathbf{x}_1 \times \mathbf{x}_2\| = $ area U^*. But we also have area $U^* = \|\mathbf{x}_1\|\|\mathbf{x}_2\| \sin\theta$, which is $\sqrt{\det G} = \sqrt{g}$, by Proposition 5.1. END OF PROOF

The next proposition shows how any parallelogram $\mathbf{v} \wedge \mathbf{w}$ can be linked to the "basic" parallelogram $\mathbf{x}_1 \wedge \mathbf{x}_2$.

Proposition 5.3 $\mathbf{v} \times \mathbf{w} = (v^1 w^2 - v^2 w^1)(\mathbf{x}_1 \times \mathbf{x}_2) = (\det R)(\mathbf{x}_1 \times \mathbf{x}_2)$. *Here (using the summation convention)*

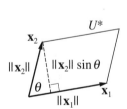

$$\mathbf{v} = v^i \mathbf{x}_i, \quad \mathbf{w} = w^j \mathbf{x}_j, \quad \text{and} \quad R = \begin{pmatrix} v^1 & w^1 \\ v^2 & w^2 \end{pmatrix}.$$

Area of an arbitrary parallelogram

Therefore, area $\mathbf{v} \wedge \mathbf{w} = \det R \cdot$ area $\mathbf{x}_1 \wedge \mathbf{x}_2 = \sqrt{g} \det R$.

PROOF: By the summation convention,

$$\mathbf{v} \times \mathbf{w} = v^i \mathbf{x}_i \times w^j \mathbf{x}_j = v^i w^j (\mathbf{x}_i \times \mathbf{x}_j).$$

But $\mathbf{x}_1 \times \mathbf{x}_1 = \mathbf{x}_2 \times \mathbf{x}_2 = \mathbf{0}$ and $\mathbf{x}_2 \times \mathbf{x}_1 = -\mathbf{x}_1 \times \mathbf{x}_2$, so

$$\mathbf{v} \times \mathbf{w} = v^1 w^2 (\mathbf{x}_1 \times \mathbf{x}_2) + v^2 w^1 (\mathbf{x}_2 \times \mathbf{x}_1)$$
$$= (v^1 w^2 - v^2 w^1)(\mathbf{x}_1 \times \mathbf{x}_2) = (\det R)(\mathbf{x}_1 \times \mathbf{x}_2).$$

Therefore, area $\mathbf{v} \wedge \mathbf{w} = \det R \cdot$ area $\mathbf{x}_1 \wedge \mathbf{x}_2 = \det R \sqrt{g}$, as claimed. END OF PROOF

Corollary 5.1 area $\mathbf{v} \wedge \mathbf{w} > 0$ *if and only if* $\det R > 0$.

To understand these relations between areas, it is helpful to look in more detail at the way we assign coordinates in TS_P. Recall that by definition, $\mathbf{R}^2$ is the set of all ordered pairs (a, b) of real numbers. A coordinate assignment in TS_P is a linear map $L : \mathbf{R}^2 \to TS_P$ defined by its effect on the standard basis:

Areas from coordinates

$$L : \begin{pmatrix} 1 \\ 0 \end{pmatrix} \mapsto \mathbf{x}_1, \quad L : \begin{pmatrix} 0 \\ 1 \end{pmatrix} \mapsto \mathbf{x}_2, \quad \text{and} \quad L : \begin{pmatrix} v^1 \\ v^2 \end{pmatrix} \mapsto \mathbf{v} \quad \text{if } \mathbf{v} = v^i \mathbf{x}_i.$$

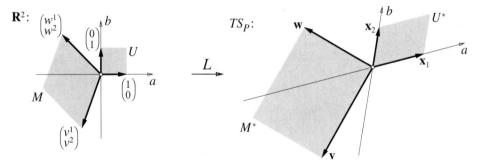

Any linear map from one 2-dimensional vector space to another magnifies areas by a certain fixed amount. The area magnification factor for L is $\sqrt{g}$, which we can see by the following argument. Let U be the unit square, spanned by $(1, 0)$ and $(0, 1)$ in $\mathbf{R}^2$. Then its image $U^* = L(U)$ is the parallelogram spanned by $\mathbf{x}_1$ and $\mathbf{x}_2$ in TS_P. Since area $U = 1$, while area $U^* = \sqrt{g}$, the linear map L magnifies areas by the factor $\sqrt{g}$.

L magnifies areas by the factor $\sqrt{g}$

Now let M be the parallelogram

$$\begin{pmatrix} v^1 \\ v^2 \end{pmatrix} \wedge \begin{pmatrix} w^1 \\ w^2 \end{pmatrix}$$

in $\mathbf{R}^2$. The exercises ask you to show that area $M = \det R$, where

$$R = \begin{pmatrix} v^1 & w^1 \\ v^2 & w^2 \end{pmatrix}.$$

Therefore, the image parallelogram $L(M) = M^* = \mathbf{v} \wedge \mathbf{w}$ has area $\sqrt{g} \det R$, as we have already demonstrated by other means. Incidentally, note that $\det R < 0$ in the figure above, so M and M^*, as drawn, have *negative* area.

Geometry on the Surface

Connecting the tangent plane to the surface

Geometric information from the tangent plane TS_P can be transferred down to the surface S itself, at least near the point P. Taylor's theorem provides the mechanism. For suppose $P = (c^1, c^2)$ (note that this is a shorthand for $P = \mathbf{x}(c^1, c^2)$), and suppose $(\Delta q^1, \Delta q^2)$ is near $(0, 0)$. Then $(c^1 + \Delta q^1, c^2 + \Delta q^2)$ is near P, and

$$\underbrace{\mathbf{x}(c^1 + \Delta q^1, c^2 + \Delta q^2)}_{\text{point on surface near } P} \approx \underbrace{\mathbf{x}(c^1, c^2) + \Delta q^1\, \mathbf{x}_1(c^1, c^2) + \Delta q^2\, \mathbf{x}_2(c^1, c^2)}_{\text{point on tangent plane to surface at } P}.$$

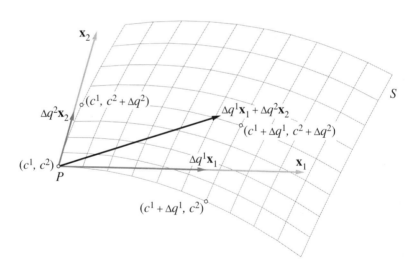

The point $\mathbf{x}(c^1 + \Delta q^1, c^2 + \Delta q^2)$, which in the figure above is shown simply as $(c^1 + \Delta q^1, c^2 + \Delta q^2)$, has coordinates $(\Delta q^1, \Delta q^2)$ relative to (c^1, c^2). The point $\mathbf{x} + \Delta q^1\, \mathbf{x}_1 + \Delta q^2\, \mathbf{x}_2$ is the vector in

§5.1 The Metric

TS_P with coordinates $(\Delta q^1, \Delta q^2)$ relative to the basis $\mathbf{x}_1$ and $\mathbf{x}_2$. Taylor's theorem says that these two points—one on the surface and one on the tangent plane—which have the same coordinates, are close in $\mathbf{R}^3$. The technical condition is that

$$\lim_{(\Delta q^1, \Delta q^2) \to (0,0)} \frac{\|\mathbf{x}(c^1 + \Delta q^1, c^2 + \Delta q^2) - (\mathbf{x} + \Delta q^1 \mathbf{x}_1 + \Delta q^2 \mathbf{x}_2)\|}{\|(\Delta q^1, \Delta q^2)\|^2} = 0.$$ exists and is finite.

(Note that where $\mathbf{x}$, $\mathbf{x}_1$, and $\mathbf{x}_2$ appear without arguments in this expression and in the figure above they are to be evaluated at (c^1, c^2).) We can rewrite the technical condition of Taylor's theorem using "big Oh" notation as

$$\mathbf{x}(c^1 + \Delta q^1, c^2 + \Delta q^2) \\ = \mathbf{x}(c^1, c^2) + \Delta q^1 \mathbf{x}_1(c^1, c^2) + \Delta q^2 \mathbf{x}_2(c^1, c^2) + O\left((\Delta q^1)^2 + (\Delta q^2)^2\right).$$

We turn now to measuring the length of a curve on the surface S, where S is the image of a parametrization $\mathbf{x} : \mathcal{R} \to \mathbf{R}^3$. If we start with a curve in the parameter plane $\mathcal{R}$, defined by a parametrization

Length of a curve on S

$$\mathbf{q} : [a, b] \to \mathcal{R} : t \mapsto (q^1(t), q^2(t)),$$

then the composition $\mathbf{z}(t) = \mathbf{x}(\mathbf{q}(t))$ gives us a parametrized curve in $\mathbf{R}^3$ that lies on S. Thus we have two curves: one in $\mathcal{R}$ and one in $S = \mathbf{x}(\mathcal{R})$.

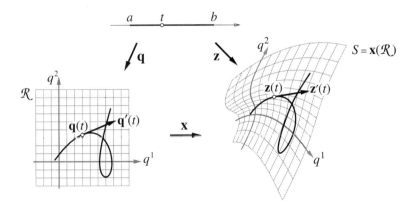

It will be instructive to calculate the lengths of both curves; for $\mathbf{z}(t)$ in S we have

$$\text{length of } \mathbf{z} = \int_a^b \|\mathbf{z}'(t)\|\, dt.$$

Since $\mathbf{z}(t) = \mathbf{x}(q^1(t), q^2(t))$, the chain rule gives

$$\mathbf{z}'(t) = \frac{d\mathbf{z}}{dt} = \frac{\partial \mathbf{x}}{\partial q^1}\frac{dq^1}{dt} + \frac{\partial \mathbf{x}}{\partial q^2}\frac{dq^2}{dt} = \frac{dq^1}{dt}\mathbf{x}_1 + \frac{dq^2}{dt}\mathbf{x}_2.$$

$\mathbf{z}' = \left(\dfrac{dq^1}{dt}, \dfrac{dq^2}{dt}\right)$
in $TS_{\mathbf{z}}$

Thus $\mathbf{z}'(t)$, which is a vector in the tangent plane at the point $\mathbf{z}(t) = \mathbf{x}(\mathbf{q}(t))$, has coordinates dq^1/dt and dq^2/dt with respect to the basis $\mathbf{x}_1$ and $\mathbf{x}_2$. We can therefore express its length in terms of the metric on $TS_{\mathbf{z}(t)}$. If we use the summation convention but otherwise write out everything in detail, the formula looks like this:

$$\|\mathbf{z}'(t)\| = \sqrt{g_{ij}(q^1(t), q^2(t))\frac{dq^i}{dt}(t)\frac{dq^j}{dt}(t)}.$$

More briefly, the length of the curve is

$$\text{length of } \mathbf{z} = \int_a^b \sqrt{g_{ij}\frac{dq^i}{dt}\frac{dq^j}{dt}}\, dt.$$

$\mathbf{q}' = \left(\dfrac{dq^1}{dt}, \dfrac{dq^2}{dt}\right)$
in $\mathcal{R}$

Now consider $\mathbf{q}(t)$ in $\mathcal{R}$. Notice that $\mathbf{q}' = (dq^1/dt, dq^2/dt)$, so $\mathbf{q}'$ and $\mathbf{z}'$ have the same components in their respective vector spaces. We have

$$\text{length of } \mathbf{q} = \int_a^b \|\mathbf{q}'(t)\|\, dt = \int_a^b \sqrt{\left(\frac{dq^1}{dt}\right)^2 + \left(\frac{dq^2}{dt}\right)^2}\, dt.$$

To make this resemble the length formula for $\mathbf{z}(t)$ even more closely, observe that $\mathcal{R}$ uses the ordinary Euclidean inner product, which is defined by the identity matrix. That is, $\mathbf{v} \cdot \mathbf{w} = \mathbf{v}^t I \mathbf{w}$, where

$$I = \begin{pmatrix} \delta_{11} & \delta_{12} \\ \delta_{21} & \delta_{22} \end{pmatrix} = \begin{pmatrix} 1 & 0 \\ 0 & 1 \end{pmatrix}.$$

In other words, we can take the view that $\mathcal{R}$ is itself a surface whose metric is everywhere equal to

$$\delta_{ij} = \begin{cases} 1 & i = j, \\ 0 & i \neq j. \end{cases}$$

Then, if $\mathbf{v} = (v^1, v^2)$, you can check that $\delta_{ij} v^i v^j = (v^1)^2 + (v^2)^2$, so $\|\mathbf{v}\| = \sqrt{\delta_{ij} v^i v^j}$. Therefore, we finally have

$$\text{length of } \mathbf{q} = \int_a^b \sqrt{\delta_{ij} \frac{dq^i}{dt} \frac{dq^j}{dt}} \, dt.$$

Certainly the formula for the length of $\mathbf{q}(t)$ depends only on quantities defined in $\mathcal{R}$. The similarity between the calculations for $\mathbf{q}(t)$ and $\mathbf{z}(t)$ draws attention to the fact that the length of $\mathbf{z}(t)$ *also* depends only on quantities defined in $\mathcal{R}$.

Theorem 5.1 *Suppose D is a region in $\mathcal{R}$ and $S = \mathbf{x}(\mathcal{R})$ is a surface. Then*

$$\text{area } D = \iint_D dq^1 dq^2, \qquad \text{area } \mathbf{x}(D) = \iint_D \sqrt{g(q^1, q^2)} \, dq^1 dq^2.$$

Area of a region in S

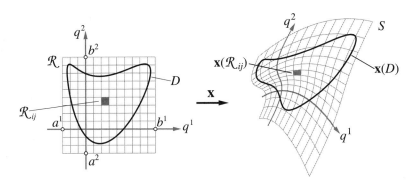

PROOF: This is a sketch of a proof. While the assertion about area D itself is immediate, we will still find it useful to start with some of the details in that case.

Suppose that D lies inside the rectangle $\mathcal{R} = [a^1, b^1] \times [a^2, b^2]$ in the (q^1, q^2)-plane. A partition of these two intervals,

Area of D

$$a^1 = q_1^1 < q_2^1 < q_3^1 < \cdots < q_{m+1}^1 = b^1,$$
$$a^2 = q_1^2 < q_2^2 < q_3^2 < \cdots < q_{n+1}^2 = b^2,$$

subdivides $\mathcal{R}$ into small rectangles $\mathcal{R}_{ij} = [q_i^1, q_{i+1}^1] \times [q_j^2, q_{j+1}^2]$. Note that area $\mathcal{R}_{ij} = \Delta q_i^1 \Delta q_j^2$, where $\Delta q_i^1 = q_{i+1}^1 - q_i^1$, $\Delta q_j^2 = q_{j+1}^2 - q_j^2$. The area of D is approximately equal to the sum of the areas of the various $\mathcal{R}_{ij}$ that intersect D. To express this more precisely, we introduce the *characteristic function* of D:

$$\chi_D(q^1, q^2) = \begin{cases} 1 & \text{if } (q^1, q^2) \text{ is in } D, \\ 0 & \text{otherwise.} \end{cases}$$

Then

$$\text{area } D \approx \sum_{i=1}^{m} \sum_{j=1}^{n} \chi_D(q_i^1, q_j^2) \text{ area } \mathcal{R}_{ij} = \sum_{i=1}^{m} \sum_{j=1}^{n} \chi_D(q_i^1, q_j^2) \Delta q_i^1 \Delta q_j^2.$$

Note that the summation convention is not appropriate here. In the limit as $(\Delta q_i^1)^2 + (\Delta q_j^2)^2 \to 0$, this sum becomes the integral

$$\iint_{\mathcal{R}} \chi_D(q^1, q^2) \, dq^1 dq^2 = \iint_D dq^1 dq^2.$$

Area of the image of a single $\mathcal{R}_{ij}$

To determine the area of $\mathbf{x}(D)$, we focus on the image of a single rectangle $\mathcal{R}_{ij}$. In fact, there are two images to consider: One is the portion $\mathbf{x}(\mathcal{R}_{ij})$ on the surface S, and the other is the parallelogram $\Delta q_i^1 \mathbf{x}_1 \wedge \Delta q_j^2 \mathbf{x}_2$ that lies in the tangent plane at $\mathbf{x}(q_i^1, q_j^2)$ and is spanned by $\Delta q_i^1 \mathbf{x}_1$ and $\Delta q_j^2 \mathbf{x}_2$. (In the figure below, the tangent plane lies transparently in front of the surface.)

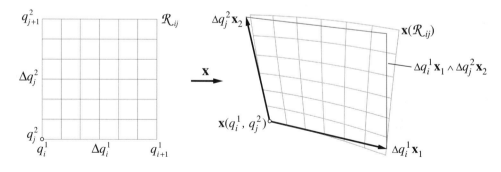

By Taylor's theorem, the parallelogram $\Delta q_i^1 \mathbf{x}_1 \wedge \Delta q_j^2 \mathbf{x}_2$ is in very close agreement with the image $\mathbf{x}(\mathcal{R}_{ij})$, so we can write

$$\text{area } \mathbf{x}(\mathcal{R}_{ij}) \approx \text{area}\left(\Delta q_i^1 \mathbf{x}_1 \wedge \Delta q_j^2 \mathbf{x}_2\right)$$
$$= \Delta q_i^1 \Delta q_j^2 \cdot \text{area } \mathbf{x}_1 \wedge \mathbf{x}_2$$
$$= \Delta q_i^1 \Delta q_j^2 \sqrt{g(q_i^1, q_j^2)}$$
$$= \sqrt{g(q_i^1, q_j^2)} \text{ area } \mathcal{R}_{ij}.$$

We already know that $\sqrt{g}$ is the area magnification factor for the map from $\mathcal{R}_{ij}$ into the tangent plane. This result shows that it is also the (approximate) local area magnification factor for the map $\mathbf{x}$ from $\mathcal{R}_{ij}$ to the surface itself.

The area of the entire image $\mathbf{x}(D)$ is therefore well-approximated by the total area of the individual images $\mathbf{x}(\mathcal{R}_{ij})$, when we sum over those small rectangles $\mathcal{R}_{ij}$ that intersect D:

Area of the entire image

$$\text{area } \mathbf{x}(D) \approx \sum_{i=1}^{m} \sum_{j=1}^{n} \chi_D(q_i^1, q_j^2) \text{ area } \mathbf{x}(\mathcal{R}_{ij})$$
$$\approx \sum_{i=1}^{m} \sum_{j=1}^{n} \chi_D(q_i^1, q_j^2) \sqrt{g(q_i^1, q_j^2)} \Delta q_i^1 \Delta q_j^2.$$

If we take the limit as $(\Delta q_i^1)^2 + (\Delta q_j^2)^2 \to 0$, then on the one hand this sum becomes the exact value of area $\mathbf{x}(D)$, while on the other it becomes the integral

$$\iint_R \chi_D(q^1, q^2) \sqrt{g(q^1, q^2)} \, dq^1 dq^2 = \iint_D \sqrt{g(q^1, q^2)} \, dq^1 dq^2.$$

END OF PROOF

Exercises

1. (a) Sketch the following surfaces in $\mathbf{R}^3$. Indicate the coordinate lines $q^1 = $ constant, $q^2 = $ constant, and describe the surface in words, where possible.

    (b) $\mathbf{x}(q^1, q^2) = (q^1, q^2, q^1 q^2)$, $-1 \leq q^1, q^2 \leq 1$.

(c) $\mathbf{x}(q^1, q^2) = (q^1, q^2, (q^1)^2 - (q^2)^2)$, $-1 \leq q^1, q^2 \leq 1$.

(d) $\mathbf{x}(q^1, q^2) = \left(q^1, q^2, \sqrt{1 - (q^1)^2 - (q^2)^2}\right)$, $(q^1)^2 + (q^2)^2 \leq 1$.

(e) $\mathbf{x}(q^1, q^2) = (\cos q^2 \cos q^1, \cos q^2 \sin q^1, \sin q^2)$, $0 \leq q^1 \leq 2\pi$, $0 \leq q^2 \leq \pi/2$.

(f) The previous surface but with the domain enlarged to $-\pi/2 \leq q^2 \leq \pi/2$.

(g) $\mathbf{x}(q^1, q^2) = (\sin q^2 \cos q^1, \sin q^2 \sin q^1, \cos q^2)$, $0 \leq q^1 \leq 2\pi$, $0 \leq q^2 \leq \pi$.

(h) $\mathbf{x}(q^1, q^2) = (\cosh q^2 \cos q^1, \cosh q^2 \sin q^1, \sinh q^2)$, $0 \leq q^1 \leq 2\pi$, $-1 \leq q^2 \leq 1$.

2. For each of the surfaces $\mathbf{x}(q^1, q^2)$ in the previous exercise, determine $\mathbf{x}_1$, $\mathbf{x}_2$, $\mathbf{x}_1 \times \mathbf{x}_2$, g_{ij}, and $\sqrt{g}$. Also determine whether the coordinate lines on the surfaces are orthogonal.

3. (a) Consider the "bowtie" surface $\mathbf{y}(q^1, q^2)$ given in the text:

$$x = q^1 + 0.9q^2, \qquad y = 0, \qquad z = \left(q^1 - 0.9q^2\right)\left(q^1 + 0.9q^2\right)^2.$$

Calculate the tangent vectors $\mathbf{y}_1$ and $\mathbf{y}_2$ and show that they are collinear at $(q^1, q^2) = (0, 0)$.

(b) Assume that the domain of $\mathbf{y}$ is the entire (q^1, q^2)-plane; show that the surface covers the entire (x, z)-plane excluding the z-axis but including the origin.

4. Show that area $M = v_1 w_2 - v_2 w_1$ when M is the parallelogram

$$M = \begin{pmatrix} v_1 \\ v^2 \end{pmatrix} \wedge \begin{pmatrix} w_1 \\ w_2 \end{pmatrix}.$$

5. For each of the surfaces $\mathbf{x}(q^1, q^2)$ given below, determine $\mathbf{x}_1$, $\mathbf{x}_2$, $\mathbf{x}_1 \times \mathbf{x}_2$, g_{ij}, and $\sqrt{g}$. Also determine whether the coordinate lines on the surfaces are orthogonal.

(a) The plane $\mathbf{x}(q^1, q^2) = (q^1, q^2, aq^1 + bq^2 + c)$.

(b) The graph of a function: $\mathbf{x}(q^1, q^2) = (q^1, q^2, f(q^1, q^2))$.

(c) A surface of revolution: $\mathbf{x}(q^1, q^2) = (r(q^2) \cos q^1, r(q^2) \sin q^1, z(q^2)$ where $0 \leq q^1 \leq 2\pi$.

(d) A sphere of radius R (a particular kind of surface of revolution):

$$\mathbf{x}(q^1, q^2) = (R\cos q^2 \cos q^1, R\cos q^2 \sin q^1, R\sin q^2),$$

where $0 \leq q^1 \leq 2\pi$, $-\pi/2 \leq q^2 \leq \pi/2$.

(e) A torus $T_{a,r}$ (a particular kind of surface of revolution):

$$\mathbf{x}(q^1, q^2) = ((a + r\cos q^2)\cos q^1, (a + r\cos q^2)\sin q^1, r\sin q^2),$$

where $0 \leq q^1 \leq 2\pi$, $-\pi \leq q^2 \leq \pi$. What geometric features of the torus do the parameters a and r determine? Sketch the tori $T_{a,3}$ for $a = 1, 2, 3, 4$.

6. How does the parametrized curve $(r, z) = (r(q), z(q))$ appear in the surface of revolution $\mathbf{x}(q^1, q^2) = (r(q^2)\cos q^1, r(q^2)\sin q^1, z(q^2))$? Illustrate your answer with the surfaces in Exercises 1d, 1f, 1g, 5d, and 5e.

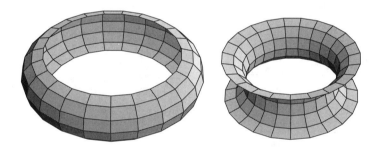

7. Consider the *outer* and *inner* halves of the torus $T_{a,r}$, as shown above. Determine their areas as functions of a and r. First identify the subdomains of $\mathcal{R}: 0 \leq q^1 \leq 2\pi$, $-\pi \leq q^2 \leq \pi$ that determine these two portions of the torus. What is the area of the whole torus?

8. (a) The coordinate lines on the torus $T_{a,r}$ consist of "meridians of longitude," given by $q^1 =$ constant, and "parallels of latitude," given by $q^2 =$ constant. Determine their lengths; in particular, show that the length of a latitude circle depends on the value of q^2 but all longitude circles have the same length.

(b) Sketch the curve $q^2 = q^1 - \pi$ on $T_{a,r}$; take $0 \leq q^1 \leq 2\pi$. Your sketch should make it clear that the curve is *closed*: Its starting and ending points are the same.

(c) Sketch the curve $\mathbf{x}(2t, 3t)$ on $T_{a,r}$; take $0 \leq t \leq 2\pi$. Your sketch should make it clear that this is a closed curve that has the form of a *trefoil knot*.

9. (a) Sketch the helix given by the arc-length parametrization

$$\mathbf{z}(s) = \left(\cos \frac{s}{2}, \sin \frac{s}{2}, \frac{\sqrt{3}}{2} s \right), \qquad 0 \leq s \leq \pi.$$

(b) Verify that the tangent $\mathbf{z}'(s)$ is indeed a unit vector for all s. On your sketch of the curve draw the individual tangents $\mathbf{z}'(s)$ for $s = 0, \pi/3, \pi/2, 2\pi/3, \pi$.

(c) For a fixed s, consider the tangent line parametrized by t as

$$\mathbf{w}(t) = \mathbf{z}(s) + t\mathbf{z}'(s).$$

Taking $-1 \leq t \leq 2$ and letting s take in turn the values specified in part (b), draw each of these tangent lines on your sketch of the helix.

(d) Now "fill in" the picture by drawing the family of tangent lines that occur when s is allowed to sweep out all values from 0 to π. This is a portion of the surface called the **tangent developable** of $\mathbf{z}$. Your drawing should show two sheets that meet in a cusp edge along the curve $\mathbf{z}$.

10. (a) Let $\mathbf{z}(s)$ be an arbitrary space curve parametrized by arc length and let $\mathbf{x}(q^1, q^2) = \mathbf{z}(q^1) + q^2 \mathbf{z}'(q^1)$ be its tangent developable surface. Determine $\mathbf{x}_1, \mathbf{x}_2, \mathbf{x}_1 \times \mathbf{x}_2, g_{ij}$, and $\sqrt{g}$ in terms of $\mathbf{z}$.

(b) At which points (q^1, q^2) is the parametrization $\mathbf{x}(q^1, q^2)$ singular? What is happening to the surface at the singular points?

(c) Show that all points on a given tangent line (that is, on a coordinate line $q^1 = $ constant) have the same the unit normal vector $\mathbf{n} = (\mathbf{x}_1 \times \mathbf{x}_2)/\|\mathbf{x}_1 \times \mathbf{x}_2\|$.

(d) Determine whether the coordinate lines on the surface are orthogonal.

(e) Consider the coordinate line $q^2 \mapsto \mathbf{x}(c^1, q^2)$; show that the segment for which $a \leq q^2 \leq b$ has length $b - a$.

5.2 Intrinsic Geometry on the Sphere

Suppose the surface S in $\mathbf{R}^3$ is defined by a parametrization $\mathbf{x} : \mathcal{R} \to \mathbf{R}^3 : \mathbf{q} \mapsto (x(\mathbf{q}), y(\mathbf{q}), z(\mathbf{q}))$. In the last section we saw that the metric tensor g_{ij} allows us to carry out all the basic geometric measurements on S without ever leaving the flat parameter plane $\mathcal{R}$. We call the study of S that we can carry out in the parameter plane **intrinsic geometry** because we need not pay attention to the way S sits in $\mathbf{R}^3$. (Of course the metric g_{ij} was constructed by using that knowledge; the point is that once we *have* the metric, we can proceed without further reference to the map $\mathbf{x}$.)

Intrinsic geometry

To illustrate how much we can learn about a surface from its intrinsic geometry, we take some time now to explore the familiar example of a sphere. We start with one of the standard parametrizations; q^1 is longitude and q^2 is latitude:

Example: a sphere

$$\mathbf{x}(q^1, q^2) = (\cos(q^1)\cos(q^2), \sin(q^1)\cos(q^2), \sin(q^2)),$$

$$\mathcal{R} : \begin{cases} -\pi \leq q^1 \leq \pi, \\ -\pi/2 \leq q^2 \leq \pi/2. \end{cases}$$

You should check that $\|\mathbf{x}(\mathbf{q})\| = 1$ for all $\mathbf{q}$, so $\mathbf{x}(\mathbf{q})$ does indeed lie on a sphere of radius 1. The basis vectors for the tangent space are

$$\mathbf{x}_1 = (-\sin(q^1)\cos(q^2), \cos(q^1)\cos(q^2), 0),$$
$$\mathbf{x}_2 = (-\cos(q^1)\sin(q^2), -\sin(q^1)\sin(q^2), \cos(q^2)).$$

They must satisfy the technical condition $\mathbf{x}_1 \times \mathbf{x}_2 \neq \mathbf{0}$. In fact,

$$\begin{aligned}\mathbf{x}_1 \times \mathbf{x}_2 &= \left(\cos(q^1)\cos^2(q^2), \sin(q^1)\cos^2(q^2), \sin q^2 \cos(q^2)\right) \\ &= \cos(q^2)\left(\cos(q^1)\cos(q^2), \sin(q^1)\cos(q^2), \sin(q^2)\right) \\ &= \cos(q^2)\,\mathbf{x}.\end{aligned}$$

Singular points of x

Thus $\mathbf{x}_1 \times \mathbf{x}_2$ points in the same direction as $\mathbf{x}$. This is not surprising: $\mathbf{x}_1 \times \mathbf{x}_2$ is always perpendicular to the surface, and since $\mathbf{x}$ is the radius vector for a sphere, it is perpendicular to the surface as well. Since $\|\mathbf{x}\| = 1$, $\|\mathbf{x}_1 \times \mathbf{x}_2\| = 0$ precisely when $\cos(q^2) = 0$, that is, when $q^2 = \pm\pi/2$. These points, which form the top and bottom edges of $\mathcal{R}$, are therefore the *singular* points of the map $\mathbf{x}$. The top edge is singular because it collapses to the north pole; the bottom edge is singular because it collapses to the south pole. The map is not locally one-to-one on these sets, but we can permit them because they are on the boundary of the domain $\mathbf{R}$.

As the figure illustrates (and you should confirm by direct calculation), $\|\mathbf{x}_2\|^2 = 1$ always, while $\|\mathbf{x}_1\|^2 = \cos^2(q^2)$, a value that drops to 0 as the point $\mathbf{q}$ approaches one of the poles. Furthermore, $\mathbf{x}_1 \cdot \mathbf{x}_2 = 0$, so the coordinate grid is *orthogonal* (latitude and longitude lines are perpendicular). Thus the metric tensor G has the following components:

$$G = \begin{pmatrix} g_{11} & g_{12} \\ g_{21} & g_{22} \end{pmatrix} = \begin{pmatrix} \cos^2(q^2) & 0 \\ 0 & 1 \end{pmatrix}.$$

Length of a parallel of latitude

Let us begin our exploration of intrinsic geometry with this question: How long is a complete circuit around a parallel of latitude $q^2 = \varphi$? We can parametrize this curve as $(q^1(t), q^2(t)) = (t, \varphi)$, so $(dq^1/dt, dq^2/dt) = (1, 0)$. Therefore,

$$g_{ij}\left(q^1(t), q^2(t)\right) \frac{dq^i}{dt}\frac{dq^j}{dt} = g_{11}(t, \varphi)\left(\frac{dq^1}{dt}\right)^2 + g_{22}(t, \varphi)\left(\frac{dq^2}{dt}\right)^2$$
$$= \cos^2(\varphi)$$

and

$$\text{length} = \int_{-\pi}^{\pi} \sqrt{g_{ij}\left(q^1(t), q^2(t)\right) \frac{dq^i}{dt} \frac{dq^j}{dt}} \, dt = \int_{-\pi}^{\pi} \cos(\varphi) \, dt = 2\pi \cos(\varphi).$$

Therefore, at the equator (where $\varphi = 0$ and $\cos\varphi = 1$), the length is 2π. Halfway to the poles, where $\varphi = \pm\pi/4$ and $\cos\varphi = \sqrt{2}/2$, the length is only $\sqrt{2}\,\pi$. At the poles, $\cos\varphi = 0$, so the length is 0.

Of course, in the (q^1, q^2)-plane, the *apparent* length of each of these paths is 2π: They all run from the left side of $\mathcal{R}$, where $q^1 = 0$, to the right side, where $q^1 = 2\pi$. Thus,

$$\frac{\text{apparent length}}{\text{true length}} = \frac{2\pi}{2\pi \cos\varphi} = \frac{1}{\cos\varphi} = \sec\varphi.$$

This is precisely the relation we noted in Section 4.1 when we were discussing the distortions in a Miller cylindrical projection of the earth.

By contrast, vertical distances (along meridians of longitude) are *exactly* as they appear. To see this, fix longitude $q^1 = \theta$ and consider the path $\mathbf{q}(t) = (\theta, t)$, $-\pi/2 \leq t \leq \pi/2$. In the exercises you are asked to confirm that the length of this path is π, independent of the value of θ.

Length of a meridian of longitude

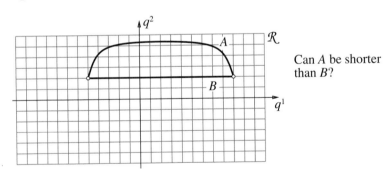

Can A be shorter than B?

We can read the metric tensor this way: Vertical distances are what they appear to be, but horizontal distances are not, the farther we go from the q^1-axis, the more severe the distortion is *along a horizontal path*. This suggests the following: If we consider two points at the same level $q^2 = c$, could a path between them that goes to more extreme q^2-values—like A, above—be shorter

Can paths that *appear* longer actually be shorter?

than one that goes "straight" from one point to the other at the same q^2-level—like B?

To pursue this question, first consider the family of curves C_α defined by the parametrizations

$$\mathbf{q}_\alpha(t) = (t, \arctan(\alpha \sin t)), \qquad 0 \le t \le \pi.$$

All these curves have the same endpoints: They start at $(q^1, q^2) = (0, 0)$ and end at $(\pi, 0)$. How do *their* lengths compare? In particular, what is the balance between additional vertical length and the advantage of having a greater portion of the path at "higher latitudes"?

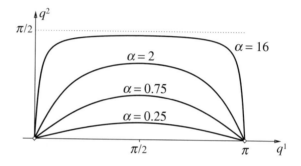

Our starting point in calculating the length is

$$\frac{dq^1}{dt} = 1, \qquad \frac{dq^2}{dt} = \frac{\alpha \cos t}{1 + \alpha^2 \sin^2 t}.$$

Therefore,

$$g_{ij}\left(q^1(t), q^2(t)\right) \frac{dq^i}{dt} \frac{dq^j}{dt} = \cos^2(\arctan(\alpha \sin t)) + \frac{\alpha^2 \cos^2 t}{\left(1 + \alpha^2 \sin^2 t\right)^2}.$$

Now, $\cos(\arctan y) = 1/\sqrt{1 + y^2}$, so

$$\cos^2(\arctan(\alpha \sin t)) = \frac{1}{1 + \alpha^2 \sin^2 t},$$

and hence

$$g_{ij} \frac{dq^i}{dt} \frac{dq^j}{dt} = \frac{1 + \alpha^2 \sin^2 t + \alpha^2 \cos^2 t}{\left(1 + \alpha^2 \sin^2 t\right)^2} = \frac{1 + \alpha^2}{\left(1 + \alpha^2 \sin^2 t\right)^2}.$$

§5.2 Intrinsic Geometry on the Sphere

Therefore, the length L_α of C_α is

$$L_\alpha = \int_0^\pi \sqrt{g_{ij}\frac{dq^i}{dt}\frac{dq^j}{dt}}\,dt = \int_0^\pi \frac{\sqrt{1+\alpha^2}}{1+\alpha^2\sin^2 t}\,dt.$$

A little rewriting and a trigonometric substitution put the integrand in a standard form:

$$\frac{\sqrt{1+\alpha^2}}{1+\alpha^2\sin^2 t} = \frac{\sqrt{1+\alpha^2}}{\cos^2 t + \sin^2 t + \alpha^2 \sin^2 t}$$

$$= \frac{\sqrt{1+\alpha^2}}{\cos^2 t + (1+\alpha^2)\sin^2 t} = \frac{\sqrt{1+\alpha^2}}{\cos^2 t} \cdot \frac{1}{1+(1+\alpha^2)\tan^2 t}.$$

Let $u = \sqrt{1+\alpha^2}\cdot\tan t$; then $du = \sqrt{1+\alpha^2}\cdot\sec^2 t\,dt$, so

$$L_\alpha(t) = \int_0^t \frac{\sqrt{1+\alpha^2}}{1+\alpha^2\sin^2 t}\,dt = \int_0^u \frac{1}{1+u^2}\,du$$
$$= \arctan(u) = \arctan\left(\sqrt{1+\alpha^2}\tan t\right).$$

The value we want is $L_\alpha(\pi)$. As you can see on the following graphs, the various L_α are different, but they all agree at $\pi/2$ and π. In fact, $L_\alpha(\pi/2) = \pi/2$ and $L_\alpha(\pi) = \pi$, for all α.

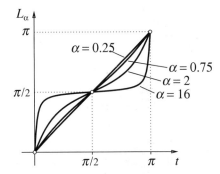

So, all the curves C_α are the same length! They are, in fact, all as long as the direct path from $(0,0)$ to $(\pi,0)$ along the equator. The arc-length function L_α tells us how length accumulates as t increases; it answers our question about the trade-off between the increase in vertical length and the savings by traveling horizontally at higher latitudes. This is most readily seen in the graph of

Every C_α has length π

$L_{16}(t)$, which has the strongest contrasts. The steep initial portion for t near 0 says that length accumulates quickly at first—while the path is climbing vertically—but then, as the graph levels off when t gets nearer to $\pi/2$, length accumulates very slowly indeed.

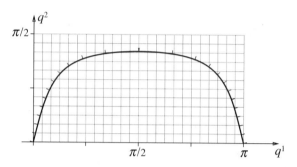

The figure above shows the curve C_4; the ticks along it mark arc length increments of $\pi/24$, exactly the same as the horizontal and vertical increments on the background grid. Note that while the path is climbing steeply, the ticks match vertical grid quite well, but when the path is more nearly horizontal, the ticks are very widely spaced in comparison to the horizontal grid. This disparity helps us see the difference between the metric as given by g_{ij} and the Euclidean metric our eye assumes. You should compare this figure to the worldcurve plots in Section 3.3 that used ticks to indicate constant increments of proper time.

Obviously, the paths C_α are quite special, and they were very carefully chosen. We can see how by going back to *extrinsic* geometry—that is, by considering how S sits in space. From that viewpoint we claim that the C_α come from great circles on the sphere. The points $(0, 0)$ and $(\pi, 0)$ are antipodal points, and any plane $z = \alpha y$ passes through them. The intersection of this plane with the sphere is a great circle through $(0, 0)$ and $(\pi, 0)$; the curve C_α is just the preimage in $\mathcal{R}$ of this great circle. See below. To determine the parametrization $\mathbf{q}_\alpha(t)$ of C_α, consider what the condition $z = \alpha y$ means for $\mathbf{x}(\mathbf{q}(t))$:

$$\sin(q^2) = z = \alpha y = \alpha \sin(q^1) \cos(q^2).$$

Hence $\tan(q^2) = \alpha \sin(q^1)$, so $q^2 = \arctan(\alpha \sin(q^1))$, which is equivalent to the parametrization we used.

The C_α are great circles

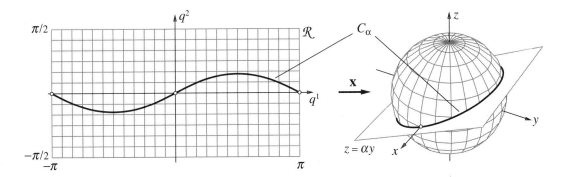

We return to the question of whether a path connecting two points at the same latitude can be *strictly* shorter than the "straight" path along the parallel of latitude that runs between them. Take the points $q^1 = \pi/4$ and $q^1 = 3\pi/4$ on the parallel $q^2 = \pi/4$. The distance between them along the parallel B is $(3\pi/4 - \pi/4)\cos(\pi/4) = \pi\sqrt{2}/4$.

Can "polar" paths be strictly shorter?

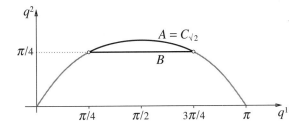

One of the great circles $A = C_\alpha$ also passes through these points. It is the one for which $q^2 = \pi/4$ when $q^1 = \pi/4$. Therefore,
$$1 = \tan(\pi/4) = \alpha \sin(\pi/4) = \alpha\sqrt{2}/2$$
so $\alpha = \sqrt{2}$. The entire length of $C_{\sqrt{2}}$ is π. The length of the initial segment, from $q^1 = 0$ to $q^1 = \pi/4$, is
$$L_{\sqrt{2}}\left(\frac{\pi}{4}\right) = \arctan\left(\sqrt{1+2}\,\tan\left(\frac{\pi}{4}\right)\right) = \arctan\left(\sqrt{3}\right) = \frac{\pi}{3}.$$
The length of the final segment, from $q^1 = 3\pi/4$ to $q^1 = \pi$, is also $\pi/3$, so the length of the middle segment A is $\pi/3$. Since $\pi/3 < \pi\sqrt{2}/4$, A is indeed strictly shorter than B.

While we have just determined that A is shorter than B by intrinsic means, we still do not have the intrinsic tools to prove

Measuring angles

that A is the shortest possible path that connects its endpoints. However, the *extrinsic* proof is immediate, since A is a portion of a great circle.

Intrinsic geometry also determines the angles between intersecting curves. Consider the line $q^2 = q^1$ in $\mathcal{R}$. It apparently makes a steady angle of 45° with the horizontal. What is the *true* angle at different points?

We can parametrize the line as $\mathbf{q}(t) = (t, t)$, $0 \leq t \leq \pi/2$. Then we want to know the angle θ between its tangent, $\mathbf{q}' = (1, 1)$, and the horizontal basis vector $\mathbf{x}_1 = (1, 0)$. We have

$$\cos \theta = \frac{\mathbf{q}' \cdot \mathbf{x}_1}{\sqrt{\mathbf{q}' \cdot \mathbf{q}'}\sqrt{\mathbf{x}_1 \cdot \mathbf{x}_1}}.$$

We use the metric tensor $g_{ij}(\mathbf{q}(t))$ to determine these inner products:

$$\mathbf{q}' \cdot \mathbf{x}_1 = (1, 1) \begin{pmatrix} \cos^2(t) & 0 \\ 0 & 1 \end{pmatrix} \begin{pmatrix} 1 \\ 0 \end{pmatrix} = \cos^2(t),$$

$$\mathbf{q}' \cdot \mathbf{q}' = (1, 1) \begin{pmatrix} \cos^2(t) & 0 \\ 0 & 1 \end{pmatrix} \begin{pmatrix} 1 \\ 1 \end{pmatrix} = \cos^2(t) + 1,$$

$$\mathbf{x}_1 \cdot \mathbf{x}_1 = (1, 0) \begin{pmatrix} \cos^2(t) & 0 \\ 0 & 1 \end{pmatrix} \begin{pmatrix} 1 \\ 0 \end{pmatrix} = \cos^2(t).$$

Therefore,

$$\cos \theta = \frac{\cos(t)}{\sqrt{\cos^2(t) + 1}}.$$

This is a monotonic decreasing function of t, so θ increases with t. At the start, when $t = 0$, θ is indeed 45°, but it rises to 90° as $t \to \pi/2$. You should confirm that this makes sense extrinsically—that is, by seeing that the path $\mathbf{q}(t)$ follows on the sphere itself implies that the angle increases.

Area of the sphere

Finally, we use the metric tensor to calculate areas. The area of the whole sphere is

$$\iint_{\mathcal{R}} \sqrt{g}\, dq^1 dq^2 = \int_{-\pi}^{\pi} \left(\int_{-\pi/2}^{\pi/2} \cos(q^2)\, dq^2 \right) dq^1 = \int_{-\pi}^{\pi} 2\, dq^1 = 4\pi.$$

§5.2 Intrinsic Geometry on the Sphere

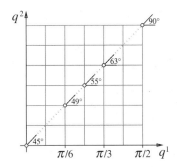

By contrast, the *apparent* area of $\mathcal{R}$, regarded simply as a rectangle in the plane $\mathbf{R}^2$, is $2\pi \times \pi = 2\pi^2$.

Exercises

For all exercises involving a sphere of unit radius, use the parametrization given in the text.

1. Adapt the parametrization of the unit sphere to a sphere of radius R in $\mathbf{R}^3$, and use it to determine the associated metric tensor g_{ij}. Compare this tensor to the one given for the unit sphere; compare the area magnification factors $\sqrt{g}$.

2. (a) Show that a meridian of longitude on the unit sphere ($q^2 =$ constant) has length π. Conclude that a circle of radius d centered at the north pole is given by $q^2 = \frac{\pi}{2} - d$.

   (b) Show that the area of a spherical cap of radius d centered at the north pole on the unit sphere is $2\pi(1 - \cos d)$.

   (c) Show that the area of a spherical cap of radius d on a sphere of radius R is $2\pi R^2(1 - \cos(d/R))$. Then explain why, when d is small in relation to R, the area of the cap is $\pi d^2 + O(d^4/R^2)$.

3. Graph the function $\theta = \arccos\left(\dfrac{\cos t}{\sqrt{1 + \cos^2 t}}\right)$ for $0 \leq t \leq \pi/2$.

4. Determine the length of the line $q^2 = q^1$, $0 \leq q^1 \leq \pi/2$, on the unit sphere.

5. (a) A **loxodrome** on the sphere is a curve that makes a constant angle with the parallels of latitude (or meridians of

longitude). Show that

$$\mathbf{q}(t) = (t, \arcsin(\tanh t)), \quad -\infty < t < \infty,$$

is a loxodrome on the unit sphere. Note: q^1 is not restricted to the interval $[0, 2\pi]$ here but is instead allowed to take arbitrarily large positive and negative values.

(b) What angle does $\mathbf{q}$ make with the parallels of latitude? Sketch $\mathbf{q}(t)$ on the sphere for $0 \leq t \leq 3\pi$.

(c) Obtain the parametrization of a loxodrome that makes an *arbitrary* fixed angle with the parallels of latitude.

(d) Near one of the poles, a loxodrome looks like an equiangular spiral; cf. the exercises in Section 3.2. Illustrate this by making a sketch of $\mathbf{q}(t)$ for $\pi \leq t \leq 4\pi$; use a view of the sphere taken over the north pole.

(e) Calculate the full length of the loxodrome $\mathbf{q}(t)$, $-\infty < t < \infty$ (that is, from the south pole to the north pole).

5.3 De Sitter Spacetime

A simple curved spacetime

The de Sitter universe is perhaps the simplest curved spacetime possible. It is the spacelike unit sphere in a $(1+4)$-dimensional Minkowski space; that is, it is the set of spacelike unit vectors. To see what this means concretely, we will take a $(1+2)$-dimensional slice of the $(1+4)$-dimensional ambient space and examine the slice of de Sitter spacetime that appears there. Our larger purpose is to see how the ideas we have begun to develop about curved surfaces in Euclidean space can be adapted to Minkowski space.

Let $\mathbb{X} = (t, x, y)$ be a point in the $(1+2)$-dimensional Minkowski space with the standard metric $Q(\mathbb{X}) = t^2 - (x^2 + y^2)$. The spacelike unit vectors $\mathbb{X}$ are those that satisfy the condition $\|\mathbb{X}\|^2 = -Q(\mathbb{X}) = x^2 + y^2 - t^2 = 1$. They lie on a surface S that in ordinary $\mathbf{R}^3$ forms a *hyperboloid of one sheet*.

This is the de Sitter spacetime. It is $(1+1)$-dimensional: In the time direction, it extends indefinitely far into the past and into the future; in the spatial direction, it is just a circle (albeit one that first contracts and then expands over time). This is new; we have

§5.3 De Sitter Spacetime

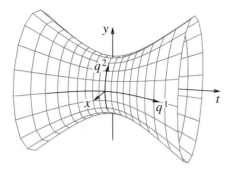

never contemplated a spacetime in which space is finite but has no boundary or "edge." When observers and objects move, they move on this circle. It is conceivable, therefore, that an observer, or a photon, could "circumnavigate" the entire space, making a journey that returns to the starting point without ever reversing direction.

To explore the new spacetime, we construct a parametrization. Since it is the analogue in Minkowski space of an ordinary Euclidean sphere, we can parametrize it simply by adapting the parametrization of the sphere that we used in the previous section:

Parametrization of de Sitter spacetime

$$\mathbb{X}(q^1, q^2) = (\sinh(q^1), \cosh(q^1)\cos(q^2), \cosh(q^1)\sin(q^2)),$$

$-\infty < q^1 < \infty$, $-\pi \leq q^2 \leq \pi$. As the figure shows, q^1 serves to label time and q^2 to label position.

The basis vectors in the tangent space are

$$\mathbb{X}_1 = (\cosh(q^1), \sinh(q^1)\cos(q^2), \sinh(q^1)\sin(q^2)),$$
$$\mathbb{X}_2 = (0, -\cosh(q^1)\sin(q^2), \cosh(q^1)\cos(q^2)).$$

The components of the metric tensor are

$$g_{11} = \mathbb{X}_1 \cdot \mathbb{X}_1 = \cosh^2(q^1) - \sinh^2(q^1)\cos^2(q^2) - \sinh^2(q^1)\sin^2(q^2) \equiv 1,$$
$$g_{12} = \mathbb{X}_1 \cdot \mathbb{X}_2 = \sinh(q^1)\cosh(q^1)\cos(q^2)\sin(q^2)$$
$$\qquad - \sinh(q^1)\cosh(q^1)\cos(q^2)\sin(q^2) \equiv 0,$$
$$g_{22} = \mathbb{X}_2 \cdot \mathbb{X}_2 = -\cosh^2(q^1)\sin^2(q^2) - \cosh^2(q^1)\cos^2(q^2) = -\cosh^2(q^1).$$

Minkowski geometry in each tangent space

The minus signs occur in these expressions because we are calculating the *Minkowski* inner product. The values of the g_{ij} show that each tangent plane is, in fact, a $(1+1)$-dimensional Minkowski space in which $\mathbb{X}_1$ is future-timelike, $\mathbb{X}_2$ is spacelike, and $\mathbb{X}_1$ and $\mathbb{X}_2$ are Minkowski-orthogonal.

In Section 3.3 we established that any curve whose tangent is always a future-timelike vector can be taken as the worldcurve of an observer. This implies that the "horizontal" coordinate lines $q^2 = $ constant are worldcurves, because their tangents $\mathbb{X}_1$ are always future-timelike. In particular, we take the q^1-axis to be the worldcurve of the observer G. In fact, q^1 is G's proper time τ, as we can see by this calculation:

$$\tau(q^1) = \int_0^{q^1} \|\mathbb{X}_1\| \, dq^1 = \int_0^{q^1} 1 \cdot dq^1 = q^1.$$

Henceforth we will use τ and q^1 interchangeably.

The radius of space

To address the question of circumnavigation, let us first measure the circumference of space. At time τ,

$$\text{circumference} = \int_{-\pi}^{\pi} \|\mathbb{X}_2\| \, dq^2 = \int_{-\pi}^{\pi} \cosh(\tau) \, dq^2 = 2\pi \cosh(\tau).$$

(Since $\mathbb{X}_2$ is a *spacelike* vector, $\|\mathbb{X}_2\| = \sqrt{-\mathbb{X}_2 \cdot \mathbb{X}_2} = +\cosh(q^1) > 0$.) It is natural, then, to say that the *radius* of space at time τ is $\cosh(\tau)$, a value that grows essentially exponentially with τ when $\tau > 0$.

Can a photon circumnavigate space?

Now consider, in the (q^1, q^2)-plane, a photon emitted by G at the event O in the positive q^2-direction and detected later by G at the event E_2. It might have a worldcurve like that shown below; E_1 appears twice on the worldcurve simply because the chart "wraps around" in the q^2-direction.

The light cone at different points

However, this does not happen. In fact, we now prove that no photon can travel more than halfway around the circle. To begin, consider the light cone determined in each tangent plane by the metric tensor

$$G = \begin{pmatrix} 1 & 0 \\ 0 & -\cosh^2(q^1) \end{pmatrix}, \qquad g = -\cosh^2(q^1) \leq -1.$$

§5.3 De Sitter Spacetime   233

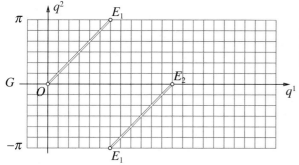

Is this a possible worldcurve for a photon?

This has the essential features of the familiar metric in the ordinary flat Minkowski plane. In fact, when $q^1 = 0$, it reduces to that metric:

$$G = \begin{pmatrix} 1 & 0 \\ 0 & -1 \end{pmatrix}, \qquad g = -1.$$

You should check that the light-like vectors that separate the time-like from the space-like vectors at the point (q^1, q^2) are multiples of

$$L_\pm = \begin{pmatrix} 1 \\ \pm \operatorname{sech}(q^1) \end{pmatrix}.$$

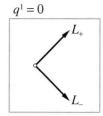
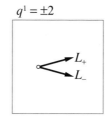

The slopes of L_+ and L_- are therefore $\pm \operatorname{sech}(q^1)$, and these slopes rapidly approach 0 as $|q^1|$ increases.

There is such a light cone at each point (q^1, q^2) in the parameter plane. We can think of the cone as a pair of vector fields, as shown below, that define the possible directions of photon worldcurves.

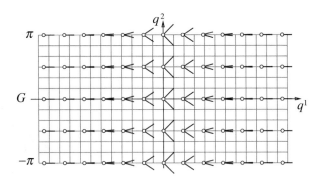

The worldcurve satisfies a differential equation

Now suppose that the worldcurve of a photon is the graph of a function $q^2 = \varphi(\tau)$ in the (q^1, q^2)-plane. This graph must be everywhere tangent to one of the light-cone fields. For the sake of illustration, let us take the L_+ field, with slope $\text{sech}(\tau)$. Then

$$\frac{d\varphi}{d\tau} = \text{sech}(\tau), \qquad q^2 = \varphi(\tau) = \int \text{sech}(\tau)\, d\tau.$$

To integrate this, we write the integrand as

$$\text{sech}(\tau) = \frac{2}{e^\tau + e^{-\tau}} = \frac{2e^\tau}{e^{2\tau} + 1}$$

and then make the substitution $u = e^\tau$, $du = e^\tau\, d\tau$:

$$\varphi(\tau) = \int \text{sech}(\tau)\, d\tau = \int \frac{2e^\tau\, d\tau}{e^{2\tau} + 1}$$

$$= \int \frac{2\, du}{u^2 + 1} = 2\arctan(u) + C$$

$$= 2\arctan(e^\tau) + C.$$

When $\tau \to -\infty$, then $e^\tau \to 0$ and $\varphi(\tau) \to C$. When $\tau \to +\infty$, then $e^\tau \to +\infty$, $\arctan(e^\tau) \to \pi/2$, and $\varphi(\tau) \to \pi + C$. One of the graphs $q^2 = \varphi(\tau)$ is shown below, superimposed on the vector fields. The others are vertical translates of this one, obtained by changing the value of C. Each graph lies in a horizontal band whose vertical width is π.

§5.3 De Sitter Spacetime

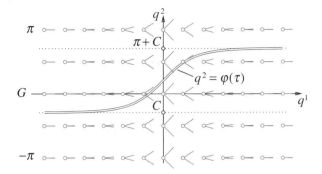

The exercises ask you to show that the worldcurves tangent to the field L_- are just the reflections of the worldcurves tangent to the field L_+. In fact, the symmetries of $q^2 = \varphi(\tau)$ are such that *either* of the reflections $\tau \mapsto -\tau$ or $q^2 \mapsto -q^2$ will do the job. Since every photon worldcurve lies in a horizontal band of vertical width $\Delta q^2 = \pi$, no photon ever travels more than halfway around the circle.

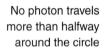
No photon travels more than halfway around the circle

The two worldcurves through the event $O = (0, 0)$ define the boundary of the future set $\mathcal{F}_O$ of O.

The future set $\mathcal{F}$

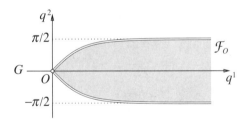

Since $\mathcal{F}_O$ is the set of events that G can possibly influence after time $q^1 = 0$, it follows that points on the far side of the circle from G are permanently beyond the reach of G—at least by the time $q^1 = 0$. The future set $\mathcal{F}_P$ of an event P that happened much earlier on G's worldline reaches considerably further. Thus, as time passes, G's "influence horizon" steadily contracts. By contrast, G's "viewing horizon," defined by the set of *timelike past* events relative to an event on G's worldcurve, steadily increases. See the exercises.

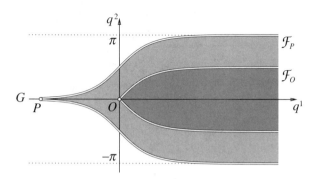

The speed of light is constant

It appears to our eyes that the velocity of a photon, as indicated by the slope of $q^2 = \varphi(\tau)$, decreases to 0 as $\tau \to \infty$. But this is not so; slope is $\Delta q^2/\Delta q^1$, while velocity is Δdistance/Δtime. Even though Δq^1 agrees with elapsed time, Δq^2 does *not* correspond to distance. The velocity of a photon is the ratio of the spacelike to the timelike component of one of the lightlike vectors $L_\pm$. Thus, since

$$L_\pm = \begin{pmatrix} 1 \\ \pm\operatorname{sech}(\tau) \end{pmatrix} = \underbrace{1 \cdot \mathbb{X}_1}_{\text{timelike}} \pm \underbrace{\operatorname{sech}(\tau) \cdot \mathbb{X}_2}_{\text{spacelike}},$$

we have

$$\text{velocity} = \frac{\pm\operatorname{sech}(\tau)\,\|\mathbb{X}_2\|}{\|\mathbb{X}_1\|} = \frac{\pm\operatorname{sech}(\tau)\cosh(\tau)}{1} = \pm 1.$$

So, in the de Sitter universe, the speed of light is constant after all: $c = 1$. The photon worldcurves look the way they do because spatial distance, as the metric g_{ij} defines it, is very different from the Euclidean distance that our eyes see.

An extrinsic analysis of the light cones

If we return to the embedding of de Sitter spacetime as a hyperboloid S in (t, x, y)-space, we can visualize the light cones in a way that is more compatible with our Euclidean/Minkowskian intuitions. To begin, consider the worldcurves of the two photons emitted at the event $O = (q^1, q^2) = (0, 0)$. You should check that they are the graphs of

$$q^2 = 2\arctan(e^\tau) - \pi/2,$$
$$q^2 = -2\arctan(e^\tau) + \pi/2;$$

we can write the pair as $q^2 = \pm 2\arctan(e^\tau) \mp \pi/2$. In (t, x, y)-space they are the curves obtained by composing these functions with the map $\mathbb{X}$. Thus they are given by $\tau \mapsto \mathbb{X}(\tau, \pm 2\arctan(e^\tau) \mp \pi/2)$:

$$t(\tau) = \sinh(\tau),$$
$$x(\tau) = \cosh(\tau) \cos\left(\pm 2\arctan(e^\tau) \mp \pi/2\right),$$
$$y(\tau) = \cosh(\tau) \sin\left(\pm 2\arctan(e^\tau) \mp \pi/2\right).$$

We claim that these curves are a pair of straight lines and that, in fact, they lie on the intersection of the hyperboloid and the plane $x = 1$. Furthermore, $x = 1$ is the tangent plane to the hyperboloid at O:

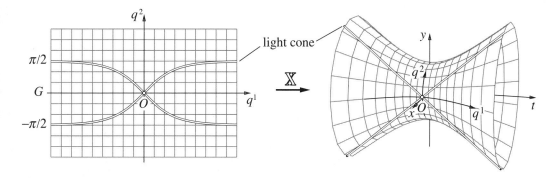

To prove the claim we first show that $x(\tau) \equiv 1$ by simplifying the cosine expression in the formula for $x(\tau)$. The relation

$$\cos(\pm A \mp \pi/2) = \pm \sin(\pm A) = \sin(A)$$

and the double angle formula for the sine function give us

$$\cos\left(\pm 2\arctan(e^\tau) \mp \pi/2\right) = \sin\left(2\arctan(e^\tau)\right)$$
$$= 2\sin\left(\arctan(e^\tau)\right) \cdot \cos\left(\arctan(e^\tau)\right).$$

Now, $\sin(\arctan(B)) = B/\sqrt{B^2 + 1}$ and $\cos(\arctan(B)) = 1/\sqrt{B^2 + 1}$, so the original cosine expression reduces to

$$\cos\left(\pm 2\arctan(e^\tau) \mp \pi/2\right) = \frac{2e^\tau}{e^{2\tau} + 1} = \frac{2}{e^\tau + e^{-\tau}} = \frac{1}{\cosh(\tau)},$$

and then finally,

$$x(\tau) = \cosh(\tau)\cos\left(\pm 2\arctan(e^\tau) \mp \pi/2\right) = \cosh(\tau)\cdot\frac{1}{\cosh(\tau)} \equiv 1,$$

as claimed. Now, $x=1$ is a plane in (t,x,y)-space, so the light cone is the intersection of the hyperboloid S with this plane. Furthermore, since $x=1$ contains two different curves in S, it must actually be the tangent plane to S at the point where the curves intersect.

Next we show that $y(\tau) = \pm\sinh(\tau) = \pm t(\tau)$. This will imply that the curves are the two straight lines $y \pm t$ in the plane $x=1$. The argument is a straightforward modification of the argument for $x(\tau)$:

$$\sin\left(\pm 2\arctan(e^\tau) \mp \pi/2\right) = \mp\cos\left(2\arctan(e^\tau)\right)$$
$$= \mp\left(\cos^2\left(\arctan(e^\tau)\right) - \sin^2\left(\arctan(e^\tau)\right)\right)$$
$$= \mp\left(\frac{1}{e^{2\tau}+1} - \frac{e^{2\tau}}{e^{2\tau}+1}\right)$$
$$= \pm\frac{e^{2\tau}-1}{e^{2\tau}+1} = \pm\tanh(\tau).$$

Therefore, $y(\tau) = \cosh(\tau)\cdot\pm\tanh(\tau) = \pm\sinh(\tau) = \pm t(\tau)$. In the plane $x=1$ these curves are straight lines with slope ± 1, and they look more like the worldlines of photons in ordinary Minkowski space.

These photons stay on the front side of the hyperboloid

Seeing the light cone on the hyperboloid S makes it clear why a photon cannot circumnavigate space: If a photon were to go all the way around the hyperboloid, its worldcurve would have to take on negative x values. But these worldcurves, at least, are stuck in the plane $x=1$.

As the last comment indicates, we have investigated the light cone only at the single event O. We can, however, exploit a special feature of the hyperboloid to determine the light cones everywhere. The first step is to note that the hyperboloid, though clearly curved, has a pair of straight lines embedded in it. Moreover, because the hyperboloid has rotational symmetry around the t-axis, the image of either of these lines under an arbitrary rotation around the t-axis must also lie in S. Arbitrary rotations of even just one of these two lines will therefore sweep out the entire hyperboloid, as you can see below.

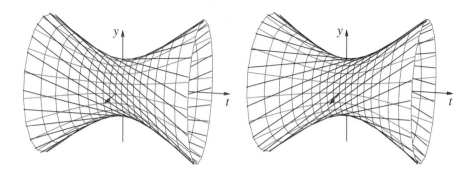

Such a line is called a *generator* of the surface, the complete family of lines is called a *ruling*, and the surface itself is called a **ruled surface**. The hyperboloid has two separate rulings, and thus can be said to be a doubly ruled surface.

The hyperboloid is a doubly ruled surface

These two separate rulings are therefore the photon worldcurves at every point of the de Sitter universe. They determine the light cone at every event, and in the figure below you can see how the apparent (Euclidean) angle between the two lines in the light cone decreases as $|t|$ increases—that is, as we move away from the narrow waist of the hyperboloid.

The rulings form the light cones

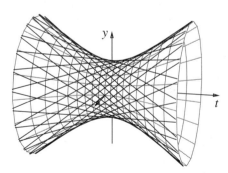

Since every worldcurve is a straight line, no worldcurve loops around the hyperboloid, so no photon circumnavigates the circular space that underlies the de Sitter universe. The network of light cones also makes it clear how space can be finite in extent and yet not completely accessible to a single observer at any moment.

Incidentally, space appears here as a 1-dimensional circle $x^2 + y^2 = 1$ only because we are looking at a 2-dimensional slice of the full de Sitter universe. The full 4-dimensional de Sitter universe U is the set of spacelike unit vectors $\mathbb{X} = (t, x, y, z, w)$ in a $(1+4)$-dimensional Minkowski space; U is defined by the equation

$$-Q(\mathbb{X}) = x^2 + y^2 + z^2 + w^2 - t^2 = 1.$$

In U, space is the 3-dimensional sphere $x^2 + y^2 + z^2 + w^2 = 1$.

Exercises

1. (a) For the parametrization $\mathbb{X}$ of de Sitter spacetime that we use, show that $\mathbb{X}_1 \times \mathbb{X}_2 = \cosh(q^1)\, \mathbb{X}$.

   (b) Show that $(\mathbb{X}_1 \times \mathbb{X}_2) \perp \mathbb{X}_1$ and $(\mathbb{X}_1 \times \mathbb{X}_2) \perp \mathbb{X}_2$, in the sense of the Minkowski norm.

   (c) Show that $\|\mathbb{X}_1 \times \mathbb{X}_2\| = \sqrt{-g} = \cosh q^1$, the area magnification factor.

2. (a) Consider the hyperboloid

   $$\mathbf{x}(q^1, q^2) = (\sinh(q^1), \cosh(q^1)\cos(q^2), \cosh(q^1)\sin(q^2))$$

   as being an embedding in *Euclidean* space instead of Minkowski space, and calculate $\mathbf{x}_1$, $\mathbf{x}_2$, $\mathbf{x}_1 \times \mathbf{x}_2$, g_{ij}, and $\sqrt{g}$. Are the coordinate lines still orthogonal?

   (b) It is still true that $\mathbf{q}(t) = (t, 2\arctan(e^t) - \pi/2)$ is a straight line on the surface. Determine the normal vector $\mathbf{n} = (\mathbf{x}_1 \times \mathbf{x}_2)/\|\mathbf{x}_1 \times \mathbf{x}_2\|$ along this line. In particular, show that $\mathbf{n}$ is *not* constant along the line.

   (c) At a representative collection of points on the hyperboloid, sketch both the Euclidean normal $\mathbf{n}$ and the Minkowski normal $\mathbb{N} = (\mathbb{X}_1 \times \mathbb{X}_2)/\|\mathbb{X}_1 \times \mathbb{X}_2\|$; $\mathbb{X}_1 \times \mathbb{X}_2$ is from the previous exercise. Compare the two normals, especially for $|t|$ large.

3. Prove that $\cos(\pm A \mp \pi/2) = \sin(A)$ and $\sin(\pm A \mp \pi/2) = \mp \cos(A)$.

4. (a) Prove that $q^2 = 2\arctan(e^\tau) - \pi/2$ has odd symmetry, that is, that the substitutions $\tau \mapsto -\tau$ and $q^2 \mapsto -q^2$ have the same effect.

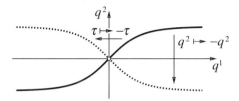

(b) Show that the reflection of any worldcurve $q^2 = 2\arctan(e^\tau)+ C$ across the q^2-axis is tangent to the field L_- and is thus the worldcurve of a photon.

5. Construct the *past timelike* sets at various points along G's worldcurve, and show that these sets grow and eventually cover the entire circle as G's proper time increases without limit.

5.4 Curvature of a Surface

Curves Reconsidered

It is natural to try to define the curvature of a surface by analogy with the curvature of a curve. The basic idea is simple enough: Curvature is the rate at which a curve changes direction. To make the idea precise, though, we used an arc-length parametrization $\mathbf{y} : [0, L] \to \mathbf{R}^n : s \mapsto \mathbf{y}(s)$. But $\mathbf{y}$ is quite special: It preserves distances. That is, if s_1 and s_2 are any two points in $[0, L]$, then the distance between $\mathbf{y}(s_1)$ and $\mathbf{y}(s_2)$, *as measured along the curve*, is the same as the distance between s_1 and s_2 in $[0, L]$, namely $\Delta s = s_2 - s_1$. We call a distance-preserving map like $\mathbf{y}$ an **isometry**.

The arc-length parametrization is an isometry

But we have seen that any map from a flat plane to a curved surface will distort distances. In other words, we cannot expect to parametrize a surface with an isometry. Therefore, if we are to use the curvature of a curve as a guide in defining the curvature of a surface, we must go back and reconsider our work on curves in a way that will be more useful for surfaces.

In general, surface parametrizations are not isometries

Fortunately, this is not difficult. For simplicity, though, we will consider only curves C that lie in the plane $\mathbf{R}^2$. We start with an arbitrary smooth parametrization $\mathbf{x} : [a, b] \to \mathbf{R}^2$ of C, and require, as always, that $\mathbf{x}$ have no singular points—that is, no points q_0 where $\mathbf{x}'(q_0) = \mathbf{0}$. The *direction* of C at $\mathbf{x}(q)$ is given by the unit tangent vector

$$\mathbf{t}(q) = \frac{\mathbf{x}'(q)}{\|\mathbf{x}'(q)\|}.$$

The unit tangent parametrization

Usually, we draw the tangent vector $\mathbf{t}(q)$ with its tail at the point of tangency $\mathbf{x}(q)$. If instead, we bring all the tangents to the origin O, we get the parametrization $\mathbf{t} : [a, b] \to \mathbf{R}^2$ of another curve C^*. Since $\|\mathbf{t}(q)\| = 1$ for all q in $[a, b]$, C^* lies on the unit circle S^1.

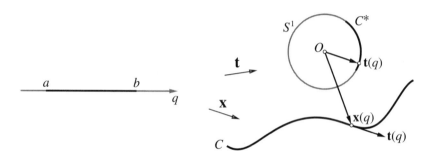

In general, the map $\mathbf{t}$ will have singularities even though $\mathbf{x}$ has none. At a typical singular point, C will have an inflection, while C^* will fold back upon itself. Notice in the figure above that $\mathbf{t}$ has three such singular points. You can explore this in the exercises. We can now define curvature in terms of the *Gauss map*, which assigns to each point on a curve its direction as a point on S^1.

The Gauss map of a curve

Definition 5.2 *Suppose the plane curve C is parametrized by $\mathbf{x} : [a, b] \to \mathbf{R}^2$. The **Gauss map** $\mathcal{G} : C \to S^1$ assigns to the point $P = \mathbf{x}(q)$ the unit tangent vector $\mathcal{G}(P) = \mathbf{t}(q) = \mathbf{x}'(q)/\|\mathbf{x}'(q)\|$ at P.*

Definition 5.3 *The **curvature** of the curve C at the point P is*

$$\kappa(P) = \lim_{\alpha \downarrow P} \frac{\text{length of } \mathcal{G}(\alpha)}{\text{length of } \alpha},$$

where the limit is taken over all segments α that contain P, as the length of α approaches 0.

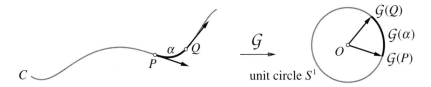

In fact, at each point of C there are *two* unit tangent vectors, pointing in opposite directions. In terms of **x**, the two choices are $\pm \mathbf{t} = \pm \mathbf{x}'/\|\mathbf{x}'\|$. Therefore, if α is any segment on C, there are two possible segments $\mathcal{G}(\alpha)$ in S^1, but these are antipodal and thus have the same length. Ultimately, then, the curvature $\kappa(P)$ is independent of the choice of unit tangent.

Theorem 5.2 *Suppose* $P = \mathbf{x}(u)$; *then* $\kappa(P) = \dfrac{\|\mathbf{t}'(u)\|}{\|\mathbf{x}'(u)\|}.$

PROOF: By the definition,

$$\kappa(P) = \lim_{\epsilon \to 0} \frac{\text{length of } \mathbf{t} \text{ from } u - \epsilon \text{ to } u + \epsilon}{\text{length of } \mathbf{x} \text{ from } u - \epsilon \text{ to } u + \epsilon} = \lim_{\epsilon \to 0} \frac{\int_{u-\epsilon}^{u+\epsilon} \|\mathbf{t}'(q)\| \, dq}{\int_{u-\epsilon}^{u+\epsilon} \|\mathbf{x}'(q)\| \, dq}.$$

According to the integral form of the mean value theorem, each integral can be replaced by the product of the length of the integration interval (2ϵ, in this case) and a suitably chosen value of the integrand. Thus there are numbers $-1 \leq \theta_1, \theta_2 \leq 1$ for which

$$\int_{u-\epsilon}^{u+\epsilon} \|\mathbf{t}'(q)\| \, dq = 2\epsilon \cdot \|\mathbf{t}'(u + \theta_1 \epsilon)\|, \quad \int_{u-\epsilon}^{u+\epsilon} \|\mathbf{x}'(q)\| \, dq = 2\epsilon \cdot \|\mathbf{x}'(u + \theta_2 \epsilon)\|.$$

Hence

$$\kappa(P) = \lim_{\epsilon \to 0} \frac{2\epsilon \cdot \|\mathbf{t}'(u + \theta_1 \epsilon)\|}{2\epsilon \cdot \|\mathbf{x}'(u + \theta_2 \epsilon)\|} = \frac{\|\mathbf{t}'(u)\|}{\|\mathbf{x}'(u)\|}. \qquad \text{END OF PROOF}$$

Corollary 5.2 $\kappa(P) = \sqrt{\dfrac{(\mathbf{x}' \cdot \mathbf{x}')(\mathbf{x}'' \cdot \mathbf{x}'') - (\mathbf{x}' \cdot \mathbf{x}'')^2}{(\mathbf{x}' \cdot \mathbf{x}')^3}}.$

PROOF: See the exercises.

The definitions are consistent

The theorem and corollary give us a way to calculate the curvature directly from any parametrization of a curve. However, if we use the arc-length parametrization $\mathbf{y}(s)$, then the unit tangent vector $\mathbf{t}$ is $\mathbf{y}'(s) = \mathbf{u}(s)$ and its derivative $\mathbf{t}'$ is $\mathbf{u}'(s) = \mathbf{k}(s)$. Therefore,

$$\kappa(P) = \frac{\|\mathbf{u}'(s)\|}{\|\mathbf{y}'(s)\|} = \|\mathbf{k}(s)\|,$$

which agrees with the original definition in Section 3.2. Our two definitions are consistent.

Since $\|\mathbf{x}'(u)\|$ is the *length magnification factor* for the map $\mathbf{x}$ at u, we have another way to view curvature:

$$\kappa(P) = \frac{\text{length magnification factor for } \mathbf{t}}{\text{length magnification factor for } \mathbf{x}}.$$

Examples

If C is a straight line, then $\mathcal{G}(C)$ is a single point, so the length of any segment $\mathcal{G}(\alpha)$ is 0, implying that $\kappa(P) = 0$ at every point of C. If C is a circle of radius r and α is a segment of C, then α and $\mathcal{G}(\alpha)$ are similar in the sense of Euclidean geometry. Specifically, each segment on C is r times larger than the similar segment on the unit circle S^1:

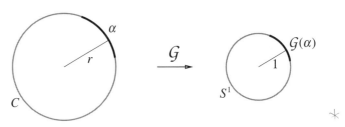

$$\frac{\text{length of } \mathcal{G}(\alpha)}{\text{length of } \alpha} = \frac{1}{r}.$$

Therefore, $\kappa(P) = 1/r$ at every point P on C, so the curvature of a circle is the reciprocal of its radius.

* $\mathcal{G}(\alpha)$ should be rotated 50° counterclockwise.

Gaussian Curvature

Our aim is to define the curvature of a surface as the rate at which it changes direction. But how do we specify the direction of a surface at a point? For a curve we use the *tangent* direction. This is adequately specified by a single vector, the unit tangent, because all the tangents to a curve at a single point lie on a 1-dimensional line. But the tangents to a surface at a point form a plane, not a line. The plane is 2-dimensional, so no single tangent vector can give its direction. However, any plane in $\mathbf{R}^3$ has a unique normal direction, and this *is* 1-dimensional, so it can be specified by a single unit vector. Thus we use the unit normal to define the direction of a surface at a point.

The direction of a surface is given by its normal

We can now implement this idea and define the Gauss map for a surface S given by a parametrization $\mathbf{x} : \mathcal{R} \to \mathbf{R}^3$:

The Gauss map

$$\mathbf{x} : (q^1, q^2) = \mathbf{q} \mapsto (x(\mathbf{q}), y(\mathbf{q}), z(\mathbf{q})).$$

Since we require $\mathbf{x}$ to be nonsingular, the normal vector $\mathbf{x}_1(\mathbf{q}) \times \mathbf{x}_2(\mathbf{q})$ is nonzero at every point $\mathbf{q}$ and the unit normal vector is defined:

$$\mathbf{n}(\mathbf{q}) = \frac{\mathbf{x}_1 \times \mathbf{x}_2}{\|\mathbf{x}_1 \times \mathbf{x}_2\|}.$$

The unit normals lie on the sphere S^2. In fact, $\mathbf{n} : \mathcal{R} \to S^2$ is itself the parametrization of a surface S^* that lies in S^2.

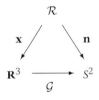

Definition 5.4 *Let S be a surface with the parametrization $\mathbf{x} : \mathcal{R} \to \mathbf{R}^3$. The **Gauss map** $\mathcal{G} : S \to S^2$ assigns to the point $P = \mathbf{x}(\mathbf{q})$ the unit normal vector $\mathcal{G}(P) = \mathbf{n}(\mathbf{q})$ at P, making the following diagram commutative.*

The figure below shows the typical relation between a surface S and its Gaussian image $S^* = \mathcal{G}(S)$. Notice that S curves more sharply one way than the other; specifically, the direction of the normal undergoes a large change as we move along S in the q^1 direction but only a small change in the q^2-direction. This means that S^* will cover a significant portion of the sphere in the q^1-direction but will be quite compressed in the q^2-direction. And this is true even though $\mathcal{R}$ is elongated in the q^2-direction when it

The relative size of the Gaussian image

Chapter 5 Surfaces and Curvature

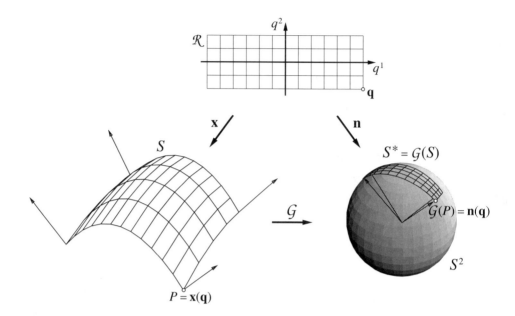

is mapped onto S by $\mathbf{x}$. Informally, then, we can say that the more sharply curved a region on the surface is, the larger its Gaussian image will be—in relation to the size of the region itself.

Definition 5.5 *The **curvature**, or **Gaussian curvature**, of the surface S at the point P is*

$$K(P) = \lim_{\Omega \downarrow P} \frac{\text{area of } \mathcal{G}(\Omega)}{\text{area of } \Omega}$$

where the limit is taken over all regions Ω of S that contain P, as the diameter of Ω approaches 0. The diameter of a region is the maximum distance between any two points in the region.

Plane and sphere

If S is a plane, then $\mathcal{G}(S)$ is a single point on S^2, so its area is 0. This implies $K(P) = 0$ at every point P in S—exactly what we would expect. If S is a sphere of radius r, then Ω and $\mathcal{G}(\Omega)$ are similar figures. Linear dimensions on S are r times what they are

§5.4 Curvature of a Surface

on the unit sphere S^2, so

$$\frac{\text{area of } \mathcal{G}(\Omega)}{\text{area of } \Omega} = \frac{1}{r^2},$$

a constant ratio, independent of Ω. Therefore, $K(P) = 1/r^2$ at every point P of S. The curvature of a given sphere is constant; if its radius increases, its curvature decreases.

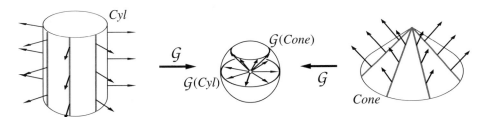

Cylinder and cone

More surprising, perhaps, are the results for a cylinder or cone. Since the normal vectors along a generator of a cylinder or cone are parallel, all the points on that generator have the same image under the Gauss map. Points on different generators do have different images, but those different images fill only a circle in the target sphere S^2. Thus area $\mathcal{G}(Cyl) = $ area $\mathcal{G}(Cone) = 0$, so $K(P) = 0$ at every point P on either surface.

Theorem 5.3 *Suppose* $\mathbf{x}$ *parametrizes the surface* S *and* $P = \mathbf{x}(\mathbf{q})$; *then*

$$K(P) = \frac{\|\mathbf{n}_1(\mathbf{q}) \times \mathbf{n}_2(\mathbf{q})\|}{\|\mathbf{x}_1(\mathbf{q}) \times \mathbf{x}_2(\mathbf{q})\|}.$$

PROOF: The theorem is analogous to the one for curves, and so is the proof. Let D_ϵ be the square of side 2ϵ centered at the point $\mathbf{q}$ in the parameter plane $\mathcal{R}$. Then $\Omega_\epsilon = \mathbf{x}(D_\epsilon)$ is a region in S containing P, and $\mathcal{G}(\Omega_\epsilon) = \mathbf{n}(D_\epsilon)$ is its Gaussian image. Therefore,

$$K(P) = \lim_{\epsilon \to 0} \frac{\text{area of } \mathcal{G}(\Omega_\epsilon)}{\text{area of } \Omega_\epsilon} = \lim_{\epsilon \to 0} \frac{\iint_{D_\epsilon} \|\mathbf{n}_1 \times \mathbf{n}_2\| \, dq^1 dq^2}{\iint_{D_\epsilon} \|\mathbf{x}_1 \times \mathbf{x}_2\| \, dq^1 dq^2}.$$

By the mean value theorem for double integrals, each integral is equal to the area of D_ϵ times the value of the integrand at some point inside D_ϵ. Thus there are points $\mathbf{q}_1$ and $\mathbf{q}_2$ in D_ϵ for which

$$K(P) = \lim_{\epsilon \to 0} \frac{4\epsilon^2 \, \|\mathbf{n}_1(\mathbf{q}_1) \times \mathbf{n}_2(\mathbf{q}_1)\|}{4\epsilon^2 \, \|\mathbf{x}_1(\mathbf{q}_2) \times \mathbf{x}_2(\mathbf{q}_2)\|}.$$

In the limit as $\epsilon \to 0$, the square D_ϵ shrinks down to the point $\mathbf{q}$, forcing $\mathbf{q}_1 \to \mathbf{q}$ and $\mathbf{q}_2 \to \mathbf{q}$. This establishes the theorem.

END OF PROOF

K is a ratio of oriented areas

The theorem gives us a concrete way to calculate the Gaussian curvature $K(P)$ in terms of the areas of two parallelograms:

$$K(P) = \frac{\text{area } \mathbf{n}_1 \wedge \mathbf{n}_2}{\text{area } \mathbf{x}_1 \wedge \mathbf{x}_2}.$$

These parallelograms lie in different tangent planes ($\mathbf{x}_1 \wedge \mathbf{x}_2$ in TS_P and $\mathbf{n}_1 \wedge \mathbf{n}_2$ in $T\mathcal{G}(S)_{\mathcal{G}(P)}$), but in fact, the two planes are parallel. That is proven in the following proposition. As a consequence, we can compare the orientations of the two parallelograms—as we did in Section 5.1—and thus reinterpret Gaussian curvature in terms of *oriented* areas. Since oriented areas can be negative as well as positive, the Gaussian curvature can then take on negative, as well as positive, values.

Proposition 5.4 *For each $\mathbf{q}$ in $\mathcal{R}$, the vectors $\mathbf{n}_1(\mathbf{q})$ and $\mathbf{n}_2(\mathbf{q})$ lie in the tangent plane $TS_{\mathbf{x}(\mathbf{q})}$ whose basis is $\mathbf{x}_1(\mathbf{q})$ and $\mathbf{x}_2(\mathbf{q})$.*

PROOF: It is sufficient to show $\mathbf{n}_1 \perp \mathbf{n}$ and $\mathbf{n}_2 \perp \mathbf{n}$, because $\mathbf{n}$ is normal to the plane spanned by $\mathbf{x}_1$ and $\mathbf{x}_2$. Since $\mathbf{n} \cdot \mathbf{n} = 1$, differentiation gives

$$\frac{\partial}{\partial q^i}(\mathbf{n} \cdot \mathbf{n}) = \frac{\partial \mathbf{n}}{\partial q^i} \cdot \mathbf{n} + \mathbf{n} \cdot \frac{\partial \mathbf{n}}{\partial q^i} = 2\mathbf{n}_i \cdot \mathbf{n} = 0. \qquad \text{END OF PROOF}$$

Expressing $\mathbf{n}_1$ and $\mathbf{n}_2$ in terms of $\mathbf{x}_1$ and $\mathbf{x}_2$

The proposition implies that for each $\mathbf{q}$, $\mathbf{n}_1(\mathbf{q})$ and $\mathbf{n}_2(\mathbf{q})$ are linear combinations of the basis vectors $\mathbf{x}_1(\mathbf{q})$ and $\mathbf{x}_2(\mathbf{q})$. Thus there are scalar functions $-b_j^i(\mathbf{q})$ that are the coordinates of $\mathbf{n}_1$ and $\mathbf{n}_2$ with respect to this basis:

$$\mathbf{n}_1 = -b_1^1 \mathbf{x}_1 - b_1^2 \mathbf{x}_2, \qquad \mathbf{n}_2 = -b_2^1 \mathbf{x}_1 - b_2^2 \mathbf{x}_2,$$

or just $\mathbf{n}_j = -b^i_j \mathbf{x}_i$, $j = 1, 2$ (summation convention). The reason for the minus signs will emerge in the next section. To connect $\mathbf{n}_1 \wedge \mathbf{n}_2$ to $\mathbf{x}_1 \wedge \mathbf{x}_2$, we use the matrix whose columns are the coordinates of $\mathbf{n}_1$ and $\mathbf{n}_2$ with respect to the basis $\{\mathbf{x}_1, \mathbf{x}_2\}$:

$$\widetilde{B}(\mathbf{q}) = \begin{pmatrix} -b^1_1 & -b^1_2 \\ -b^2_1 & -b^2_2 \end{pmatrix}.$$

The connection between the areas is then provided by Proposition 5.3:

$$\text{area } \mathbf{n}_1 \wedge \mathbf{n}_2 = \det \widetilde{B} \cdot \text{area } \mathbf{x}_1 \wedge \mathbf{x}_2 = \det \widetilde{B} \sqrt{g},$$

where $g = \det g_{ij}$. (Note that $\det(-\widetilde{B}) = \det \widetilde{B}$ because $\widetilde{B}$ has an even number of rows and columns.)

We are now in a position to reinterpret the theorem so that $K(P)$ is expressed in terms of the areas of *oriented* parallelograms:

Theorem 5.4 $K(P) = \dfrac{\text{area } \mathbf{n}_1 \wedge \mathbf{n}_2}{\text{area } \mathbf{x}_1 \wedge \mathbf{x}_2} = \det \widetilde{B} = b^1_1 b^2_2 - b^2_1 b^1_2.$

We can even adapt our ideas about oriented areas to regions in S and $S^* = \mathcal{G}(S)$. Consider an arbitrary region Ω in S, where $\Omega = \mathbf{x}(D)$. Then

Oriented areas on S^*

$$\text{area } \Omega = \iint_D \|\mathbf{x}_1 \times \mathbf{x}_2\| \, dq^1 dq^2 = \iint_D \sqrt{g} \, dq^1 dq^2 \geq 0.$$

Up to this point we would calculate the area of its Gaussian image $\mathcal{G}(\Omega) = \mathbf{n}(D)$ by

$$\text{area } \mathcal{G}(\Omega) = \iint_D \|\mathbf{n}_1 \times \mathbf{n}_2\| \, dq^1 dq^2 \geq 0.$$

Note that the area can never be negative because the integrand $\|\mathbf{n}_1 \times \mathbf{n}_2\|$ is itself nonnegative. However, since this integrand represents the *un*oriented area of the parallelogram $\mathbf{n}_1 \wedge \mathbf{n}_2$, we can simply replace it by the oriented version,

$$\text{area } \mathbf{n}_1 \wedge \mathbf{n}_2 = \det \widetilde{B} \sqrt{g},$$

in the integral:

$$\text{area}\,\mathcal{G}(\Omega) = \iint_D \det \widetilde{B}\,\sqrt{g}\,dq^1 dq^2.$$

Now, of course, the area of $\mathcal{G}(\Omega)$ *can* be negative. The new definition is a consequence of the fact that we can compare the relative orientations of Ω and $\mathcal{G}(\Omega)$; area $\mathcal{G}(\Omega)$ will be negative precisely when the Gauss map $\mathcal{G}$ reverses orientation.

Negative Curvature: An Example

A saddle has negative curvature

To see how a surface comes to have negative Gaussian curvature, and to see how the Gauss map reverses orientation, we will work through an example in detail. Consider the graph of $z = xy$; this is a saddle-shaped surface S that we can parametrize as

$$x = q^1, \qquad y = q^2, \qquad z = q^1 q^2.$$

However, the normal map $\mathbf{n}$ has rotational symmetry that becomes more apparent when we use polar coordinates $q^1 = r\cos\theta$ and $q^2 = r\sin\theta$ in the (q^1, q^2)-plane. In terms of r and θ the parametrization is now

$$\mathbf{x}(r, \theta) = \left(r\cos\theta,\, r\sin\theta,\, \tfrac{1}{2}r^2 \sin 2\theta\right).$$

Before we calculate $K(P)$, first note the shape of S in the figure below. It curves upward in one direction (in the first and third quadrants, in this case) and downward in the other. This pattern is characteristic for a negatively curved surface.

$\mathcal{G}: S \to S^*$ reverses orientation

Next, compare the orientations of S and S^*. Viewed from above, the coordinate axes have a counterclockwise orientation on S but clockwise on S^*. In particular, you should convince yourself that normal vectors along the positive q^1-axis on S point roughly forward (and over your left shoulder), arranging themselves in S^2 so that their tips will lie along the positive q^1-axis as marked on S^*. Similarly, the normal vectors along the positive

§5.4 Curvature of a Surface

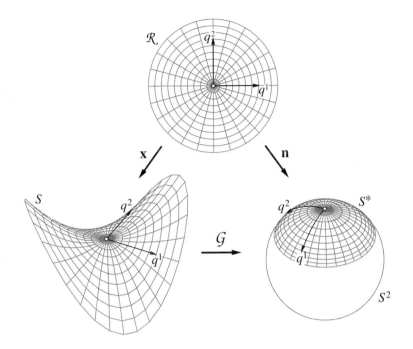

q^2-axis on S point roughly to the left, so their tips will lie along the positive q^2-axis on S^*. The fact that $\mathcal{G}$ is orientation-reversing is also characteristic for a negatively curved surface.

Finally, note the special symmetry of the normal map $\mathbf{n}:\mathcal{R}\to S^2$ in this example. Concentric circles in $\mathcal{R}$ have concentric images in S^2, though their spacing is not uniform: As r increases, the circles are packed more and more closely together. We will see that the entire (q^1, q^2)-plane is compressed by $\mathbf{n}$ into the upper hemisphere of the target S^2.

We now begin the calculation of $K(P)$. Since the summation convention will play no role, we abandon numerical subscripts and use r and θ instead to denote partial derivatives.

Calculating $K(P)$

$$\mathbf{x}_r = (\cos\theta, \sin\theta, r\sin 2\theta), \qquad \mathbf{x}_\theta = (-r\sin\theta, r\cos\theta, r^2\cos 2\theta),$$
$$\mathbf{x}_r \times \mathbf{x}_\theta = (-r^2\sin\theta, -r^2\cos\theta, r).$$

Therefore, $g = \|\mathbf{x}_r \times \mathbf{x}_\theta\|^2 = r^2(1+r^2)$, so we require $r > 0$ to make **x** nonsingular. If we set $\Delta = \sqrt{g}/r = \sqrt{1+r^2}$, then

$$\mathbf{n} = \frac{1}{\Delta}(-r\sin\theta, -r\cos\theta, 1),$$

$$\mathbf{n}_r = -\frac{1}{\Delta^3}(\sin\theta, \cos\theta, r), \qquad \mathbf{n}_\theta = \frac{r}{\Delta}(-\cos\theta, \sin\theta, 0).$$

According to the last proposition, we can express $\mathbf{n}_r$ and $\mathbf{n}_\theta$ in terms of the basis $\mathbf{x}_r$, $\mathbf{x}_\theta$. You can check that the result is

$$\mathbf{n}_r = -\frac{\sin 2\theta}{\Delta^3}\mathbf{x}_r - \frac{\cos 2\theta}{r\Delta^3}\mathbf{x}_\theta, \qquad \mathbf{n}_\theta = -\frac{r\cos 2\theta}{\Delta}\mathbf{x}_r + \frac{\sin 2\theta}{\Delta}\mathbf{x}_\theta.$$

Therefore,

$$\widetilde{B} = \begin{pmatrix} \dfrac{\sin 2\theta}{\Delta^3} & \dfrac{r\cos 2\theta}{\Delta} \\ \dfrac{\cos 2\theta}{r\Delta^3} & -\dfrac{\sin 2\theta}{\Delta} \end{pmatrix}.$$

By Theorem 5.4,

$$K(P) = \det \widetilde{B} = -\frac{1}{\Delta^4} = -\frac{1}{(1+r^2)^2}.$$

Thus $K(P)$ depends only on distance from $P = \mathbf{x}(\mathbf{q})$ to the origin $\mathbf{q} = \mathbf{0}$. Furthermore, K is always negative; in fact, $-1 \leq K < 0$. In particular, $K = -1$ only at the origin, while $K \to 0$ as $P \to \infty$.

We can now verify that the image of **n** lies entirely in the upper hemisphere of S^2; it is sufficient to note that the z-coordinate of **n** is positive for any r and θ:

$$z(r,\theta) = \frac{1}{\sqrt{1+r^2}} > 0.$$

Thus $z \to 0$ as $r \to \infty$, so the entire upper hemisphere is covered by the map $\mathbf{n}: \mathcal{R} \to S^2$.

Note: Strictly speaking, the polar parametrization we have been using is not valid at the origin—it is singular there. In the exercises you are asked to use a Cartesian parametrization—which is nonsingular everywhere—and confirm that $K = -1$ at the origin.

§5.4 Curvature of a Surface

Exercises

1. (a) Show that the curvature $\kappa(q)$ of the curve $\mathbf{x}(q)$ can be expressed directly in terms of $\mathbf{x}$ and its derivatives by the formula

$$\kappa = \sqrt{\frac{(\mathbf{x}' \cdot \mathbf{x}')(\mathbf{x}'' \cdot \mathbf{x}'') - (\mathbf{x}' \cdot \mathbf{x}'')^2}{(\mathbf{x}' \cdot \mathbf{x}')^3}}.$$

Here is one approach you can take using Theorem 5.2, which says

$$\kappa(q) = \frac{\|\mathbf{t}'(q)\|}{\|\mathbf{x}'(q)\|} \quad \text{when} \quad \mathbf{t}(q) = \frac{\mathbf{x}'(q)}{\|\mathbf{x}'(q)\|}.$$

(b) Show that $\mathbf{t}' = A\mathbf{x}'' - B\mathbf{x}'$, where $A = \dfrac{1}{\sqrt{\mathbf{x}' \cdot \mathbf{x}'}}$ and $B = \dfrac{\mathbf{x}' \cdot \mathbf{x}''}{(\mathbf{x}' \cdot \mathbf{x}')^{3/2}}$.

(c) Now calculate $\|\mathbf{t}'\|^2 = \mathbf{t}' \cdot \mathbf{t}'$ and then $\kappa = \|\mathbf{t}'(q)\|/\|\mathbf{x}'(q)\|$.

2. Calculate the unit tangent map $\mathbf{t}(q) = \mathbf{x}'(q)/\|\mathbf{x}'(q)\|$ for each of the curves $\mathbf{x}(q)$. Sketch each pair of maps and note how the inflections of $\mathbf{x}$ correspond to the points where the image of $\mathbf{t}$ folds back on itself.

   (a) $\mathbf{x}(q) = (q, q^3)$, $-1 < q < 2$.
   (b) $\mathbf{x}(q) = (q, \sin q)$, $0 < q < 4\pi$.
   (c) $\mathbf{x}(q) = (q^2, q - q^3)$, $-1 < q < 1$.

3. Show that the Gauss map of a curve does not depend on the parametrization chosen for the curve. (However, if the parametrizations have opposite orientation, the Gauss maps will be antipodal to each other.)

4. Calculate K for the cylinder

$$\mathbf{x}(q^1, q^2) = \left(f(q^1), g(q^1), q^2\right).$$

Sketch S, the image of $\mathbf{x}$, and S^*, the image of the unit normal map $\mathbf{n}$.

5. Calculate K for the "Cartesian" parametrization of the the saddle:
$$\mathbf{x}(q^1, q^2) = (q^1, q^2, q^1 q^2).$$
Show that this is nonsingular everywhere (unlike the "polar" parametrization, which is singular at the origin), and show that $K = -1$ at the origin.

6. (a) Let C be the plane curve defined by $\mathbf{w}(u) = (f(u), g(u))$, where $f(u) = u^2$, $g(u) = u^3 - u$. Let S be the surface parametrized as
$$\mathbf{x}(q^1, q^2) = (f(q^1), g(q^1), q^2).$$
Sketch C and S. Make it clear that S is the vertical cylinder over C; that is, S is the union of vertical lines (generators) that pass through C.

(b) Calculate the unit normal map $\mathbf{n}$ defined by $\mathbf{x}$ and sketch the image S^* of $\mathbf{n}$ in the unit sphere S^2. Why does this demonstrate that the curvature of S is identically zero?

(c) Calculate the matrix $\widetilde{B}(\mathbf{q})$ associated with $\mathbf{n}(\mathbf{q})$ and $\mathbf{x}(\mathbf{q})$ and show that the curvature of S is $K = \det \widetilde{B}$.

7. Suppose $\mathbf{w}(u) = (f(u), g(u))$ is a nonsingular smooth plane curve; that is, $\mathbf{w}' = (f', g')$ is never zero. Let S be the vertical cylinder over this curve:
$$\mathbf{x}(q^1, q^2) = (f(q^1), g(q^1), q^2).$$

(a) Show that S is nonsingular and calculate the unit normal map $\mathbf{n}$.

(b) Calculate the matrix $\widetilde{B}(\mathbf{q})$ and show that the curvature of S is $K = \det \widetilde{B}$.

8. Sketch by eye the Gauss map of an ellipsoid, a bell-shaped surface, and a torus. (For the bell, you can use the surface obtained by rotating the graph of $z = 1/(1 + x^2)$ around the z-axis.)

9. Compute the Gauss map of the plane $\mathbf{x} = (q^1, q^2, aq^1 + bq^2 + c)$.

10. (a) Compute the Gauss map of the torus $T_{a,r}$; use the parametrization

$$\mathbf{x}(q^1, q^2) = \left((a + r\cos q^2)\cos q^1, (a + r\cos q^2)\sin q^1, r\sin q^2\right).$$

Sketch the Gauss map $\mathcal{G}$, making it clear how $\mathcal{G}$ affects a representative collection of regions on $T_{a,r}$.

(b) Determine the curvature function $K(P)$ on $T_{a,r}$.

(c) Verify that $K(P) > 0$ when P is on the outer half of $T_{a,r}$ while $K(P) < 0$ on the inner half. Confirm that the Gauss map $\mathcal{G}: T_{a,r} \to S^2$ *reverses* orientation on the inner half of $T_{a,r}$.

(d) Identify the points P for which $K(P) = 0$. Describe the image of these points under the Gauss map $\mathcal{G}$.

Suppose D is a region in the domain of the surface $\mathbf{x}(q^1, q^2)$. The **total curvature over** D is the integral

$$\iint_D K(q^1, q^2)\sqrt{g(q^1, q^2)}\, dq^1\, dq^2,$$

where $g(q^1, q^2)$ is the determinant of the metric tensor and $K(q^1, q^2)$ is the curvature of the surface, both at the parameter point (q^1, q^2).

11. (a) Determine the total curvature of the upper hemisphere of a sphere of radius R. Determine the total curvature of the whole sphere. Does either of these values depend on the radius R?

(b) How are the total curvatures of the hemisphere and the whole sphere related to the areas of their images in S^2 under the Gauss map? Does this explain the way total curvature depends on R?

12. (a) Determine the total curvature over the following three regions of the torus $T_{a,r}$: the inner half, the outer half, and the entire torus. Do the values depend on the parameters a or r?

(b) How are the total curvatures of the these three regions on $T_{a,r}$ related to the areas of their images in S^2 under the Gauss map? Does this explain the way total curvature depends on the parameters a and r?

13. Show that the total curvature of a region D on an arbitrary surface is equal to the net oriented area of the image of D under the Gauss map of the surface.

Further Reading for Chapter 5

The treatment and notation we use for the differential geometry of surfaces follows Kreyszig [16] most closely. However, the subject is treated in a wide range of texts, a reasonable sample of which includes those by Crampin and Pirani [5], Dodson and Poston [6], Dubrovnin et al. [7], Klingenberg [15], and McCleary [22]. Naber [25] discusses the full $(1 + 3)$-dimensional de Sitter spacetime.

6
CHAPTER

Intrinsic Geometry

When Gauss defined the curvature of a surface as the rate of change of its normal direction, he made explicit use of the way the surface sits in space. Evidently, this is the extrinsic "curvature as bending" rather than the intrinsic "curvature as stretching" that we argued in Section 4.2 must be the basis of general relativity. It is altogether remarkable, then, that Gauss was able to prove that curvature *is* intrinsic. We begin this chapter by analyzing Gauss's famous argument, the *theorema egregium*, that curvature can be determined from a knowledge of the metric tensor alone, without reference to the surface's embedding in space.

The theorema egregium is the key to intrinsic differential geometry. Once it is established we can focus on developing the tools of intrinsic geometry—geodesics, tensors, and covariant derivatives—that will allow us to incorporate a theory of gravity into relativity.

6.1 Theorema Egregium

A flat plane can be wrapped around a cone or cylinder without stretching or tearing; we say that the plane can be *developed* on the

Developing one surface on another

cone or cylinder. In fact, any one of these three surfaces can be developed on any one of the others. More generally, we say that a surface S_1 can be **developed** on another surface S_2 if there is a map $f : S_1 \to S_2$ that preserves distances. A distance-preserving map is called an **isometry**.

Under what conditions can one surface be developed on another? Gauss took up this question in his long paper "Disquisitiones Generales circa Superficies Curvas" (General Investigations of Curved Surfaces, 1827) and determined that the surfaces must have the same curvature at corresponding points. The crucial step in Gauss's proof is a formula that expresses the curvature function of a surface entirely in terms of the metric tensor and its derivatives. Gauss, who wrote in Latin, called this result the *theorema egregium*. "Egregious theorem" is not a good translation, because "egregious" now has a pejorative meaning. The etymological roots are *e(x)*- "out of" and *grex*, "herd"; in Gauss's view, this theorem "stands out from the herd."

The Theorem

Theorem 6.1 (Theorema egregium) *Let* $\mathbf{x} : \mathcal{R} \to \mathbf{R}^3$ *be a parametrization of the surface S. Then the Gaussian curvature K can be expressed entirely in terms of derivatives of the metric tensor* $g_{ij} = \mathbf{x}_i \cdot \mathbf{x}_j$ *and thus is an intrinsic feature of S.*

PROOF: Our starting point is Theorem 5.4: $K = \det \widetilde{B} = \det b^i_j$, where the b^i_j are defined by $\mathbf{n}_j = -b^i_j \mathbf{x}_i$. Our goal is to prove that $\det \widetilde{B}$ depends only on the g_{ij} and their derivatives. We do this in a sequence of steps that will introduce a rather bewildering number of new functions. After we finish the proof we will pause to organize these functions and to have a systematic look at the process of "raising or lowering an index" that generates a number of them. We begin with

$$\mathbf{x}_{jk} = \frac{\partial^2 \mathbf{x}}{\partial q^k \partial q^j},$$

where $j = 1, 2$, $k = 1, 2$. These functions are defined at each point $\mathbf{q}$ in $\mathcal{R}$. Furthermore, we can carry out the differentiations in either order, so $\mathbf{x}_{jk} = \mathbf{x}_{kj}$.

STEP 1. Since $\mathbf{x}_1$, $\mathbf{x}_2$ form a basis of the tangent plane and $\mathbf{n}$ is a nonzero vector orthogonal to that plane, each $\mathbf{x}_{jk}$ can be written as a linear combination of $\mathbf{x}_1$, $\mathbf{x}_2$, and $\mathbf{n}$. Therefore, there are functions Γ^i_{jk} and b_{jk} for which

Γ^i_{jk} and b_{jk}

$$\mathbf{x}_{jk} = \Gamma^i_{jk}\mathbf{x}_i + b_{jk}\mathbf{n}.$$

Since $\mathbf{x}_{jk} = \mathbf{x}_{kj}$, we have

$$\mathbf{0} = \mathbf{x}_{kj} - \mathbf{x}_{jk} = \left(\Gamma^i_{kj} - \Gamma^i_{jk}\right)\mathbf{x}_i + \left(b_{kj} - b_{jk}\right)\mathbf{n},$$

and since $\mathbf{x}_1$, $\mathbf{x}_2$, and $\mathbf{n}$ are linearly independent, each coefficient on the right must be zero. This gives us the symmetries

$$\Gamma^i_{jk} = \Gamma^i_{kj} \quad \text{and} \quad b_{jk} = b_{kj}$$

for $i = 1, 2$.

STEP 2. We show that the new b_{jk} are connected to the b^i_j that appear in the matrix $\widetilde{B}$. First, $b_{jk} = \mathbf{x}_{jk} \cdot \mathbf{n}$ because

$$\mathbf{x}_{jk} \cdot \mathbf{n} = \Gamma^i_{jk}\mathbf{x}_i \cdot \mathbf{n} + b_{jk}\mathbf{n} \cdot \mathbf{n} = \Gamma^i_{jk} \cdot 0 + b_{jk} \cdot 1.$$

Next, $b_{jk} = g_{ji}b^i_k$; this follows by differentiating $\mathbf{x}_j \cdot \mathbf{n} = 0$:

$$0 = \frac{\partial}{\partial q^k}\left(\mathbf{x}_j \cdot \mathbf{n}\right) = \mathbf{x}_{jk} \cdot \mathbf{n} + \mathbf{x}_j \cdot \mathbf{n}_k = b_{jk} + \mathbf{x}_j \cdot \left(-b^i_k\mathbf{x}_i\right) = b_{jk} - b^i_k g_{ji}.$$

We can interpret $b_{jk} = g_{ji}b^i_k$ as a matrix product:

$$B = \begin{pmatrix} b_{11} & b_{12} \\ b_{21} & b_{22} \end{pmatrix} = \begin{pmatrix} g_{11} & g_{12} \\ g_{21} & g_{22} \end{pmatrix} \begin{pmatrix} b^1_1 & b^1_2 \\ b^2_1 & b^2_2 \end{pmatrix} = -G\widetilde{B}$$

It follows that $\det B = \det G \det \widetilde{B}$, so $K = \det \widetilde{B} = \det B / \det G = \det B/g$, and K will be expressible entirely in terms of the metric tensor and its derivatives if and only if $\det B = \det b_{jk}$ is.

$K = \dfrac{\det B}{g}$

STEP 3. Let

$$\mathbf{x}_{jkl} = \frac{\partial \mathbf{x}_{jk}}{\partial q^l} = \frac{\partial \Gamma^i_{jk}}{\partial q^l}\mathbf{x}_i + \Gamma^i_{jk}\mathbf{x}_{il} + \frac{\partial b_{jk}}{\partial q^l}\mathbf{n} + b_{jk}\mathbf{n}_l.$$

This vector must also be a linear combination of $\mathbf{x}_1$, $\mathbf{x}_2$, and $\mathbf{n}$. To find it, we substitute $\mathbf{n}_l = -b^i_l\mathbf{x}_i$ in the fourth term and $\mathbf{x}_{pl} = \Gamma^i_{pl}\mathbf{x}_i + b_{pl}\mathbf{n}$ in the second, first changing the dummy summation index there from i to p:

$$\begin{aligned}\mathbf{x}_{jkl} &= \frac{\partial \Gamma^i_{jk}}{\partial q^l}\mathbf{x}_i + \Gamma^p_{jk}\mathbf{x}_{pl} + \frac{\partial b_{jk}}{\partial q^l}\mathbf{n} - b_{jk}b^i_l\mathbf{x}_i \\ &= \frac{\partial \Gamma^i_{jk}}{\partial q^l}\mathbf{x}_i + \Gamma^p_{jk}\left(\Gamma^i_{pl}\mathbf{x}_i + b_{pl}\mathbf{n}\right) + \frac{\partial b_{jk}}{\partial q^l}\mathbf{n} - b_{jk}b^i_l\mathbf{x}_i \\ &= \left(\frac{\partial \Gamma^i_{jk}}{\partial q^l} + \Gamma^p_{jk}\Gamma^i_{pl} - b_{jk}b^i_l\right)\mathbf{x}_i + \left(\frac{\partial b_{jk}}{\partial q^l} + \Gamma^p_{jk}b_{pl}\right)\mathbf{n}.\end{aligned}$$

We get a similar expression for $\mathbf{x}_{jlk}$ by interchanging k and l:

$$\mathbf{x}_{jlk} = \left(\frac{\partial \Gamma^i_{jl}}{\partial q^k} + \Gamma^p_{jl}\Gamma^i_{pk} - b_{jl}b^i_k\right)\mathbf{x}_i + \left(\frac{\partial b_{jl}}{\partial q^k} + \Gamma^p_{jl}b_{pk}\right)\mathbf{n}.$$

$\mathbf{0} = \mathbf{x}_{jlk} - \mathbf{x}_{jkl}$

STEP 4. Since we can carry out the differentiations in either order,

$$\begin{aligned}\mathbf{0} &= \frac{\partial^2 \mathbf{x}_j}{\partial q^k \partial q^l} - \frac{\partial^2 \mathbf{x}_j}{\partial q^l \partial q^k} = \mathbf{x}_{jlk} - \mathbf{x}_{jkl} \\ &= \left[\left(\frac{\partial \Gamma^i_{jl}}{\partial q^k} - \frac{\partial \Gamma^i_{jk}}{\partial q^l} + \Gamma^p_{jl}\Gamma^i_{pk} - \Gamma^p_{jk}\Gamma^i_{pl}\right) - \left(b_{jl}b^i_k - b_{jk}b^i_l\right)\right]\mathbf{x}_i + A\,\mathbf{n}.\end{aligned}$$

We have written the coefficient of $\mathbf{n}$ simply as A because it will play no further role. Since $\mathbf{x}_1$, $\mathbf{x}_2$, and $\mathbf{n}$ are linearly independent and the sum above is zero, each coefficient must equal zero. If we introduce the abbreviation

R^i_{jkl}

$$R^i_{jkl} = \frac{\partial \Gamma^i_{jl}}{\partial q^k} - \frac{\partial \Gamma^i_{jk}}{\partial q^l} + \Gamma^p_{jl}\Gamma^i_{pk} - \Gamma^p_{jk}\Gamma^i_{pl},$$

§6.1 Theorema Egregium

then the fact that the coefficient of $\mathbf{x}_i$ is 0 implies

$$R^i_{jkl} = b_{jl}b^i_k - b_{jk}b^i_l.$$

Note that this represents $2^4 = 16$ equations, because each of the four indices i, j, k, and l can take two values.

STEP 5. We define $R_{hjkl} = g_{ih}R^i_{jkl} = b_{jl}b^i_k g_{ih} - b_{jk}b^i_l g_{ih} = b_{jl}b_{kh} - b_{jk}b_{lh}$. Then

$$R_{1212} = b_{22}b_{11} - b_{21}b_{21} = b_{22}b_{11} - b_{21}b_{12} = \det B,$$

so we can write the curvature as

$$K = \frac{R_{1212}}{g}.$$

$K = \dfrac{R_{1212}}{g}$

STEP 6. Our goal is now to show that R_{1212} depends only on the g_{ij} and their derivatives. But since

$$R_{1212} = g_{11}R^1_{212} + g_{12}R^2_{212}$$

and

$$R^i_{jkl} = \frac{\partial \Gamma^i_{jl}}{\partial q^k} - \frac{\partial \Gamma^i_{jk}}{\partial q^l} + \Gamma^p_{jl}\Gamma^i_{pk} - \Gamma^p_{jk}\Gamma^i_{pl},$$

to prove the theorem it now suffices to show that each Γ^i_{jk} depends only on the g_{ij} and their derivatives.

STEP 7. Let $\Gamma_{jk,h} = \mathbf{x}_{jk} \cdot \mathbf{x}_h$. Then $\Gamma_{jk,h} = \Gamma^i_{jk} g_{ih}$ because

$\Gamma_{jk,h}$ and g^{hm}

$$\Gamma_{jk,h} = \left(\Gamma^i_{jk}\mathbf{x}_i + b_{jk}\mathbf{n}\right) \cdot \mathbf{x}_h = \Gamma^i_{jk}\mathbf{x}_i \cdot \mathbf{x}_h + b_{jk}\mathbf{n} \cdot \mathbf{x}_h = \Gamma^i_{jk} g_{ih}.$$

We would like to invert this equation and thereby express Γ^i_{jk} in terms of $\Gamma_{jk,h}$. But $\Gamma_{jk,h} = \Gamma^i_{jk} g_{ih}$ is a matrix equation,

$$\Gamma_j = \begin{pmatrix} \Gamma_{j1,1} & \Gamma_{j1,2} \\ \Gamma_{j2,1} & \Gamma_{j2,2} \end{pmatrix} = \begin{pmatrix} \Gamma^1_{j1} & \Gamma^2_{j1} \\ \Gamma^1_{j2} & \Gamma^2_{j2} \end{pmatrix} \begin{pmatrix} g_{11} & g_{12} \\ g_{21} & g_{22} \end{pmatrix} = \widetilde{\Gamma}_j G,$$

so it can be solved for $\widetilde{\Gamma}_j$ using the inverse of G: $\Gamma_j G^{-1} = \widetilde{\Gamma}_j G G^{-1} = \widetilde{\Gamma}_j$. If we write the inverse with superscripts,

$$G^{-1} = (g^{hm}) = \begin{pmatrix} g^{11} & g^{12} \\ g^{21} & g^{22} \end{pmatrix},$$

then the summation convention works nicely:

$$g_{ih} g^{hm} = g^{mh} g_{hi} = \delta_i^m = \begin{cases} 1, & m = i, \\ 0, & m \neq i. \end{cases}$$

Thus $\Gamma_{jk,h} g^{hm} = \Gamma_{jk}^i g_{ih} g^{hm} = \Gamma_{jk}^i \delta_i^m = \Gamma_{jk}^m$. To finish our proof of the theorem, it suffices to show that each g^{hm} and each $\Gamma_{jk,h}$ depends only on the g_{ij} and their derivatives.

STEP 8. Standard facts about the inverse of a 2×2 matrix give us

$$g^{11} = \frac{g_{22}}{g}, \qquad g^{12} = g^{21} = -\frac{g_{12}}{g}, \qquad g^{22} = \frac{g_{11}}{g}.$$

STEP 9. The proof of the theorem will be complete when we show

$$\Gamma_{ij,k} = \frac{1}{2} \left(\frac{\partial g_{ik}}{\partial q^j} + \frac{\partial g_{jk}}{\partial q^i} - \frac{\partial g_{ij}}{\partial q^k} \right).$$

From $g_{ik} = \mathbf{x}_i \cdot \mathbf{x}_k$ we get

$$\frac{\partial g_{ik}}{\partial q^j} = \mathbf{x}_{ij} \cdot \mathbf{x}_k + \mathbf{x}_i \cdot \mathbf{x}_{kj} = \Gamma_{ij,k} + \Gamma_{jk,i}.$$

Similarly,

$$\frac{\partial g_{jk}}{\partial q^i} = \Gamma_{ij,k} + \Gamma_{ik,j}, \qquad -\frac{\partial g_{ij}}{\partial q^k} = -\Gamma_{ik,j} - \Gamma_{jk,i}.$$

Adding these three expressions gives the result.   END OF PROOF

Multi-index Quantities

Matrix multiplication

We can think of the summation convention essentially as a shorthand for matrix multiplication. For example, if A is the $p \times n$ matrix that has the entry a_i^j in the ith row and jth column,

$$A = (a_i^j) = \begin{pmatrix} a_1^1 & a_1^2 & \cdots & a_1^j & \cdots & a_1^n \\ a_2^1 & a_2^2 & \cdots & a_2^j & \cdots & a_2^n \\ \vdots & \vdots & & \vdots & & \vdots \\ a_i^1 & a_i^2 & \cdots & a_i^j & \cdots & a_i^n \\ \vdots & \vdots & & \vdots & & \vdots \\ a_p^1 & a_p^2 & \cdots & a_p^j & \cdots & a_p^n \end{pmatrix},$$

and B is the $n \times q$ matrix that has the entry b_j^k in the jth row and kth columns, then then their product $C = AB$ is the $p \times q$ matrix that has the entry $c_i^k = a_i^j b_j^k$ in the ith row and kth column.

Note that $b_j^k a_i^j = a_i^j b_j^k$, so $b_j^k a_i^j$ still represents an entry in AB, *not* an entry in BA. In fact, the product BA does not exist unless $p = q$, and then the element in the lth row and mth column of $D = BA$ is $d_l^m = b_l^k a_k^m = a_k^m b_l^k$.

The last example indicates the kind of freedom we have in altering index labels. In particular, a summation index is a *dummy* index: It can be replaced by any other symbol not already in use without altering the result; thus, for fixed i and k,

Dummy index of summation

$$a_i^j b_j^k = \sum_j a_i^j b_j^k = \sum_l a_i^l b_l^k = a_i^l b_l^k.$$

Since we will have further use for the multi-index quantities that are introduced in the proof of the theorema egregium, we list them now with the names they are usually given.

Catalog of quantities

They are all expressed in terms of a parametrization $\mathbf{x}(q^1, q^2)$ of a surface S; initially, a solitary subscript i represents partial differentiation with respect to q^i.

First fundamental form: $g_{ij} = \mathbf{x}_i \cdot \mathbf{x}_j$.

Second fundamental form: $b_{ij} = \mathbf{x}_{ij} \cdot \mathbf{n}$,

b_j^i defined by $\mathbf{n}_j = -b_j^i \mathbf{x}_i$,

$b_{jk} = b_j^i g_{ik}$.

Christoffel symbols, first kind: $\Gamma_{jk,l} = \mathbf{x}_{jk} \cdot \mathbf{x}_l$.

Christoffel symbols, second kind:

Γ^i_{jk} defined by $\mathbf{x}_{jk} = \Gamma^i_{jk}\mathbf{x}_i + b_{jk}\mathbf{n}$,

$\Gamma_{jk,l} = \Gamma^i_{jk} g_{il}$.

Riemann curvature tensor, mixed:

$$R^i_{jkl} = \frac{\partial \Gamma^i_{jl}}{\partial q^k} - \frac{\partial \Gamma^i_{jk}}{\partial q^l} + \Gamma^p_{jl}\Gamma^i_{pk} - \Gamma^p_{jk}\Gamma^i_{pl}.$$

Riemann curvature tensor, covariant:

$$R_{hjkl} = R^i_{jkl} g_{ih}, \quad \text{by definition.}$$

Einstein, among others, uses an older notation for the Christoffel symbols due to Christoffel himself:

$$\Gamma_{jk,l} = [jk, l], \qquad \Gamma^i_{jk} = \{jk, i\}.$$

In all cases, we call a subscript a **covariant index**, and a superscript a **contravariant index**. The meaning of these terms, and the term "tensor," will become clear in Sections 4 and 5 when we take up *relativity* on surfaces—that is, the relation between analogous multi-index quantities in two different parametrizations of the same surface. For the moment, just think of the terms as convenient labels.

Lowering an index

The multi-index quantities come in pairs; one of each pair is mixed in the sense that it has a single superscript, while the other is purely covariant. Moreover, the two forms are linked in exactly the same way; roughly speaking, this is the pattern:

$$B_{j*} = g_{ij} B^i_*.$$

(The asterisk stands for any number of indices.) We call multiplication by the metric tensor, as it is done here, **lowering an index**; we shall use it often in what follows.

Raising an index

By analogy, we define **raising an index** to be multiplication by the inverse g^{jk} of the metric tensor:

$$B^k_* = g^{jk} B_{j*}.$$

Since the inverse g^{jk} is defined by the condition

$$g^{jk}g_{ij} = \delta_i^k = \begin{cases} 1 & \text{if } i=k, \\ 0 & \text{otherwise,} \end{cases}$$

we have

$$B_*^k = g^{jk}B_{j*} = g^{jk}\left(g_{ij}B_*^i\right) = \left(g^{jk}g_{ij}\right)B_*^i = \delta_i^k B_*^i = B_*^k.$$

Therefore, raising and lowering indices are inverse processes. We first used this fact in the final steps of the proof of the theorema egregium. We had an expression for the Christoffel symbols of the first kind in terms of those of the second kind,

$$\Gamma_{jk,h} = \Gamma_{jk}^i g_{ih},$$

and we wanted to invert the relation, to express the symbols of the second kind in terms of those of the first kind. We did it like this:

$$\Gamma_{jk}^m = \Gamma_{jk}^i \delta_i^m = \Gamma_{jk}^i g_{ih} g^{hm} = \Gamma_{jk,h} g^{hm}.$$

More generally, if an arbitrary multi-index quantity B_*^{i*} has $k+1$ upper indices and l lower indices, we can lower an index to get the following new quantity with k upper indices and $l+1$ lower indices:

$$B_{j*}^* = g_{ij}B_*^{i*}.$$

Likewise, we can raise an index

$$B_*^{h*} = g^{hm}B_{m*}^*$$

to convert a quantity with k upper and $l+1$ lower indices into one with $k+1$ upper and l lower indices.

Another operation that we will carry out on multi-index quantities is **contraction**. Computing the trace of a matrix is an example: if $A = \left(a_i^j\right)$, then

$$\text{tr}\, A = a_1^1 + a_2^2 + \cdots + a_n^n = a_i^i = \text{a scalar}.$$

We have set the upper and lower indices equal to each other and then added, according to the summation convention. Note that

Contraction

the result in this example is a scalar; it has no indices at all. In general, to contract any multi-index quantity, set one upper and one lower index equal and sum. The result is a new quantity that has one index fewer in both the upper and the lower positions.

The standard computation of the length of a vector $\mathbf{v} = v^i$ is a contraction. We have

$$\|\mathbf{v}\|^2 = g_{ij}v^iv^j = (g_{ij}v^i)v^j = v_jv^j = \text{a scalar}.$$

The inner product of two vectors is similar:

$$\mathbf{v} \cdot \mathbf{w} = g_{ij}v^iw^j = (g_{ij}v^i)w^j = v_jw^j = \text{a scalar}.$$

Our final example is the Ricci tensor, which we will use in the sequel. It is obtained from the Riemann curvature tensor by contraction:

Ricci tensor: $R_{ik} = R^h_{ihk}$.

Exercises

Carry out the calculations in Exercises 1–6 for each of these four surfaces:

- the plane $\mathbf{x} = (q^1, q^2, aq^1 + bq^2 + c)$;
- the cylinder $\mathbf{x} = (f(q^1), g(q^1), q^2)$;
- the sphere of radius R, $\mathbf{x} = (R\cos q^1 \sin q^2, R\sin q^1 \sin q^2, R\cos q^2)$;
- the torus $T_{a,r}$, $\mathbf{x} = ((a + r\cos q^2)\cos q^1, (a + r\cos q^2)\sin q^1, r\sin q^2)$

1. Calculate the components g_{ij} of the metric tensor and the components g^{il} of its inverse.

2. Calculate the components of second fundamental forms b_{jk} and b^i_j. Verify that $b_{jk} = g_{ik}b^i_j$ and $b^i_j = g^{ik}b_{kj}$.

3. Calculate the Christoffel symbols of the second kind using both

$$\Gamma_{jk,l} = \mathbf{x}_{jk} \cdot \mathbf{x}_l \quad \text{and} \quad \Gamma_{jk,l} = \frac{1}{2}\left(\frac{\partial g_{jl}}{\partial q^k} + \frac{\partial g_{kl}}{\partial q^j} - \frac{\partial g_{jk}}{\partial q^l}\right).$$

4. Calculate $\Gamma^i_{jk} = g^{il}\Gamma_{jk,l}$ and verify that $\mathbf{x}_{jk} = \Gamma^i_{jk}\mathbf{x}_i + b_{jk}\mathbf{n}$.

5. Calculate the partial derivatives $\partial \Gamma^i_{jk}/\partial q^l$ and the mixed Riemann tensor

$$R^i_{jkl} = \frac{\partial \Gamma^i_{jl}}{\partial q^k} - \frac{\partial \Gamma^i_{jk}}{\partial q^l} + \Gamma^p_{jl}\Gamma^i_{pk} - \Gamma^p_{jk}\Gamma^i_{pl}.$$

6. Calculate $R_{hjkl} = g_{ih}R^i_{jkl}$ and $K = R_{1212}/g$.

7. Consider $n \times n$ matrices $\Gamma = (\gamma_{pq})$, $G = (g_{kl})$, and $M = (m^j_i)$, where the first or upper index specifies the row and the second or lower index specifies the column.

   (a) If the transpose of M is $M^t = (a^j_i)$, express a^j_i in terms of the components of M.

   (b) If $\Gamma = M^t G M$, express the component γ_{pq} in terms of the components of M and G using the summation convention.

8. (a) Show that the mixed Riemann curvature tensor R^i_{jkl} for any surface $\mathbf{x}$ satisfies the condition $R^i_{jlk} = -R^i_{jkl}$ for every i, j, k, l.

   (b) Conclude that at least 8 of the 16 components of the mixed Riemann tensor R^i_{jkl} must equal 0. List the components that are automatically 0.

9. The metric tensor $g_{ij}(q^1, q^2)$ of a general surface of revolution has the following form:

$$g_{11} = \left[r(q^2)\right]^2, \quad g_{12} = 0,$$
$$g_{21} = 0, \quad g_{22} = \left[r'(q^2)\right]^2 + \left[z'(q^2)\right]^2.$$

   Note: The g_{ij} do not depend explicitly on q^1, and $r(q^2) > 0$ for all q^2.

   (a) Using only the metric tensor, calculate g^{kl}, $\Gamma_{jk,l}$, Γ^i_{jk}, R^i_{jkl}, R_{1212}, and $K = R_{1212}/g$.

   (b) Show that at a point (q^1, q^2) where $K = 0$, either $z'(q^2) = 0$ or $r'(q^2)z''(q^2) = r''(q^2)z'(q^2)$.

   (c) Determine the points where $K < 0$ in terms of conditions on r and z and their derivatives.

6.2 Geodesics

Return to acceleration-free observers

Special relativity concerns observers who move without accelerating, along straight paths at constant speed. But in a curved space or spacetime, there may be no straight paths at all. Nevertheless, there are still paths that are "acceleration-free"; these are the *geodesics*. We will define geodesics on ordinary 2-dimensional surfaces and determine some of their properties.

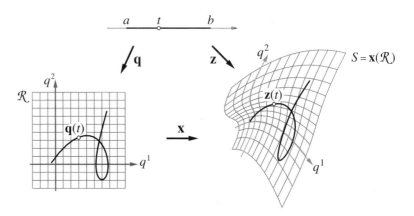

Let S be a surface parametrized by $\mathbf{x} : \mathcal{R} \to \mathbf{R}^3$, and let $\mathbf{q} : [a, b] \to \mathcal{R}$ be a curve in the parameter plane. Then $\mathbf{z}(t) = \mathbf{x}(\mathbf{q}(t))$ is a curve in S. The velocity and acceleration of $\mathbf{z}$ are

$$\text{velocity:} \quad \mathbf{z}'(t) = \mathbf{x}_i \frac{dq^i}{dt};$$

$$\text{acceleration:} \quad \mathbf{z}''(t) = \mathbf{x}_{ij} \frac{dq^i}{dt} \frac{dq^j}{dt} + \mathbf{x}_i \frac{d^2 q^i}{dt^2}$$

$$= \left(\Gamma_{ij}^k \mathbf{x}_k + b_{ij} \mathbf{n} \right) \frac{dq^i}{dt} \frac{dq^j}{dt} + \mathbf{x}_k \frac{d^2 q^k}{dt^2}$$

$$= \left(\frac{d^2 q^k}{dt^2} + \Gamma_{ij}^k \frac{dq^i}{dt} \frac{dq^j}{dt} \right) \mathbf{x}_k + \underline{b_{ij} \mathbf{n}}. \quad *$$

The normal and tangential components of acceleration

The normal component $\mathbf{N} = \underline{b_{ij} \mathbf{n}}$ of acceleration is independent of the curve $\mathbf{q}(t)$; it depends only on the second fundamental form b_{ij} of the surface *and it cannot be removed*. In other words, any curve $\mathbf{z}(t)$ has to accelerate to stay on the surface. The tan-

$\quad \tilde{N} = b_{ij} \dfrac{dq^i}{dt} \dfrac{dq^j}{dt} \vec{n}$

THE LAST TERM IS THE NORMAL COMPONENT N OF ACCELERATION
IT DEPENDS ON THE SECOND FUNDAMENTAL FORM b_{ij} AND THE
VELOCITY COMPONENTS dq^i/dt OF THE CURVE $z(t)$, BUT IS IN
GENERAL DIFFERENT FROM ZERO.

gential component

$$\mathbf{T} = \left(\frac{d^2 q^k}{dt^2} + \Gamma_{ij}^k \frac{dq^i}{dt} \frac{dq^j}{dt} \right) \mathbf{x}_k$$

is a different matter; it clearly depends on the curve and will equal zero if $\mathbf{q}(t)$ is properly chosen.

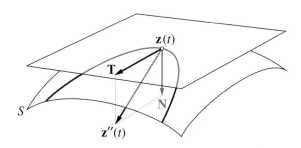

Definition 6.1 *The curve* $\mathbf{q}(t) = \left(q^1(t), q^2(t) \right)$ *in* $\mathcal{R}$ *is a* **geodesic** *on the surface* $\mathbf{x} : \mathcal{R} \to \mathbf{R}^3$ *if*

$$\frac{d^2 q^k(t)}{dt^2} + \Gamma_{ij}^k \left(q^1(t), q^2(t) \right) \frac{dq^i(t)}{dt} \frac{dq^j(t)}{dt} = 0, \qquad k = 1, 2.$$

These are called the **geodesic equations**. They are second-order nonlinear ordinary differential equations, *but they depend only on the metric tensor* g_{ij}. On an ordinary surface there are just two equations ($k = 1, 2$), but in curved spacetime that number will become four.

Let us determine the geodesics in some very simple cases. First, let S be the (x, y)-plane, parametrized by

Example: a flat plane

$$\mathbf{x} \left(q^1, q^2 \right) = \left(q^1, q^2, 0 \right).$$

Then $g_{ij} = \delta_{ij} = 1$ if $i = j$ and 0 otherwise. Therefore, all partial derivatives $\partial g_{ij} / \partial q^k$ are identically zero, implying

$$\Gamma_{ij,k} \equiv 0 \qquad \text{and} \qquad \Gamma_{ij}^h \equiv 0$$

for all i, j, k, and h. Therefore, the geodesic equations reduce to

$$\frac{d^2 q^1(t)}{dt^2} \equiv 0, \qquad \frac{d^2 q^2(t)}{dt^2} \equiv 0.$$

The solutions are the straight lines

$$\mathbf{q}(t) = \left(q^1(t), q^2(t)\right) = (at + b, ct + d);$$

a, b, c, and d are arbitrary constants of integration. The parameter moves along this line with constant speed $\|(a, c)\| = \sqrt{a^2 + c^2}$.

This last point is crucial. Consider, for example, the straight ray

$$q^2 = q^1 > 0$$

with the variable-speed parametrization $q^1(t) = q^2(t) = t^2$, $t > 0$. Since

$$\frac{d^2 q^1}{dt^2} = \frac{d^2 q^2}{dt^2} \equiv 2 \neq 0,$$

this is not a geodesic. Although the path itself is straight, the point (t^2, t^2) accelerates as it moves along that path. Thus, in the plane, every geodesic is a straight line, but not every straight line is a geodesic. The parametrization matters.

A geodesic is more than a "straightest" path

This is an important lesson: We customarily think of geodesics as certain *paths* on a surface, independent of any parametrization. But this is untenable; only constant-speed parametrizations work, as the next theorem shows.

Theorem 6.2 *If $\mathbf{q}(t)$ is a geodesic, then $\|\mathbf{q}'(t)\|^2 =$ constant.*

PROOF: In terms of the metric tensor,

$$\|\mathbf{q}'(t)\|^2 = g_{ij}\left(q^1(t), q^2(t)\right) \frac{dq^i(t)}{dt} \frac{dq^j(t)}{dt}.$$

To prove that this is constant we show that its derivative with respect to t is identically zero. Thus,

$$\frac{d}{dt}\left(\|\mathbf{q}'(t)\|^2\right) = \frac{\partial g_{ij}}{\partial q^k} \frac{dq^k}{dt} \frac{dq^i}{dt} \frac{dq^j}{dt} + g_{ij} \frac{d^2 q^i}{dt^2} \frac{dq^j}{dt} + g_{ij} \frac{dq^i}{dt} \frac{d^2 q^j}{dt^2}$$

$$= \frac{\partial g_{ij}}{\partial q^k} \frac{dq^k}{dt} \frac{dq^i}{dt} \frac{dq^j}{dt} + 2 g_{ij} \frac{d^2 q^i}{dt^2} \frac{dq^j}{dt}.$$

By assumption, $\mathbf{q}(t)$ is a geodesic,

$$\frac{d^2 q^i}{dt^2} = -\Gamma^i_{lm} \frac{dq^l}{dt} \frac{dq^m}{dt},$$

so

$$\begin{aligned}
\frac{d}{dt}\left(\|\mathbf{q}'(t)\|^2\right) &= \frac{\partial g_{ij}}{\partial q^k} \frac{dq^k}{dt} \frac{dq^i}{dt} \frac{dq^j}{dt} - 2 g_{ij} \Gamma^i_{lm} \frac{dq^l}{dt} \frac{dq^m}{dt} \frac{dq^j}{dt} \\
&= \frac{\partial g_{ij}}{\partial q^k} \frac{dq^k}{dt} \frac{dq^i}{dt} \frac{dq^j}{dt} - 2 \Gamma_{lm,j} \frac{dq^l}{dt} \frac{dq^m}{dt} \frac{dq^j}{dt} \\
&= \frac{\partial g_{ij}}{\partial q^k} \frac{dq^k}{dt} \frac{dq^i}{dt} \frac{dq^j}{dt} - \left(\frac{\partial g_{lj}}{\partial q^m} + \frac{\partial g_{mj}}{\partial q^l} - \frac{\partial g_{lm}}{\partial q^j}\right) \frac{dq^l}{dt} \frac{dq^m}{dt} \frac{dq^j}{dt}.
\end{aligned}$$

Now write this as four separate terms and relabel the dummy summation indices in each of the last three so that that term becomes

$$\frac{\partial g_{ij}}{\partial q^k} \frac{dq^k}{dt} \frac{dq^i}{dt} \frac{dq^j}{dt}.$$

This shows that the four terms are identical; since two occur with a plus sign and two with a minus, their sum is zero. END OF PROOF

It is possible to have different parametrizations of the same geodesic path, but the theorem forces the parameters of any two to be affinely related—that is, connected by an equation of the form $t = au + b$.

Corollary 6.1 *Suppose $\mathbf{q}(t)$ and $\mathbf{Q}(u) = \mathbf{q}(\varphi(u))$ are two geodesic parametrizations of the same path. Then $t = \varphi(u) = au + b$ for some constants a and b.*

PROOF: By the chain rule, $\mathbf{Q}'(u) = \mathbf{q}'(\varphi(u))\varphi'(u) = \mathbf{q}'(t)\varphi'(u)$. Therefore, since $\mathbf{Q}$ and $\mathbf{q}$ are geodesics,

$$\underbrace{\|\mathbf{Q}'(u)\|}_{\text{constant}} = \underbrace{\|\mathbf{q}'(t)\|}_{\text{constant}} \cdot |\varphi'(u)|,$$

so $\varphi'(u) = a$, where a is one of the constants $\pm \|\mathbf{Q}'(u)\|/\|\mathbf{q}'(t)\|$.

Example: the sphere

Hence $\varphi(u) = au + b$ for some constant of integration b.

END OF PROOF

The sphere gives us a more complex example on which to test these observations. Since geodesics are determined solely by the metric tensor, we start there (cf. Section 5.2):

$$\begin{pmatrix} g_{11} & g_{12} \\ g_{21} & g_{22} \end{pmatrix} = \begin{pmatrix} \cos^2(q^2) & 0 \\ 0 & 1 \end{pmatrix}, \quad \begin{pmatrix} g^{11} & g^{12} \\ g^{21} & g^{22} \end{pmatrix} = \begin{pmatrix} \cos^{-2}(q^2) & 0 \\ 0 & 1 \end{pmatrix}.$$

The Christoffel symbols of the first kind are sums of partial derivatives of the g_{ij}. But the only nonzero derivative is

$$\frac{\partial g_{11}}{\partial q^2} = -2\sin(q^2)\cos(q^2),$$

so only the Christoffel symbols whose indices are $\{1, 1, 2\}$ in some order will be nonzero:

$$\Gamma_{11,2} = \sin(q^2)\cos(q^2),$$
$$\Gamma_{21,1} = \Gamma_{12,1} = -\sin(q^2)\cos(q^2).$$

For the Christoffel symbols of the second kind we have

$$\Gamma^2_{11} = g^{21}\Gamma_{11,1} + g^{22}\Gamma_{11,2} = \sin(q^2)\cos(q^2),$$
$$\Gamma^1_{21} = \Gamma^1_{12} = g^{11}\Gamma_{12,1} + g^{12}\Gamma_{12,2} = \frac{-\sin(q^2)\cos(q^2)}{\cos^2(q^2)} = -\tan(q^2).$$

In all other cases $\Gamma_{ij,k} = \Gamma^h_{ij} \equiv 0$. The geodesic equations are

$$\frac{d^2q^1}{dt^2} - 2\tan(q^2)\frac{dq^1}{dt}\frac{dq^2}{dt} = 0, \quad \frac{d^2q^2}{dt^2} + \sin(q^2)\cos(q^2)\left(\frac{dq^1}{dt}\right)^2 = 0.$$

These equations are strongly coupled—the equation for q^1 involves q^2 and vice versa—and it is not obvious what their solutions might be.

Rather than attempt to solve the differential equations, we shall just check whether certain familiar curves are geodesics. The simplest to check are the coordinate curves $q^1 = $ constant (the meridians of longitude) and $q^2 = $ constant (the parallels of latitude).

Meridians of longitude: Let $q^1 = a$, $q^2 = t$, $-\pi/2 < t < \pi/2$. Here a is the longitude, so $0 \le a < 2\pi$:

$$\frac{d^2 q^1}{dt^2} - 2\tan(q^2)\frac{dq^1}{dt}\frac{dq^2}{dt} = 0 - 2\tan(t) \cdot 0 \cdot 1 = 0,$$

$$\frac{d^2 q^2}{dt^2} + \sin(q^2)\cos(q^2)\left(\frac{dq^1}{dt}\right)^2 = 0 + \sin(t)\cos(t) \cdot 0^2 = 0,$$

so these are all geodesics.

Parallels of latitude: Let $q^1 = t$, $q^2 = b$, $0 \le t \le 2\pi$. Here b is the latitude, so $-\pi/2 < b < \pi/2$:

$$\frac{d^2 q^1}{dt^2} - 2\tan(q^2)\frac{dq^1}{dt}\frac{dq^2}{dt} = 0 - 2\tan(b) \cdot 1 \cdot 0 = 0,$$

$$\frac{d^2 q^2}{dt^2} + \sin(q^2)\cos(q^2)\left(\frac{dq^1}{dt}\right)^2 = 0 + \sin(b)\cos(b) \cdot 1^2 = \tfrac{1}{2}\sin(2b).$$

The second equation is 0 only if $2b$ is an integer multiple of π; in the allowable range for b, this happens only when $b = 0$. This is the equator, so among the parallels of latitude only the equator is a geodesic.

Great circles: In Section 5.2 we defined a collection of great circles by the parametrizations

$$q^1(t) = t, \qquad q^2(t) = \arctan(\alpha \sin t), \qquad 0 \le t \le \pi.$$

Here α can take any real value. Since great circles on the sphere are the paradigmatic example of geodesics, it is natural to expect these curves to solve the geodesic equations. However, they are not even constant-speed parametrizations, because in Section 5.2 we established that their speed is

$$\sqrt{g_{ij}\frac{dq^i}{dt}\frac{dq^j}{dt}} = \frac{\sqrt{1+\alpha^2}}{1+\alpha^2 \sin^2 t}.$$

Therefore, they are not geodesics. Only after we reparametrize them—for example, by arc length—will they prove to be geodesics. Arc length is given by

$$s = \int_0^t \frac{\sqrt{1+\alpha^2}}{1+\alpha^2 \sin^2 t}\, dt = \arctan\left(\sqrt{1+\alpha^2}\, \tan t\right)$$

(cf. Section 5.2), implying

$$t = \arctan\left(\frac{\tan s}{\sqrt{1+\alpha^2}}\right).$$

The unit speed (i.e., arc-length) parametrization is therefore

$$q^1 = \arctan\left(\frac{\tan s}{\sqrt{1+\alpha^2}}\right), \qquad q^2 = \arctan\left(\frac{\alpha \tan s}{\sqrt{\alpha^2 + \sec^2 s}}\right).$$

In the exercises you are asked to work through the details and verify that these are indeed geodesics. The main ingredients of the proof are

$$\tan(q^2) = \frac{\alpha \tan s}{\sqrt{\alpha^2 + \sec^2 s}}, \qquad \sin(q^2)\cos(q^2) = \frac{\alpha \tan s \sqrt{\alpha^2 + \sec^2 s}}{(1+\alpha^2)\sec^2 s},$$

and

$$\frac{dq^1}{ds} = \frac{\sqrt{1+\alpha^2}\,\sec^2 s}{\alpha^2 + \sec^2 s}, \qquad \frac{dq^2}{ds} = \frac{\alpha}{\sqrt{\alpha^2 + \sec^2 s}},$$

$$\frac{d^2 q^1}{ds^2} = \frac{2\alpha^2 \sqrt{1+\alpha^2}\,\tan s \sec^2 s}{(\alpha^2 + \sec^2 s)^2}, \qquad \frac{d^2 q^2}{ds^2} = -\frac{\alpha \tan s \sec^2 s}{(\alpha^2 + \sec^2 s)^{3/2}}.$$

The first geodesic equation is

$$\frac{d^2 q^1}{ds^2} - 2\tan(q^2)\frac{dq^1}{ds}\frac{dq^2}{ds} = \frac{2\alpha^2 \sqrt{1+\alpha^2}\,\tan s \sec^2 s}{(\alpha^2 + \sec^2 s)^2}$$

$$- 2\frac{\alpha \tan s}{\sqrt{\alpha^2 + \sec^2 s}} \cdot \frac{\sqrt{1+\alpha^2}\,\sec^2 s}{\alpha^2 + \sec^2 s} \cdot \frac{\alpha}{\sqrt{\alpha^2 + \sec^2 s}} \equiv 0.$$

The second is

$$\frac{d^2 q^2}{ds^2} + \sin(q^2)\cos(q^2)\left(\frac{dq^1}{ds}\right)^2$$

$$= -\frac{\alpha \tan s \sec^2 s}{(\alpha^2 + \sec^2 s)^{3/2}} + \frac{\alpha \tan s \sqrt{\alpha^2 + \sec^2 s}}{(1+\alpha^2)\sec^2 s} \cdot \frac{(1+\alpha^2)\sec^4 s}{(\alpha^2 + \sec^2 s)^2} \equiv 0.$$

§6.2 Geodesics

Geodesics occur everywhere

Finally, given any point on a surface and given any direction at that point, we can construct a geodesic that passes through the point in the given direction.

Theorem 6.3 *Suppose P is a point on the surface S and* $\mathbf{v}$ *is a vector in the tangent plane TS_P. Then there is a geodesic* $\mathbf{q}(t)$ *on S with* $\mathbf{q}(0) = P$ *and* $\mathbf{q}'(0) = \mathbf{v}$.

PROOF: Notice first that the statement of the theorem is abbreviated. For a more complete statement we need an explicit parametrization $\mathbf{x} : \mathcal{R} \to \mathbf{R}^3$ of S. Then, if

$$P = \mathbf{x}(q_0^1, q_0^2) \quad \text{and} \quad \mathbf{v} = v^i \mathbf{x}_i(q_0^1, q_0^2),$$

the theorem asserts that there is a curve $\mathbf{q} : [-\varepsilon, \varepsilon] \to \mathcal{R}$ in the parameter plane that satisfies the geodesic equations and for which

$$\mathbf{q}(0) = (q_0^1, q_0^2) \quad \text{and} \quad \mathbf{q}'(0) = \left(\frac{dq^1}{dt}(0), \frac{dq^2}{dt}(0)\right) = (v^1, v^2).$$

The proof then follows from the general theorem on the existence and uniqueness of solutions to systems of ordinary differential equations. We have a pair of second-order differential equations (the geodesic equations) for the functions $q^1(t)$ and $q^2(t)$ and initial conditions on q^1, q^2, dq^1/dt, and dq^2/dt at $t = 0$ that are displayed above. Under the assumption that the functions $\Gamma_{ij}^k(q^1, q^2)$ that appear in the geodesic equations are sufficiently differentiable, the theorem guarantees that there are unique functions $q^1(t)$ and $q^2(t)$ defined on an interval $[-\varepsilon, \varepsilon]$ containing the initial point $t = 0$ that satisfy the differential equations and all the initial conditions. END OF PROOF

Exercises

1. Obtain the geodesic equations for the plane $\mathbf{x}(q^1, q^2) = (q^1, q^2, aq^1 + bq^2 + c)$ and find all solutions to those equations.
2. (a) Obtain the geodesic equations for the surface of revolution
$$\mathbf{x}(q^1, q^2) = (r(q^2)\cos q^1, r(q^2)\sin q^1, q^2).$$

(b) Show that a meridian of longitude q^1 = constant is a geodesic; you must first obtain a constant-speed parametrization for the meridian.

(c) Show that a parallel of latitude q^2 = constant is a geodesic if and only if $r'(q^2) = 0$. Explain this result geometrically.

3. (a) Obtain the geodesic equations for the cylinder

$$\mathbf{x}(q^1, q^2) = (f(q^1), g(q^1), q^2).$$

(b) Assuming that q^1 is arc length on the plane curve $q^1 \mapsto (f(q^1), g(q^1))$, show that any straight line $(q^1, q^2) = (at + b, ct + d)$ in the coordinate plane is a geodesic on the cylinder.

(c) Suppose instead that q^1 is *not* arc length on the curve $(f(q^1), g(q^1))$. Let $s = s(q^1)$ be the arc length function and let $q^1 = \varphi(s)$ be its inverse. Now use φ to obtain all solutions to the geodesic equations.

4. Obtain the geodesic equations for the torus $T_{a,r}$,

$$\mathbf{x}(q^1, q^2) = \left((a + r\cos q^2)\cos q^1, (a + r\cos q^2)\sin q^1, r\sin q^2\right),$$

and show that the meridians of longitude q^1 = constant are geodesics.

5. (a) Verify that each curve C_α defined by

$$q^1 = \arctan\left(\frac{\tan s}{\sqrt{1+\alpha^2}}\right), \quad q^2 = \arctan\left(\frac{\alpha \tan s}{\sqrt{\alpha^2 + \sec^2 s}}\right)$$

is a geodesic on the unit sphere

$$\mathbf{x} = (\cos q^1 \cos q^2, \sin q^1 \cos q^2, \sin q^2).$$

(b) Make a careful sketch of C_α in the (q^1, q^2)-plane for $\alpha = -2, 0, 1, 2, 25$.

(c) Let $C_{\alpha, \beta}$ be the curve obtained from C_α by translating q^1 by β:

$$q^1 = \beta + \arctan\left(\frac{\tan s}{\sqrt{1+\alpha^2}}\right), \quad q^2 = \arctan\left(\frac{\alpha \tan s}{\sqrt{\alpha^2 + \sec^2 s}}\right).$$

Show that each $C_{\alpha,\beta}$ is a geodesic. What is the geometric relation between the various $C_{\alpha,\beta}$ and C_α on the sphere itself? Why does this relation immediately imply that every $C_{\alpha,\beta}$ is a geodesic?

(d) Prove that if $\mathbf{q}(t) = (q^1(t), q^2(t))$ is a geodesic on the sphere, then so is $\mathbf{q}_\beta(t) = (\beta + q^1(t), q^2(t))$ for every β.

(e) According to the existence and uniqueness theorem for solutions to differential equations, there is a unique geodesic on the sphere passing through a given point in a given direction. That is, there is a unique geodesic $\mathbf{q}(t)$ satisfying

$$\mathbf{q}(0) = (a^1, a^2), \qquad \mathbf{q}'(0) = (b^1, b^2).$$

Find an explicit formula for $\mathbf{q}(t)$ in terms of the given a^1, a^2, b^1, and b^2.

6. Determine which coordinate lines $\mathbf{q}(t) = (t, k^2)$ and $\mathbf{q}(t) = (k^1, t)$ are geodesics in the $(1+1)$-dimensional de Sitter spacetime given by the metric

$$(g_{ij}) = \begin{pmatrix} 1 & 0 \\ 0 & -\cosh^2(q^1) \end{pmatrix}, \qquad -\infty < q^1 < \infty, \quad 0 \le q^2 \le 2\pi.$$

6.3 Curved Spacetime

In this section we shall define a curved spacetime to be a portion of $\mathbf{R}^4$ whose differential geometry is defined intrinsically by a metric in which each tangent space is a linear $(1+3)$-dimensional spacetime. To see how this should be done we consider first an ordinary 2-dimensional surface.

Ordinary Surfaces

Henceforth, we consider a surface patch S to be an open set $\mathcal{R}$ in the (q^1, q^2)-coordinate plane supplied with a 2×2 matrix of functions

A surface with an intrinsically defined geometry

$$G(q^1, q^2) = \left(g_{ij}(q^1, q^2)\right), \qquad i, j = 1, 2,$$

satisfying the following conditions:

1. Each $g_{ij}(q^1, q^2)$ is a smooth function of (q^1, q^2) on $\mathcal{R}$;
2. at each point (q^1, q^2), G is a symmetric matrix, $g_{ji} = g_{ij}$;
3. at each point (q^1, q^2), the two eigenvalues of G are positive.

We call $G = (g_{ij})$ the **metric tensor**, or **first fundamental form**, of S. Except that it must satisfy these three conditions, the metric tensor is arbitrary. In particular, it is *not* defined in terms of an embedding $\mathbf{x} : \mathcal{R} \to \mathbf{R}^3$. Indeed, no such map is involved in the definition of S at all; there is no larger ambient space in which S resides. It is in this sense that the geometry of S is *intrinsically* defined. To indicate that S depends on $\mathcal{R}$ and g_{ij} and on these ingredients alone, we write $S = \{\mathcal{R}, g_{ij}\}$.

$S = \{\mathcal{R}, g_{ij}\}$

Geometry in S

The metric tensor provides a full set of geometric multi-index quantities defined at each point of S:

- $g = \det G$;
- $G^{-1} = (g^{jk})$; that is, $g_{ij}g^{jk} = g^{kj}g_{ji} = \delta_i^k$;
- Christoffel symbols, first kind: $\Gamma_{ij,k} = \dfrac{1}{2}\left(\dfrac{\partial g_{ik}}{\partial q^j} + \dfrac{\partial g_{kj}}{\partial q^i} - \dfrac{\partial g_{ij}}{\partial q^k}\right)$;
- Christoffel symbols, second kind: $\Gamma^i_{jk} = g^{il}\Gamma_{jk,l}$;
- Riemann curvature tensor, mixed: $R^i_{jkl} = \dfrac{\partial \Gamma^i_{jl}}{\partial q^k} - \dfrac{\partial \Gamma^i_{jk}}{\partial q^l} + \Gamma^p_{jl}\Gamma^i_{pk} - \Gamma^p_{jk}\Gamma^i_{pl}$;
- Riemann curvature tensor, covariant: $R_{ijkl} = g_{im}R^m_{jkl}$;
- Gaussian curvature: $K = R_{1212}/g$;

With these quantities we can determine all the relevant geometric information about S. For example, suppose $\mathbf{q}(t) = (q^1(t), q^2(t))$ is a path in S and $a \le t \le b$. Then,

$$\text{length of } \mathbf{q} = \int_a^b \sqrt{g_{ij}\left(q^1(t), q^2(t)\right)\dfrac{dq^i}{dt}\dfrac{dq^j}{dt}}\, dt,$$

while $\mathbf{q}(t)$ is a geodesic if and only if

$$\frac{d^2 q^k}{dt^2} + \Gamma^k_{ij}(q^1, q^2) \frac{dq^i}{dt} \frac{dq^j}{dt} = 0.$$

The metric g_{ij} defines an inner product on each of the tangent planes of S, just as it does for an embedded surface. To make this point, though, we need to clarify what we mean by a tangent plane of a surface whose geometry is defined intrinsically rather than by an embedding.

The tangent plane TS_P

Essentially, a tangent vector is just the velocity vector of a curve at a particular point. Suppose P is a point in $S = \{\mathcal{R}, g_{ij}\}$, with coordinates (c^1, c^2) in $\mathcal{R}$. Let

$$\mathbf{q} : [a, b] \to \mathcal{R} : t \mapsto (q^1(t), q^2(t))$$

be a smooth curve in S, and suppose $\mathbf{q}(c) = (c^1, c^2) = P$. Then $\mathbf{q}'(c)$ is a tangent vector to S at P. We get all the tangent vectors to S at P by allowing $\mathbf{q}$ to vary over all smooth curves that pass through P. The set of all tangent vectors at P constitutes the tangent plane TS_P. Note that we draw the tangent vectors in their own plane, and not in $\mathcal{R}$ itself.

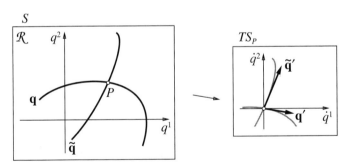

Each tangent plane inherits coordinates from S in a natural way. We shall write these coordinates as $(\dot{q}^1, \dot{q}^2)$ to indicate that they arise from the (q^1, q^2)-coordinates in S by differentiation with respect to a "time" parameter. (A dot is sometimes used to denote a time derivative.) As the following proposition asserts, $\dot{q}^1$ and $\dot{q}^2$ are the coordinates with respect to the basis $\{\mathbf{e}_1, \mathbf{e}_2\}$ defined by the curves that trace out the horizontal and vertical coordinate

Induced coordinates in TS_P

lines through (c^1, c^2) in $\mathcal{R}$:

$$\mathbf{e}_1 = \mathbf{q}'_1, \quad \text{where} \quad \mathbf{q}_1(s) = (c^1 + s, c^2) = P + s(1, 0),$$
$$\mathbf{e}_2 = \mathbf{q}'_2, \quad \text{where} \quad \mathbf{q}_2(s) = (c^1, c^2 + s) = P + s(0, 1).$$

Proposition 6.1 *The numbers*

$$\dot{q}^1 = \frac{dq^1}{dt}(c), \qquad \dot{q}^2 = \frac{dq^2}{dt}(c)$$

are the coordinates of the tangent vector $\mathbf{q}'(c)$ *with respect to the basis* $\{\mathbf{e}_1, \mathbf{e}_2\}$ *in the tangent plane* TS_P.

PROOF: Use Taylor's theorem to write

$$\mathbf{q}(t) = \left(q^1(t), q^2(t)\right)$$
$$= \left(c^1 + \dot{q}^1 \cdot (t - c) + O\left((t - c)^2\right), c^2 + \dot{q}^2 \cdot (t - c) + O\left((t - c)^2\right)\right)$$
$$= P + (t - c)\left(\dot{q}^1, \dot{q}^2\right) + O\left((t - c)^2\right)$$

for t near c. Since

$$\mathbf{q}_1(t - c) - P = (t - c)(1, 0), \qquad \mathbf{q}_2(t - c) - P = (t - c)(0, 1),$$

we can at least express $\mathbf{q}$ itself in terms of the "basis" curves in $\mathcal{R}$ for t near c:

$$\mathbf{q}(t) - P = \dot{q}^1 \left(\mathbf{q}_1(t - c) - P\right) + \dot{q}^2 \left(\mathbf{q}_2(t - c) - P\right) + O\left((t - c)^2\right).$$

Now differentiate with respect to t and set $t = c$:

$$\mathbf{q}'(c) = \dot{q}^1 \mathbf{q}'_1 + \dot{q}^2 \mathbf{q}'_2 = \dot{q}^1 \mathbf{e}_1 + \dot{q}^2 \mathbf{e}_2 = \dot{q}^i \mathbf{e}_i. \qquad \text{END OF PROOF}$$

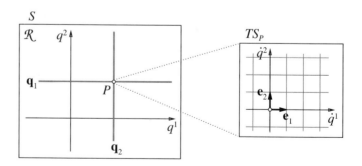

Although the individual tangent planes TS_P are not visually distinct—either from each other or from S itself—in the way that they are for an embedded surface, you should nonetheless think of them as *conceptually* distinct. From this point of view a surface S is supplied with an indexed collection of 2-dimensional inner product spaces TS_P; the index is the point P in S, and the inner product is the metric $g_{ij}(P)$. In particular, tangent vectors based at different points are in different vector spaces; we can add two vectors or compute their inner product only if they are in the same space.

The tangent planes TS_P are distinct

The simplest example of a surface S with an intrinsically defined geometry is one with a *constant* metric. The Christoffel symbols and Riemann tensors are zero everywhere, so the Gaussian curvature K is identically zero. The differential equations for geodesics take the simple form

$$\frac{d^2 q^1}{dt^2} = \frac{d^2 q^2}{dt^2} = 0,$$

so their solutions are the straight lines

$$q^1(t) = at + b, \qquad q^2(t) = ct + d.$$

Finally, the tangent planes all have the same inner product.

Example: The Hyperbolic Plane

One of the most familiar examples of a surface whose geometry is usually defined intrinsically by a metric—rather than by an embedding in space—is the hyperbolic plane H. It arises in the study of classical non-Euclidean geometry, which follows all the axioms of Euclid except the famous parallel postulate: "Given a line and a point not on that line, there exists precisely one line through the point that is parallel to the line." In non-Euclidean geometry this postulate is modified in two different ways, leading to two different geometries: Either assume that there is *no* parallel (elliptic geometry) or assume that there are *many* parallels (hyperbolic geometry).

A model for non-Euclidean geometry

Geodesics are lines

The domain $\mathcal{R}$ of the hyperbolic plane H is the upper half-plane $q^2 > 0$. The geodesics in the model H play the role of "straight lines" in the geometry. Once we define the metric in H we shall show that any vertical straight line (of the ordinary sort) is a geodesic, and so is any ordinary semicircle whose center lies on the q^1-axis. Then the figure below illustrates how H is a model for non-Euclidean hyperbolic geometry. Each of the "lines" M passing through the point P (and shown in gray) is parallel to the line L because it never intersects L.

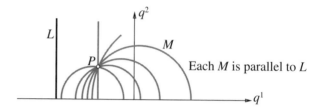

Each M is parallel to L

The metric on H

Let us now analyze H and show that these lines and circles are indeed geodesics. The metric for H is defined by

$$g_{11} = g_{22} = \frac{1}{(q^2)^2}, \qquad g_{12} = g_{21} = 0.$$

Thus $g^{11} = g^{22} = (q^2)^2$, $g^{12} = g^{21} = 0$, while

$$\frac{\partial g_{11}}{\partial q^2} = \frac{\partial g_{22}}{\partial q^2} = \frac{-2}{(q^2)^3},$$

and all other partial derivatives are zero. This implies that the only nonzero Christoffel symbols of the first kind will have indices $\{1, 1, 2\}$ in some order or $2, 2, 2$; these are

$$\Gamma_{12,1} = \Gamma_{21,1} = -\frac{1}{(q^2)^3}, \qquad \Gamma_{11,2} = \frac{1}{(q^2)^3}, \qquad \Gamma_{22,2} = -\frac{1}{(q^2)^3}.$$

Here is a complete reckoning of the Christoffel symbols of the second kind:

$$\Gamma^1_{11} = g^{11}\Gamma_{11,1} + g^{12}\Gamma_{11,2} = 0,$$

$$\Gamma^1_{12} = \Gamma^1_{21} = g^{11}\Gamma_{12,1} + g^{12}\Gamma_{12,2} = (q^2)^2 \cdot \frac{-1}{(q^2)^3} = -\frac{1}{q^2},$$

$$\Gamma_{22}^1 = g^{11}\Gamma_{22,1} + g^{12}\Gamma_{22,2} = 0,$$

$$\Gamma_{11}^2 = g^{21}\Gamma_{11,1} + g^{22}\Gamma_{11,2} = (q^2)^2 \cdot \frac{1}{(q^2)^3} = \frac{1}{q^2},$$

$$\Gamma_{12}^2 = \Gamma_{21}^2 = g^{21}\Gamma_{12,1} + g^{22}\Gamma_{12,2} = 0,$$

$$\Gamma_{22}^2 = g^{21}\Gamma_{22,1} + g^{22}\Gamma_{22,2} = (q^2)^2 \cdot \frac{-1}{(q^2)^3} = -\frac{1}{q^2}.$$

The geodesic equations are

$$0 = \frac{d^2 q^1}{dt^2} - \frac{2}{q^2} \frac{dq^1}{dt} \frac{dq^2}{dt},$$

$$0 = \frac{d^2 q^2}{dt^2} + \frac{1}{q^2}\left(\frac{dq^1}{dt}\right)^2 - \frac{1}{q^2}\left(\frac{dq^2}{dt}\right)^2.$$

Recall now that a path can satisfy the geodesic equations only if the parameter moves at a constant speed, implying that it is (a constant multiple of) the arc length. Therefore, if we are to show that vertical lines and semicircles centered on the q^1-axis are geodesic, we must first obtain their arc-length parametrizations.

Geodesics always have constant-speed parametrizations

The vertical lines. Consider the vertical line at $q^1 = c$; to start, we parametrize it as $q^1(t) = c$, $q^2(t) = t$, $t > 0$. Its speed is then

$$\|\mathbf{q}'(t)\| = \sqrt{g_{ij}(c,t) \frac{dq^i}{dt} \frac{dq^j}{dt}} = \sqrt{g_{22}(c,t)\left(\frac{dq^2}{dt}\right)^2} = \sqrt{\frac{1}{t^2} \cdot 1} = \frac{1}{t}.$$

If we measure arc length from the point where $t = a$, then the arc-length function is

$$s(t) = \int_a^t \|\mathbf{q}'(t)\|\, dt = \int_a^t \frac{dt}{t} = \ln t - \ln a.$$

This becomes simply $s = \ln t$ if we take $a = 1$, so $t = e^s$ and the arc-length parametrization of the vertical line is

$$q^1 = c, \qquad q^2 = e^s.$$

It is now straightforward to verify that these paths satisfy the geodesic equations; see the exercises.

Chapter 6 Intrinsic Geometry

The boundary of H is infinitely far from the interior

Any vertical line segment from an interior point to the boundary on the q^1-axis is infinitely long, as measured in H. Here is why. In terms of the parameter t, the boundary is at $t = 0$; at an interior point, t has some positive value b, for example. The length is therefore $s(b) - s(0) = \ln b + \infty = \infty$, as claimed. The visible boundary of H (the q^1-axis) is therefore infinitely far from any interior point. This is one reason why H is presented intrinsically, rather than as an embedded surface.

The semicircles. Suppose we want a clockwise parametrization of the semicircle of radius r centered at the point $(q^1, q^2) = (c, 0)$. The most common way is to use the sine and cosine functions:

$$q^1 = c - r\cos t, \quad q^2 = r\sin t, \quad 0 < t < \pi.$$

However, these are not constant-speed parametrizations on H; see the exercises.

Arc-length parametrization

It is the identity $\sin^2 t + \cos^2 t = 1$ that leads us to use sines and cosines to parametrize circles. But there is an analogous identity for hyperbolic functions,

$$\tanh^2 s + \operatorname{sech}^2 s = 1,$$

so we can use them for the same purpose. In fact, you are asked to prove in the exercises that

$$q^1 = c + r\tanh s, \quad q^2 = r\operatorname{sech} s, \quad -\infty < s < \infty,$$

is the arc-length parametrization we seek, and this path satisfies the geodesic equations on H. Notice, in particular, that the path is infinitely long and that the parameter values $s = \pm\infty$ give us the points $q^1 = c \pm r$ on the boundary $q^2 = 0$.

Incidentally, these paths are "semicircles" only from the Euclidean point of view. From the point of view of the hyperbolic geometry that prevails in H, they are *straight*. On the two geodesics in the figure below, the ticks mark units of arc length.

$K \equiv -1$ on H

Another aspect of H that makes it difficult to represent as a surface embedded in ordinary space is its Gaussian curvature; in the exercises you are asked to show that $K = R_{1212}/g \equiv -1$.

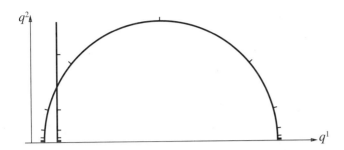

The nature of the geodesics and the fact that the curvature is negative everywhere make H very different from the Euclidean plane. Nevertheless, an individual tangent plane TH_P is remarkably similar to a Euclidean plane, as the following proposition shows.

Comparing TH_P to the Euclidean plane

Proposition 6.2 *The measure of an angle in H is the same as its Euclidean measure.*

PROOF: Let $\mathbf{q}' = (\dot{q}^1, \dot{q}^2)$ and $\mathbf{r}' = (\dot{r}^1, \dot{r}^2)$ be two vectors tangent to H at the same point $P = (c^1, c^2)$, and let θ be the angle between them. The hyperbolic measure of θ is

$$\theta = \arccos \frac{\mathbf{q}' \cdot \mathbf{r}'}{\|\mathbf{q}'\|\|\mathbf{r}'\|} = \arccos \frac{\dfrac{\dot{q}^1 \dot{r}^1 + \dot{q}^2 \dot{r}^2}{\cancel{(c^1)^2} + (c^2)^2}}{\sqrt{\dfrac{(\dot{q}^1)^2 + (\dot{q}^2)^2}{\cancel{(c^1)^2} + (c^2)^2}}\sqrt{\dfrac{(\dot{r}^1)^2 + (\dot{r}^2)^2}{\cancel{(c^1)^2} + (c^2)^2}}}$$

$(c^1)^2$ doesn't belong here.

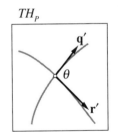

$$= \arccos \frac{\dot{q}^1 \dot{r}^1 + \dot{q}^2 \dot{r}^2}{\sqrt{(\dot{q}^1)^2 + (\dot{q}^2)^2}\sqrt{(\dot{r}^1)^2 + (\dot{r}^2)^2}}.$$

The final expression is just the Euclidean measure of θ.

END OF PROOF

The reason for this similarity between the Euclidean and hyperbolic planes—at the level of their tangent planes—becomes transparent when we write their metrics in matrix form. The Euclidean metric is given by I, the identity matrix, while the hyperbolic metric is a scalar multiple of I:

$$G = \frac{1}{(q^2)^2} I.$$

Although this scalar varies from point to point in H, at a given point it has a fixed value. Therefore, in the tangent plane at that point, every distance is that fixed multiple of its Euclidean distance, so each TH_P is strictly similar, in the sense of Euclidean geometry, to the Euclidean plane.

Spacetime

Gravity curves spacetime

In Chapter 4 we saw that, in the spacetime coordinate frame of an accelerating observer, distances between events will inevitably be distorted in a way we now attribute to curvature. In other words, an accelerating frame is curved. Furthermore, by Einstein's principle of equivalence, there is no difference, at least locally, between an observer who is stationary in a gravitational field and one undergoing acceleration with respect to an inertial frame. Therefore, the spacetime frame of an observer in a gravitational field will be curved, too: Gravity curves spacetime.

A spacetime frame and its metric

We are now in a position to deal with an arbitrary coordinate frame in spacetime—in particular, one that includes a gravitational field. Such a frame S consists of an open set $\mathcal{R}$ in the $\mathbf{q} = (q^1, q^2, q^3, q^4)$-space together with a 4×4 matrix of functions

$$G(\mathbf{q}) = \big(g_{ij}(\mathbf{q})\big), \qquad i, j = 1, 2, 3, 4,$$

satisfying the following conditions:

1. Each $g_{ij}(\mathbf{q})$ is a smooth function of $\mathbf{q}$ on $\mathcal{R}$;
2. at each point $\mathbf{q}$, G is a symmetric matrix, $g_{ji} = g_{ij}$;
3. at each point $\mathbf{q}$, G has one positive and three negative eigenvalues.

We call $G = (g_{ij})$ the **metric tensor**, or **first fundamental form**, of S. Because G is symmetric, only 10 of its 16 components g_{ij} can be distinct. According to the observations we made in Chapter 4, it is impossible to identify the quantities q^i with measurements of space and time made by an observer using rulers and clocks. On the contrary, the coordinates (q^1, q^2, q^3, q^4) are merely labels an observer uses to distinguish one spacetime event from another. However, we shall be able to do more in the tangent spaces.

§6.3 Curved Spacetime

Given a point P of S, we define the **tangent space of S at P** to be the set of velocity vectors of all curves in S that pass through P. The coordinates on S induce "dotted" coordinates $(\dot{q}^1, \dot{q}^2, \dot{q}^3, \dot{q}^4)$ on TS_P the same way they do on an ordinary surface. These are the coordinates with respect to the basis $\{\mathbf{e}_1, \mathbf{e}_2, \mathbf{e}_3, \mathbf{e}_4\}$ consisting of the tangent vectors of the coordinate curves.

The tangent spaces

Since it is impossible to visualize directly these 4-dimensional spaces, we replace them, for the sake of illustration and for most discussion, with their 2-dimensional counterparts exactly as we did when we first took "slices" of spacetime in Chapter 1. Thus, we modify the definition of a spacetime frame S in the following two ways:

Reduce the dimension for visual clarity

- The domain $\mathcal{R}$ of spacetime S is an open set in $\mathbf{R}^2$ rather than $\mathbf{R}^4$;
- the metric tensor $G(P) = g_{ij}(P)$ is a symmetric 2×2 matrix with one positive and one negative eigenvalue.

As we have in the past, we shall return to the full 4-dimensional spacetime when it is important to do so.

Each tangent space TS_P is now 2-dimensional, with coordinates $(\dot{q}^1, \dot{q}^2)$ induced from the coordinates (q^1, q^2) on S. If $\mathbf{q}'$ is tangent to S at P and $\mathbf{q}' = (\dot{q}^1, \dot{q}^2)$ in terms of the induced coordinates in TS_P, then by definition

Each TS_P is a $(1+1)$-dimensional Minkowski plane

$$\mathbf{q}' \cdot \mathbf{q}' = g_{ij}(P)\dot{q}^i \dot{q}^j.$$

Theorem 6.4 *We can choose a basis $\mathcal{B} = \{\mathbf{T}, \mathbf{X}\}$ for TS_P with the following property: If (t, x) are the coordinates of $\mathbf{q}'$ with respect to the basis $\mathcal{B}$ (i.e., $\mathbf{q}' = t\mathbf{T} + x\mathbf{X}$), then*

$$\mathbf{q}' \cdot \mathbf{q}' = t^2 - x^2.$$

PROOF: The basis $\mathcal{B}$ will consist of the eigenvectors of $G(P)$, properly scaled. Let $\mathbf{T}^*$ be an eigenvector associated with the positive eigenvalue (which we denote by $\lambda_\mathbf{T}$), and let $\mathbf{X}^*$ be an eigenvector associated with the negative eigenvalue $\lambda_\mathbf{X}$.

We claim that $\mathbf{T}^*$ and $\mathbf{X}^*$ are G-orthogonal. Let

$$\mathbf{T}^* = \dot{q}_\mathbf{T}^i \mathbf{e}_i, \quad T^* = \begin{pmatrix} \dot{q}_\mathbf{T}^1 \\ \dot{q}_\mathbf{T}^2 \end{pmatrix}, \quad \mathbf{X}^* = \dot{q}_\mathbf{X}^i \mathbf{e}_i, \quad X^* = \begin{pmatrix} \dot{q}_\mathbf{X}^1 \\ \dot{q}_\mathbf{X}^2 \end{pmatrix};$$

here we take the trouble to distinguish between the abstract vectors $\mathbf{T}^*$, $\mathbf{X}^*$ and the columns of coordinates T^*, X^* because it helps to make clear that we are converting abstract inner products to matrix multiplications in the calculations that follow. We now calculate $\mathbf{T}^* \cdot \mathbf{X}^*$ two ways:

(1) $\quad \mathbf{T}^* \cdot \mathbf{X}^* = T^{*t} G X^* = T^{*t}\left(\lambda_\mathbf{X} X^*\right) = \lambda_\mathbf{X} T^{*t} X^*;$

(2) $\quad \mathbf{T}^* \cdot \mathbf{X}^* = (T^{*t} G)^{tt} X^* = (G^t T^*)^t X^* = (G T^*)^t X^*$
$\qquad = (\lambda_\mathbf{T} T^*)^t X^* = \lambda_\mathbf{T} T^{*t} X^*.$

Therefore, $(\lambda_\mathbf{X} - \lambda_\mathbf{T}) T^{*t} X^* = 0$; since $\lambda_\mathbf{X} - \lambda_\mathbf{T} \neq 0$, we have $T^{*t} X^* = 0$ and thus $\mathbf{T}^* \cdot \mathbf{X}^* = 0$, proving the claim.

Next, consider $\mathbf{X}^* \cdot \mathbf{X}^* = X^{*t} G X^* = \lambda_\mathbf{X} X^{*t} X^*$. Noting that $\lambda_\mathbf{X}$ is negative while $X^{*t} X^* = (\dot{q}_\mathbf{X}^1)^2 + (\dot{q}_\mathbf{X}^2)^2$ is positive, we rescale $\mathbf{X}^*$ as

$$\mathbf{X} = \frac{1}{\sqrt{-\lambda_\mathbf{X}}\sqrt{X^{*t} X^*}} \mathbf{X}^*.$$

Then $\mathbf{X} \cdot \mathbf{X} = -1$. Similarly, we rescale $\mathbf{T}^*$ as

$$\mathbf{T} = \frac{1}{\sqrt{\lambda_\mathbf{T}}\sqrt{T^{*t} T^*}} \mathbf{T}^*;$$

then $\mathbf{T} \cdot \mathbf{T} = 1$. Since $\mathbf{T}$ and $\mathbf{X}$ are just scalar multiples of $\mathbf{T}^*$ and $\mathbf{X}^*$, $\mathbf{T} \cdot \mathbf{X} = 0$. Therefore, if $\mathbf{q}' = t\mathbf{T} + x\mathbf{X}$, then

$\mathbf{q}' \cdot \mathbf{q}' = (t\mathbf{T} + x\mathbf{X}) \cdot (t\mathbf{T} + x\mathbf{X})$
$\qquad = t^2 \mathbf{T} \cdot \mathbf{T} + 2tx \mathbf{T} \cdot \mathbf{X} + x^2 \mathbf{X} \cdot \mathbf{X} = t^2 - x^2.$ END OF PROOF

The Minkowski norm in TS_P

As we did in Chapter 2, we shall distinguish between timelike, spacelike, and lightlike vectors $\mathbf{q}'$ in TS_P and define the Minkowski norm of a vector:

$\mathbf{q}' \cdot \mathbf{q}' > 0 \quad \mathbf{q}'$ is **timelike** $\quad \|\mathbf{q}'\| = \sqrt{\mathbf{q}' \cdot \mathbf{q}'}$

$\mathbf{q}' \cdot \mathbf{q}' < 0 \quad \mathbf{q}'$ is **spacelike** $\quad \|\mathbf{q}'\| = \sqrt{-\mathbf{q}' \cdot \mathbf{q}'}$

$\mathbf{q}' \cdot \mathbf{q}' = 0 \quad \mathbf{q}'$ is **lightlike** $\quad \|\mathbf{q}'\| = 0$

§6.3 Curved Spacetime

The figure below illustrates how the {**T**, **X**} basis and the future light cone can vary from one tangent plane to another. Although **T** and **X** are indeed unit vectors in the G-metric, their lengths appear to us (from the Euclidean point of view) to be $1/\sqrt{\lambda_T}$ and $1/\sqrt{-\lambda_X}$. This difference is reflected, furthermore, in the slope of the light cone (i.e., the speed of light). The slope is $\pm\|\mathbf{X}\|/\|\mathbf{T}\|$; its actual value, as determined by the G-metric, is ± 1, but its apparent value is $\pm\sqrt{-\lambda_T/\lambda_X}$ (which explains why the sides of the cone do not make Euclidean 45° angles with the t- and x-axes).

The new basis in each tangent plane

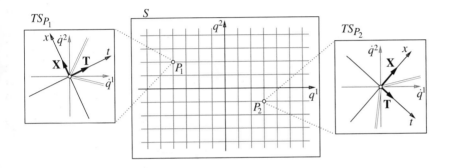

Since we can find, in each tangent plane, axes that record ordinary clock times and ruler lengths, that plane is a classical inertial frame. Since the tangent plane TS_P provides a magnified version of the microscopic environments at P, roughly speaking, we can introduce in S itself coordinates that approximately agree with clock rulers and ruler lengths, at least in a sufficiently small neighborhood of P. To this extent, at least, there are physically meaningful coordinates in S.

Each tangent plane carries an inertial frame

We have already seen, in the de Sitter spacetime, an illustration of some aspects of the theorem. The de Sitter metric is

Example: de Sitter spacetime

$$G(q^1, q^2) = \begin{pmatrix} 1 & 0 \\ 0 & \cosh^2 q^1 \end{pmatrix},$$

so the eigenvalues are $\lambda_T = 1$, $\lambda_X = -\cosh^2 q^1$. Because G is already a diagonal matrix (in terms of the induced coordinates $(\dot{q}^1, \dot{q}^2)$), the eigenvectors lie on the axes of those coordinates.

290   Chapter 6   Intrinsic Geometry

In terms of those coordinates, the basis eigenvectors are

$$\mathbf{T} = \begin{pmatrix} 1 \\ 0 \end{pmatrix} \quad \text{and} \quad \mathbf{X} = \frac{1}{\cosh q^1} \begin{pmatrix} 0 \\ 1 \end{pmatrix}.$$

With respect to the induced coordinates, the slope of the light cone is

$$\pm \sqrt{-\frac{\lambda_T}{\lambda_X}} = \frac{\pm 1}{\cosh q^1} = \pm \operatorname{sech} q^1;$$

this is precisely what we noticed in Section 5.3.

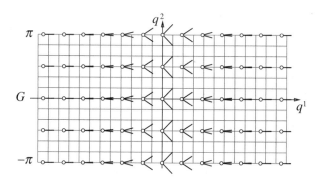

4-dimensional spacetime

A theorem analogous to Theorem 6.4 holds in the full 4-dimensional spacetime S. It says that in each tangent space TS_P there are coordinates (t, x, y, z) for which

$$\mathbf{q}' \cdot \mathbf{q}' = t^2 - x^2 - y^2 - z^2.$$

This is a consequence of the principal axes theorem of linear algebra, which says that given any quadratic form $Q(\mathbf{y})$ on $\mathbf{R}^n$, there are coordinates $(y_1, y_2, \ldots, y_n)$ for $\mathbf{y}$ for which

SPECTRAL THEOREM

$$Q(\mathbf{y}) = \lambda_1 y_1^2 + \lambda_2 y_2^2 + \cdots + \lambda_n y_n^2;$$

$\lambda_1, \lambda_2, \ldots, \lambda_n$ are the eigenvalues of Q, listing repeated eigenvalues as often as they occur. Then, if $(z_1, z_2, \ldots, z_n)$ are a suitable rescaling of the $(y_1, y_2, \ldots, y_n)$ coordinates, then

$$Q(\mathbf{y}) = \pm z_1^2 \pm z_2^2 \pm \cdots \pm z_n^2,$$

where the sign of z_i^2 is chosen equal to the sign of λ_i. For a proof of the principal axes theorem, consult a text on algebra or linear algebra.

Exercises

1. Show that the line $q^1 = c$, $q^2 = e^s$ *does* satisfy the geodesic equations for the hyperbolic plane H, while the same line parametrized simply as $q^1 = c$, $q^2 = t$ does not.

2. Compute the speed $\|\mathbf{q}'\|$ of the curve $\mathbf{q}(t) = (c - r\sin t, r\cos t)$ in the hyperbolic plane H to show that it cannot be a geodesic.

3. (a) Show that the path $\mathbf{q}(s) = (c + r\tanh s, r\,\text{sech}\, s)$ satisfies the geodesic equations in the hyperbolic plane H, for any c and $r > 0$.

   (b) Show that the length of $\mathbf{q}(s)$ from $s = s_1$ to $s = s_2$ is $|s_2 - s_1|$. Explain, then, why the whole geodesic has infinite length in H.

4. Prove that if $\mathbf{q}(t) = (q^1(t), q^2(t))$ is a geodesic on H, then so is $\mathbf{q}_\beta(t) = (\beta + q^1(t), q^2(t))$ for every β.

5. According to the existence and uniqueness theorem for solutions to differential equations, there is a unique geodesic on H passing through a given point in a given direction. That is, there is a unique geodesic $\mathbf{q}(t)$ satisfying

$$\mathbf{q}(0) = (a^1, a^2), \qquad \mathbf{q}'(0) = (b^1, b^2).$$

   Find an explicit formula for $\mathbf{q}(t)$ in terms of the given a^1, a^2, b^1, and b^2.

6. (a) Calculate the mixed Riemann curvature tensor

$$R^i_{jkl} = \frac{\partial \Gamma^i_{jl}}{\partial q^k} - \frac{\partial \Gamma^i_{jk}}{\partial q^l} + \Gamma^p_{jl}\Gamma^i_{pk} - \Gamma^p_{jk}\Gamma^i_{pl}$$

   for the hyperbolic plane H, and calculate $R_{hjkl} = g_{ih}R^i_{jkl}$.

   (b) Show that the Gaussian curvature $K = R_{1212}/g$ is identically -1 on H.

7. This is an extension of Theorem 6.4 to a 4-dimensional spacetime $\mathbf{R}^4$, using the arguments found in the proof. Suppose G is a symmetric 4×4 real matrix with one positive eigenvalue λ_0 and three different negative eigenvalues $\lambda_1, \lambda_2,$ and λ_3. Let $\mathbf{T}^* = \mathbf{X}_0^*, \mathbf{X}_1^*, \mathbf{X}_2^*, \mathbf{X}_3^*$, respectively, be eigenvectors associated with these eigenvalues. Rescale the eigenvectors as follows:

$$\mathbf{X}_0 = \frac{1}{\sqrt{\lambda_0}\sqrt{(X_0^*)^t X_0^*}} \mathbf{X}_0^*, \qquad \mathbf{X}_\alpha = \frac{1}{\sqrt{-\lambda_\alpha}\sqrt{(X_\alpha^*)^t X_\alpha^*}} \mathbf{X}_\alpha^*, \quad \alpha = 1, 2, 3.$$

Here X_j^* denotes the column vector of coordinates of $\mathbf{X}_j^*$ with respect to the standard basis in $\mathbf{R}^4$.

(a) Show that each pair of rescaled eigenvectors is G-orthogonal in the sense of the proof of Theorem 6.4: $\mathbf{X}_i \cdot \mathbf{X}_j = (X_i^*)^t G X_j = 0$, $i, j = 0, 1, 2, 3$.

(b) Show that $\mathbf{X}_0 \cdot \mathbf{X}_0 = 1$, while $\mathbf{X}_\alpha \cdot \mathbf{X}_\alpha = -1$, $\alpha = 1, 2, 3$.

(c) Show that $\mathcal{B} = \{\mathbf{X}_0, \mathbf{X}_1, \mathbf{X}_2, \mathbf{X}_3\}$ is a basis for $\mathbf{R}^4$.

(d) Suppose (y^0, y^1, y^2, y^3) are the coordinates of an arbitrary vector $\mathbf{Y}$ with respect to the basis $\mathcal{B}$ in $\mathbf{R}^4$. Show that

$$\mathbf{Y} \cdot \mathbf{Y} = (y^0)^2 - (y^1)^2 - (y^2)^2 - (y^3)^2.$$

6.4 Mappings

A pair of observers

At this point we understand how an individual observer will describe and analyze events in spacetime using an arbitrary coordinate frame. But relativity is about synthesizing a coherent objective reality from the views of numerous observers who have *different* views of the same events. So let us once again consider two observers G and R who view the same portion of spacetime. No longer, though, shall we assume that these are Galilean observers with inertial coordinate frames. Instead, we simply assume that each labels events using an arbitrary frame with a curved metric:

domain	coordinates	metric
G	$\boldsymbol{\xi} = (\xi^1, \xi^2, \xi^3, \xi^4)$	$\gamma_{ij}(\boldsymbol{\xi})$
R	$\mathbf{x} = (x^1, x^2, x^3, x^4)$	$g_{ij}(\mathbf{x})$

§6.4 Mappings

In the metaphor we introduced in Chapter 1, the coordinate frames G and R are two languages for describing events in spacetime. Because G and R describe the same events, there must be a pair of maps

$$M : G \to R, \qquad M^{-1} : R \to G$$

that serve as dictionaries to translate one language to the other. Each point $\boldsymbol{\xi} = (\xi^i)$ in G labels a certain event E in spacetime. Let $\mathbf{x} = (x^k)$ be the point in R that labels the same event. Then $\mathbf{x} = M(\boldsymbol{\xi})$ and $\boldsymbol{\xi} = M^{-1}(\mathbf{x})$ are the two translations.

The maps $M : G \to R$ and $M^{-1} : R \to G$

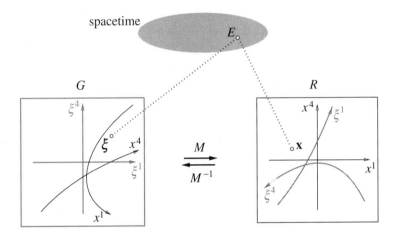

To make M easier to visualize, we have drawn G and R as planes even though they are really 4-dimensional. As the figure indicates, the relation between the two frames can be quite arbitrary and nonlinear.

Under the map M, each target coordinate x^k is a real-valued function of the source coordinates $(\xi^1, \xi^2, \xi^3, \xi^4)$; likewise, under M^{-1} each ξ^i is a function of (x^1, x^2, x^3, x^4):

Coordinate functions are smooth

$$M : \begin{cases} x^1 = f^1(\xi^1, \xi^2, \xi^3, \xi^4), \\ x^2 = f^2(\xi^1, \xi^2, \xi^3, \xi^4), \\ x^3 = f^3(\xi^1, \xi^2, \xi^3, \xi^4), \\ x^4 = f^4(\xi^1, \xi^2, \xi^3, \xi^4); \end{cases} \qquad M^{-1} : \begin{cases} \xi^1 = \varphi^1(x^1, x^2, x^3, x^4), \\ \xi^2 = \varphi^2(x^1, x^2, x^3, x^4), \\ \xi^3 = \varphi^3(x^1, x^2, x^3, x^4), \\ \xi^4 = \varphi^4(x^1, x^2, x^3, x^4). \end{cases}$$

294 Chapter 6 Intrinsic Geometry

We can write these more compactly as

$$x^k = f^k(\xi^i), \qquad \xi^i = \varphi^i(x^k).$$

We assume that M and M^{-1} are smooth maps; this means that each f^k is a smooth function of the ξ^i, and each φ^i is a smooth function of the x^k.

Tangents and Differentials

Coordinates of tangent vectors

Besides the coordinates ξ^i that G uses to label events, there are induced coordinates $\dot\xi^i$ in each tangent space that label vectors; similarly, the x^k coordinates induce coordinates $\dot x^k$ in each tangent space in R. How do the induced coordinates get translated when we use $M : G \to R$ to translate the main coordinates?

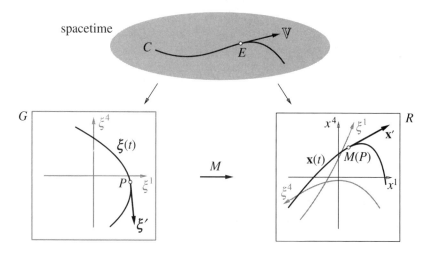

To address this question, consider how G and R describe the same tangent vector $\mathbb{V}$ to a curve C in spacetime. Suppose C is represented in G as

$$\boldsymbol{\xi}(t) = \xi^i(t);$$

then it can be represented in the Roman frame as

$$\mathbf{x}(t) = M(\boldsymbol{\xi}(t)) = f^k(\xi^i(t)) = x^k(t).$$

In G the tangent vector is

$$\xi' = \left(\frac{d\xi^i}{dt}\right),$$

while in R it is

$$\mathbf{x}' = \left(\frac{dx^k}{dt}\right) = \left(\frac{\partial f^k}{\partial \xi^1}\frac{d\xi^1}{dt} + \frac{\partial f^k}{\partial \xi^2}\frac{d\xi^2}{dt} + \frac{\partial f^k}{\partial \xi^3}\frac{d\xi^3}{dt} + \frac{\partial f^k}{\partial \xi^4}\frac{d\xi^4}{dt}\right).$$

We have used the multivariable chain rule here to differentiate $x^k = f^k(\xi^i(t))$ with respect to t. This expression represents four separate equations, for $k = 1, 2, 3, 4$; we can rewrite them as a single matrix multiplication

$$\begin{pmatrix} \dfrac{dx^1}{dt} \\ \vdots \\ \dfrac{dx^4}{dt} \end{pmatrix} = \begin{pmatrix} \dfrac{\partial f^1}{\partial \xi^1} & \dfrac{\partial f^1}{\partial \xi^2} & \dfrac{\partial f^1}{\partial \xi^3} & \dfrac{\partial f^1}{\partial \xi^4} \\ \vdots & \vdots & \vdots & \vdots \\ \dfrac{\partial f^4}{\partial \xi^1} & \dfrac{\partial f^4}{\partial \xi^2} & \dfrac{\partial f^4}{\partial \xi^3} & \dfrac{\partial f^4}{\partial \xi^4} \end{pmatrix} \begin{pmatrix} \dfrac{d\xi^1}{dt} \\ \vdots \\ \dfrac{d\xi^4}{dt} \end{pmatrix}.$$

This equation answers our question about translating induced coordinates from one tangent space to the other: Translation is by the linear map

The differential of M maps induced coordinates

$$dM_P : TG_P \to TR_{M(P)}$$

given by the matrix

$$dM_P = \left(\frac{\partial f^k}{\partial \xi^i}\right)_P.$$

The matrix is clearly related to the original map M, because its components are the partial derivatives of the coordinate functions that define M. The components are functions of ξ, and the subscript indicates that we must evaluate these functions at the point where $\xi = P$ in order to get the particular matrix that multiplies the tangent vectors in TG_P. We call both the matrix and the linear map that it defines the **differential of M at P**.

Here is a summary of what we have found so far:

Summary

- Associated with each main coordinate frame G is a family of tangent vector spaces TG_P, indexed by the points P in G.
- The coordinates ξ^i in G induce new coordinates $\dot\xi^i$ in each tangent space TG_P.
- Any map $M : G \to R$ that translates coordinates between frames induces a family of *linear* maps

$$dM_P : TG_P \to TR_{M(P)}$$

that translate the corresponding induced coordinates. The family dM_P is also indexed by the points P in G.

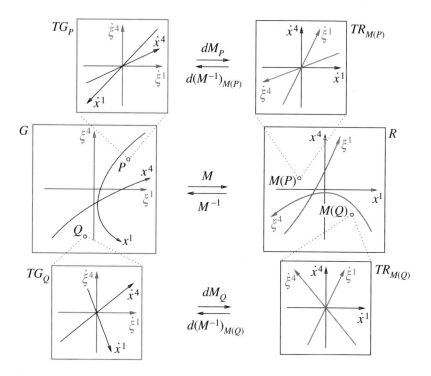

The figure above illustrates the points of the summary, and also indicates what is asserted by the following proposition: Each differential of M is invertible, and its inverse is given by the differential of the inverse of M.

§6.4 Mappings

Proposition 6.3 $(dM_P)^{-1} = d(M^{-1})_{M(P)}$.

PROOF: We shall make use of the fact that $M^{-1}(M(\xi)) = \xi$. In coordinates, this identity becomes

$$\varphi^j(f^k(\xi^i)) = \xi^j.$$

The partial derivative of this equation with respect to ξ^i is

$$\frac{\partial \varphi^j}{\partial x^1}\frac{\partial f^1}{\partial \xi^i} + \frac{\partial \varphi^j}{\partial x^2}\frac{\partial f^1}{\partial \xi^i} + \frac{\partial \varphi^j}{\partial x^3}\frac{\partial f^1}{\partial \xi^i} + \frac{\partial \varphi^j}{\partial x^4}\frac{\partial f^1}{\partial \xi^i} = \frac{\partial \xi^j}{\partial \xi^i} = \delta_i^j.$$

The result is δ_i^j (i.e., 1 or 0 depending on whether i and j are equal or not) because the variables ξ^1, ξ^2, ξ^3, and ξ^4 are independent. If we rewrite this using the summation convention, we get

$$\left(\frac{\partial \varphi^j}{\partial x^k}\right)_{M(P)} \left(\frac{\partial f^k}{\partial \xi^i}\right)_P = \delta_i^j.$$

Here we have evaluated each $\partial f^k/\partial \xi^i$ at $(\xi^i) = P$ and each $\partial \varphi^j/\partial x^k$ at $(x^k) = M(P)$. The result is an ordinary matrix multiplication, and since the numbers δ_i^j are the entries of the identity matrix I, we have the matrix equation

$$d(M^{-1})_{M(P)} \cdot dM_P = I. \qquad \text{END OF PROOF}$$

If we rewrite the coordinate equations that define M and M^{-1}, replacing the function names (f^k and φ^i) by the names of the target variables (x^k and ξ^i, respectively),

$$x^k = x^k(\xi^i), \qquad \xi^j = \xi^j(x^l),$$

then the differentials become

$$dM_P = \left(\frac{\partial x^k}{\partial \xi^i}\right), \qquad d(M^{-1})_{M(P)} = \left(\frac{\partial \xi^j}{\partial x^l}\right).$$

By the proposition, these matrices are inverses of each other (when evaluated at corresponding points), so

$$\frac{\partial x^k}{\partial \xi^i}\frac{\partial \xi^i}{\partial x^l} = \delta_l^k, \qquad \frac{\partial \xi^j}{\partial x^k}\frac{\partial x^k}{\partial \xi^i} = \delta_i^j.$$

The inverse relations are so visually compelling when expressed in this form that from now on we will usually give functions the same names as their target variables—even at the risk of losing clarity. There is certainly precedent: These relations are the multivariable equivalents of the more familiar equation

$$\frac{dx}{dy}\frac{dy}{dx} = 1$$

that applies to single-variable inverse functions like $x = \ln y$ and $y = e^x$.

Metrics

Special relativity holds in the tangent spaces

We begin, as always, with the map $M : G \to R$ that tells us how to translate G's coordinates of an event to R's. For each P in G, the differential dM_P then tells us how to translate the coordinates of tangent vectors. According to the theorem in the previous section, the tangent spaces have *inertial* frames of the sort that appear in special relativity (even when G and R themselves are quite arbitrary). We therefore make the following assumption.

Basic assumption: All the laws of special relativity hold for the inertial frames TG_P and $TR_{M(P)}$ when they are connected by the linear map

$$dM_P : TG_P \to TR_{M(P)}.$$

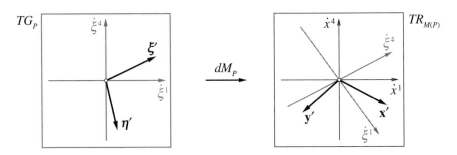

But all the laws of special relativity are consequences of a single fact: dM_P preserves inner products. That is, if $\boldsymbol{\xi}'$ and $\boldsymbol{\eta}'$ are two tangent vectors in TG_P, while $\mathbf{x}' = dM_P\boldsymbol{\xi}'$ and $\mathbf{y}' = dM_P\boldsymbol{\eta}'$ are the corresponding vectors in $TR_{M(P)}$, then

$$\underset{\text{in } TG_P}{\boldsymbol{\xi}' \cdot \boldsymbol{\eta}'} = \underset{\text{in } TR_{M(P)}}{\mathbf{x}' \cdot \mathbf{y}'}.$$

The inner product in TG_P is defined by the metric $\Gamma_P = (\gamma_{ij}(P))$, while in $TR_{M(P)}$ it is defined by $G_{M(P)} = (g_{kl}(M(P)))$. (Yes, notation is bumping into itself here: The same letter G stands for the Greek coordinate frame and the Roman metric. The context should make it clear which is which.) Therefore, we can write the inner products as matrix multiplications:

$$\begin{aligned}(\boldsymbol{\xi}')^t \Gamma_P \boldsymbol{\eta}' &= (\mathbf{x}')^t G_{M(P)} \mathbf{y}' \\ &= (dM_P \boldsymbol{\xi}')^t G_{M(P)} dM_P \boldsymbol{\eta}' \\ &= (\boldsymbol{\xi}')^t (dM_P)^t G_{M(P)} dM_P \boldsymbol{\eta}'.\end{aligned}$$

Hence

$$(\boldsymbol{\xi}')^t \left[\Gamma_P - (dM_P)^t G_{M(P)} dM_P\right] \boldsymbol{\eta}' = 0.$$

Since $\boldsymbol{\xi}'$ and $\boldsymbol{\eta}'$ are arbitrary, the expression in square brackets must be zero; thus

$$\Gamma_P = (dM_P)^t G_{M(P)} dM_P.$$

This is the relation between metrics that is induced by the transformation $M : G \to R$. Compare this with the simple case of two inertial frames G and R that each use the standard Minkowski metric

$$J_{1,3} = \begin{pmatrix} 1 & 0 & 0 & 0 \\ 0 & -1 & 0 & 0 \\ 0 & 0 & -1 & 0 \\ 0 & 0 & 0 & -1 \end{pmatrix}.$$

If $L : G \to R$, then

$$J_{1,3} = L^t J_{1,3} L.$$

The equation $\Gamma_P = (dM_P)^t G_{M(P)} dM_P$ we just found expresses the metric on TG_P in terms of the metric on $TR_{M(P)}$. This is the reverse of the way dM_P itself acts, but it is a simple matter to "invert" the relation and express the metric in $TR_{M(P)}$ in terms of

The relation between metrics: matrix form

the one in TG_P:

$$G_{M(P)} = \left(dM_P^{-1}\right)^t \Gamma_P \, dM_P^{-1}.$$

(We have used the fact that the transpose of the inverse of a matrix equals the inverse of its transpose.)

The relation between metrics: coordinate form

Now consider how the relation between Γ_P and $G_{M(P)}$ is expressed in terms of coordinates. Suppose

$$\xi' = \frac{d\xi^i}{dt}, \qquad \eta' = \frac{d\eta^j}{dt}$$

in TG_P and

$$\mathbf{x}' = \frac{dx^k}{dt} = \frac{\partial x^k}{\partial \xi^i}\frac{d\xi^i}{dt}, \qquad \mathbf{y}' = \frac{dy^l}{dt} = \frac{\partial x^l}{\partial \xi^j}\frac{d\eta^j}{dt}$$

in $TR_{M(P)}$. (The expression for $\mathbf{y}'$ is derived by the multivariable chain rule from the equation $\mathbf{y}(t) = (x^l(\eta^j(t)))$; $\mathbf{x}'$ is similar.) Our basic assumption, that the inner products are equal, then takes this form:

$$\gamma_{ij}\frac{d\xi^i}{dt}\frac{d\eta^j}{dt} = g_{kl}\frac{\partial x^k}{\partial \xi^i}\frac{d\xi^i}{dt}\frac{\partial x^l}{\partial \xi^j}\frac{d\eta^j}{dt}.$$

By rearranging factors we get

$$\left[\gamma_{ij} - g_{kl}\frac{\partial x^k}{\partial \xi^i}\frac{\partial x^l}{\partial \xi^j}\right]\frac{d\xi^i}{dt}\frac{d\eta^j}{dt} = 0.$$

Again, since the vectors $d\xi^i/dt$ and $d\eta^j/dt$ are arbitrary, we conclude that the expression in square brackets must be identically zero; thus,

$$\gamma_{ij} = g_{kl}\frac{\partial x^k}{\partial \xi^i}\frac{\partial x^l}{\partial \xi^j}.$$

This is the coordinate form of $\Gamma = dM^t G \, dM$; it involves the matrix $dM = (\partial x/\partial \xi)$. The inverse relation involves the inverse matrix $dM^{-1} = (\partial \xi/\partial x)$:

$$g_{kl} = \gamma_{ij}\frac{\partial \xi^i}{\partial x^k}\frac{\partial \xi^j}{\partial x^l}.$$

Covariance and Contravariance

At this point we pause to recall the *principle of covariance* introduced in Section 3.1 as an aspect of special relativity. The principle says that meaningful physical laws will be expressed in terms of covariant quantities. The value of a covariant quantity depends on the coordinate frame in which it is measured, and when we move from one frame to another, the value transforms in the same way as the coordinates themselves. (Roughly speaking, we can take *covariant* to mean "varies the same way as the coordinates.")

Covariant quantities

The principle becomes even more important in general relativity, where we allow arbitrary coordinate frames. However, we need to explain carefully what "varies with the coordinates" means here, because the coordinate transformations are generally nonlinear. In fact, most of the physical quantities we shall consider are defined on the tangent spaces TG_P and $TR_{M(P)}$, whose coordinates are related by *linear* transformations $dM_P : TG_P \to TR_{M(P)}$. This is the key: A *generally* covariant quantity is one whose transformation law shall be expressed in terms of dM_P.

Generally covariant quantities

A full discussion of generally covariant quantities involves tensors, which we consider in the following section. In the meantime we focus on just a single aspect of what we have already found: the difference between the ways that vectors and metrics transform. One uses dM_P; the other, dM_P^{-1}:

$$\text{transformation of vectors:} \quad \frac{d\xi^i}{dt} \xrightarrow{\left(\frac{\partial x^k}{\partial \xi^i}\right)} \frac{dx^k}{dt},$$

$$\text{transformation of the metric:} \quad \gamma_{ij} \xrightarrow{\left(\frac{\partial \xi^i}{\partial x^k}\right)} g_{kl}.$$

Note that we are not attempting here to indicate *how* γ_{ij} is transformed into g_{kl}, just that the matrix $dM_P^{-1} = (\partial \xi^i / \partial x^k)$ is involved.

All quantities that can be expressed in terms of the induced coordinates in TG_P and $TR_{M(P)}$ are ultimately expressible in terms of the induced bases for these spaces. Therefore, the most funda-

How do bases transform?

Chapter 6 Intrinsic Geometry

mental transformation is the one between the bases themselves, so we ask, *How do basis elements transform?*

We first need to make this question precise. Suppose we denote the bases as follows:

$$TG_P : \{\varepsilon_1, \varepsilon_2, \varepsilon_3, \varepsilon_4\}, \qquad TR_{M(P)} : \{\mathbf{e}_1, \mathbf{e}_2, \mathbf{e}_3, \mathbf{e}_4\}.$$

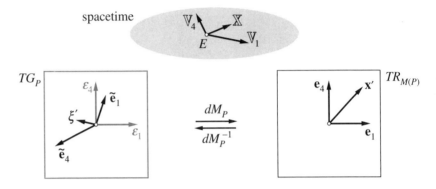

Since these bases lie in different vector spaces, we cannot connect them directly by a matrix. However, each $\mathbf{e}_k$ represents some vector $\mathbb{V}_k$ in spacetime, and this vector has a realization $\widetilde{\mathbf{e}}_k = dM_P^{-1}(\mathbf{e}_k)$ in TG_P. The vectors $\{\widetilde{\mathbf{e}}_1, \widetilde{\mathbf{e}}_2, \widetilde{\mathbf{e}}_3, \widetilde{\mathbf{e}}_4\}$ constitute a second basis for TG_P (because dM_P^{-1} is invertible), and they *are* directly comparable to $\{\varepsilon_1, \varepsilon_2, \varepsilon_3, \varepsilon_4\}$. In other words, there is a matrix α_k^i that converts the old basis to the new:

$$\widetilde{\mathbf{e}}_k = \alpha_k^i \varepsilon_i.$$

This is the transformation we seek.

In preparation, it is helpful to compare the coordinates of an arbitrary vector $\boldsymbol{\xi}'$ in TG_P with respect to the two bases. We assume that $(\dot{\xi}^i)$ are the coordinates of $\boldsymbol{\xi}'$ with respect to the basis ε_i. Let $\mathbf{x}' = dM_P(\boldsymbol{\xi}')$, as usual, and suppose $\dot{x}^k$ are the coordinates of $\mathbf{x}'$ with respect to $\mathbf{e}_k$. Then

$$\dot{\xi}^i \varepsilon_i = \boldsymbol{\xi}' = dM_P^{-1}(\mathbf{x}') = dM_P^{-1}(\dot{x}^k \mathbf{e}_k) = \dot{x}^k dM_P^{-1}(\mathbf{e}_k) = \dot{x}^k \widetilde{\mathbf{e}}_k.$$

Therefore, the coordinates of $\boldsymbol{\xi}'$ with respect to $\widetilde{\mathbf{e}}_k$ are the same as the coordinates of $\mathbf{x}' = dM_P(\boldsymbol{\xi}')$ with respect to $\mathbf{e}_k$. This is not surprising if we think of $\boldsymbol{\xi}'$ and $\mathbf{x}'$ as the realizations of the same spacetime vector $\mathbb{X}$ in TG_P and $TR_{M(P)}$, respectively.

Proposition 6.4 $\quad \widetilde{\mathbf{e}}_k = \dfrac{\partial \xi^i}{\partial x^k} \boldsymbol{\varepsilon}_i.$

PROOF: In the equation $\dot{\xi}^i \boldsymbol{\varepsilon}_i = \dot{x}^k \widetilde{\mathbf{e}}_k$, use dM_P^{-1} to express $\dot{\xi}^i$ in terms of $\dot{x}^k$:

$$\dot{x}^k \frac{\partial \xi^i}{\partial x^k} \boldsymbol{\varepsilon}_i = \dot{x}^k \widetilde{\mathbf{e}}_k.$$

Since the coordinates $\dot{x}^k$ are arbitrary, the result follows.

END OF PROOF

If we blur the distinction between $\widetilde{\mathbf{e}}_k$ and $\mathbf{e}_k$, then we can add a third transformation pattern to the two we already have:

Covariant and contravariant quantities

$$\text{basis:} \quad \boldsymbol{\varepsilon}_i \xrightarrow{\left(\frac{\partial \xi^i}{\partial x^k}\right)} \mathbf{e}_k,$$

$$\text{vectors:} \quad \frac{d\xi^i}{dt} \xrightarrow{\left(\frac{\partial x^k}{\partial \xi^i}\right)} \frac{dx^k}{dt},$$

$$\text{metric:} \quad \gamma_{ij} \xrightarrow{\left(\frac{\partial \xi^i}{\partial x^k}\right)} g_{kl}.$$

We say that the metric is *covariant* because it transforms in the same sense as the basis (that is, with the same matrix). By contrast, we say that the tangent vectors are *contravariant* because they transform in the opposite sense (that is, with the inverse of that matrix). We take the transformation of the basis as the standard; we shall call any quantity that transforms in the same sense as the basis **covariant** and any that transforms in the opposite sense **contravariant**.

Notice that the covariant quantities have *sub*scripts, while the contravariant have *super*scripts. This is invariably true and is a consequence of two things. First, we write the objects that set the standard—namely, basis elements $\mathbf{e}_k$—with subscripts. Second,

Covariant and contravariant indices

the summation convention then forces the coordinates of a vector (the paradigmatic example of a contravariant object) to be written with superscripts: $\mathbf{x}' = x^k \mathbf{e}_k$.

Exercises

1. Suppose $M : G \to R$ is a linear map: $x^k = a^k_i \xi^i$. Show that $dM_P = M$ at every point P in G.

2. Carry out the following tasks for each of the maps $M : G \to R$ given below. The first defines polar coordinates $M : (r, \theta) \to (x, y)$ in the plane, and the second defines spherical coordinates $M : (r, \varphi, \theta) \to (x, y, z)$ in space.

$$M : \begin{cases} x = r\cos\theta, \\ y = r\sin\theta; \end{cases} \qquad M : \begin{cases} x = r\sin\varphi\cos\theta, \\ y = r\sin\varphi\sin\theta, \\ z = r\cos\varphi. \end{cases}$$

   (a) Determine the inverse $M^{-1} : R \to G$ of M explicitly, as a pair of functions $r = r(x, y)$, $\theta = \theta(x, y)$ in the first case and as a triple $r = r(x, y, z)$, $\varphi = \varphi(x, y, z)$, $\theta = \theta(x, y, z)$ in the second.

   (b) Find the differential dM_P of M and calculate its inverse $(dM_P)^{-1}$ directly. What is $\det dM_P$?

   (c) Find the differential dM_Q^{-1} of M^{-1} and show that dM_Q^{-1} equals the inverse $(dM_P)^{-1}$ you calculated in part (b) when $Q = M(P)$. (This is not immediately obvious, because the two matrices are expressed in terms of the two different sets of variables. You must use M or M^{-1}, as appropriate, to convert one set of variables to the other.)

3. Let $M : G \to R$ be the polar-coordinate map defined in the previous exercise, and let $P = (r, \theta) = (2, \pi/3)$.

   (a) Calculate dM_P and sketch the image of the $(\dot{r}, \dot{\theta})$ grid under the linear map $dM_P : (\dot{r}, \dot{\theta}) \to (\dot{x}, \dot{y})$. Your sketch should make it clear that dM_P stretches the $(\dot{r}, \dot{\theta})$-plane to double its size in the $\dot{\theta}$-direction and rotates it by $\pi/3$ radians.

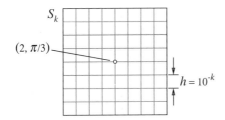

(b) Now consider a sequence of small squares S_k ($k = 1, 2, 3, 4$) centered at $(r, \theta) = (2, \pi/3)$ in the (r, θ)-plane. Let S_k measure $8h$ units, where $h = 10^{-k}$, and let it be partitioned by a grid of $h \times h$ subsquares. Sketch the image $M(S_k)$ of this grid in the (x, y)-plane, rescaling the various S_k so they all are the same size in your sketches. Call these *microscopic views* of the action of M near $P = (2, \pi/3)$ under the different magnifications $1/h = 10^k$. Compare the microscopic views to the image of dM_P; your sketches should show that the view of the differential resembles the microscopic view more and more closely as the magnification increases.

The differential gives a microscopic view of a map

4. Suppose $M : G \to R$ is the polar coordinate map and the point $P = (r, \theta)$ is arbitrary. Prove that the differential $dM_P : (\dot{r}, \dot{\theta}) \to (\dot{x}, \dot{y})$ stretches the $(\dot{r}, \dot{\theta})$-plane by the factor r in the $\dot{\theta}$-direction and rotates it by θ radians.

The remaining exercises concern the maps $M : G \to R$, $M^{-1} : R \to G$ that relate G and R when G undergoes constant linear acceleration α in the direction of R's z-axis (cf. Section 4.2):

$$M : \begin{cases} t = \dfrac{e^{\alpha \zeta}}{\alpha} \sinh \alpha \tau, \\ z = \dfrac{e^{\alpha \zeta}}{\alpha} \cosh \alpha \tau; \end{cases} \qquad M^{-1} : \begin{cases} \tau = \dfrac{1}{\alpha} \tanh^{-1} \left(\dfrac{t}{z} \right), \\ \zeta = \dfrac{1}{2\alpha} \ln \left(\alpha^2 (z^2 - t^2) \right). \end{cases}$$

5. (a) Show that $dM_P = e^{\alpha \zeta} H_{\alpha \tau}$ at an arbitrary point $P = (\tau, \zeta)$; H_u is hyperbolic rotation by the hyperbolic angle u in the Minkowski plane.

   (b) Calculate $\det dM_P$ and the inverse $(dM_P)^{-1}$.

(c) Calculate dM_Q^{-1} and $\det dM_Q^{-1}$ at an arbitrary point $Q = (t, z)$. Show that $dM_Q^{-1} = (dM_P)^{-1}$ when $Q = M(P)$.

6. (a) Take $\alpha = 1$ and sketch the image of the $(\dot{\tau}, \dot{\zeta})$ coordinate grid under the map $dM_P : (\dot{\tau}, \dot{\zeta}) \to (\dot{t}, \dot{z})$ when $P = (\tau, \zeta) = (-2, 1)$.

(b) Construct the small grids S_k ($k = 1, 2, 3, 4$) as in Exercise 3(b), but center them at $(\tau, \zeta) = (-2, 1)$. Still using $\alpha = 1$, sketch the image $M(S_k)$ of this grid in the (t, z)-plane, rescaling the various S_k so they all are the same size in your sketches. Compare these microscopic views of the action of M near $P = (-2, 1)$ under the different magnifications $1/h = 10^k$ to the image of the linear map dM_P.

7. Assume that R is an inertial frame whose metric is given by the standard matrix $J_{1,1}$; that is, $g_{11} = 1$, $g_{12} = g_{21} = 0$, $g_{22} = -1$. Use the map M to define the metric γ_{ij} on G; thus if $\Gamma = (\gamma_{ij})$, then $\Gamma_P = dM_P^t J_{1,1} dM_P$. (Note: Define indexed coordinates as follows: $x^1 = t$, $x^2 = z$, $\xi^1 = \tau$, $\xi^2 = \zeta$.)

8. (a) Show that the hyperbolic angle between two vectors $\boldsymbol{\xi}$ and $\boldsymbol{\eta}$ on TG_P is the same whether you use the induced metric Γ_P or the ordinary Minkowski metric $J_{1,1}$. In this way G resembles the hyperbolic plane (Proposition 6.2), and the explanation is the same. The induced metric is a scalar multiple of the Minkowski metric: $\Gamma_P = \lambda J_{1,1}$, where $\lambda = \lambda(\tau, \zeta)$ depends on the point $P = (\tau, \zeta)$. What is the function λ?

(b) Use your result in part (a) to prove that the lightlike vectors $\boldsymbol{\xi} = (\dot{\tau}, \dot{\zeta})$ in TG_P are the same as for the Minkowski metric—namely, $\dot{\zeta} = \pm\dot{\tau}$.

9. Using the induced metric γ_{ij} for G from the previous exercise, determine the inverse γ^{kl}, the Christoffel symbols $\Gamma_{ij,h}$ and Γ_{ij}^k, the Riemann tensors R_{ijk}^l and R_{hijk}, and $K = R_{1212}/\gamma$.

10. (a) Show that the geodesic equations in G are

$$\frac{d^2\tau}{du^2} + 2\alpha \frac{d\tau}{du}\frac{d\zeta}{du} = 0, \quad \frac{d^2\zeta}{du^2} + \alpha \left(\frac{d\tau}{du}\right)^2 + \alpha \left(\frac{d\zeta}{du}\right)^2 = 0.$$

(b) Prove that horizontal and vertical translates of a geodesic are also geodesics. That is, if $\boldsymbol{\xi}(u) = (\tau(u), \zeta(u))$ is a geodesic, then so is

$$\boldsymbol{\xi}_{p,q}(u) = (p + \tau(u), q + \zeta(u)) \quad \text{for any } p \text{ and } q.$$

(c) Explain why the image $M(\boldsymbol{\xi}(u))$ of any geodesic $\boldsymbol{\xi}(u)$ in G is a geodesic in R.

11. (a) Show that the following curves are geodesics in G. Sketch them and several representative horizontal and vertical translates:

$$\boldsymbol{\xi}_{\text{ti}}(u) = \left(\frac{1}{\alpha} \tanh^{-1} \alpha u, \frac{1}{2\alpha} \ln(1 - \alpha^2 u^2)\right),$$

$$\boldsymbol{\xi}_{\text{sp}}(u) = \left(\frac{1}{\alpha} \tanh^{-1} \frac{1}{\alpha u}, \frac{1}{2\alpha} \ln(\alpha^2 u^2 - 1)\right).$$

(b) Show that $\boldsymbol{\xi}_{\text{ti}}$ and $\boldsymbol{\xi}_{\text{sp}}$ are timelike and spacelike curves, respectively. Determine proper time along $\boldsymbol{\xi}_{\text{ti}}$.

(c) To which geodesics in R do $\boldsymbol{\xi}_{\text{ti}}$ and $\boldsymbol{\xi}_{\text{sp}}$ and their translates correspond?

12. (a) Determine the arc-length parametrization of the coordinate line $\tau = k$; that is, determine $s(u)$ such that $\boldsymbol{\eta}(s) = (k, s(u))$ has $\|\boldsymbol{\eta}'\| \equiv 1$ in G.

(b) Show that $\boldsymbol{\eta}(s)$ is a geodesic in G.

(c) Can the coordinate line $\zeta = $ constant be given a parametrization that makes it a geodesic? Explain your position.

6.5 Tensors

We come back to the central question: How can observers synthesize their different views of events in spacetime into a coherent physical theory? Einstein says that observers should express themselves in terms of generally covariant quantities, which, for Einstein, are *tensors*. He devotes nearly half of his fundamental paper on general relativity to defining tensors and developing

Tensors are generally covariant quantities

their properties, in the part called "Mathematical Aids to the Formulation of Generally Covariant Equations" ([10], pages 120–142).

Basically, a tensor is a multi-index quantity that is defined in each coordinate frame and transforms linearly in a contra- or covariant way as we move from one frame to another. To make this more precise, we consider in the usual way a pair of coordinate frames for the same portion of spacetime:

$$G \underset{M^{-1}}{\overset{M}{\rightleftarrows}} R, \qquad \xi^i = \xi^i(x^k), \quad x^k = x^k(\xi^i),$$

$$TG_P \underset{dM_P^{-1}}{\overset{dM_P}{\rightleftarrows}} TR_{M(P)}, \qquad \dot{\xi}^i = \frac{\partial \xi^i}{\partial x^k}\dot{x}^k, \quad \dot{x}^k = \frac{\partial x^k}{\partial \xi^i}\dot{\xi}^i.$$

Definition 6.2 A **tensor of type** (p, q), denoted by

$$\overset{G}{T}{}^{i_1 \cdots i_p}_{j_1 \cdots j_q} \longleftrightarrow \overset{R}{T}{}^{k_1 \cdots k_p}_{l_1 \cdots l_q},$$

is a multi-index quantity defined in each coordinate system that transforms according to the rule

$$\overset{R}{T}{}^{k_1 \cdots k_p}_{l_1 \cdots l_q} = \overset{G}{T}{}^{i_1 \cdots i_p}_{j_1 \cdots j_q} \frac{\partial x^{k_1}}{\partial \xi^{i_1}} \cdots \frac{\partial x^{k_p}}{\partial \xi^{i_p}} \frac{\partial \xi^{j_1}}{\partial x^{l_1}} \cdots \frac{\partial \xi^{j_q}}{\partial x^{l_q}}.$$

*This tensor is **contravariant of rank** p and **covariant of rank** q. Tensors of type $(1, 0)$ and $(0, 1)$ are called vectors.*

We say that a multi-index quantity is **tensorial** if it transforms by this rule. The pattern is simple: For each covariant index, multiply by a copy of the same matrix that transforms bases; for each contravariant index, multiply by a copy of the inverse matrix.

Language and notation

When we want to call attention to the fact that a tensor is defined on an open set and its value varies from point to point, we shall refer to it as a **tensor field**. Sometimes we shall use corresponding letters in the Greek and Roman alphabets for the names of a tensor in the two coordinate frames; other times we shall follow the practice shown above and use the same letter but put the name of the coordinate frame over it. Also, we can make the notation clearer by collapsing the various multi-indices into single capital letters. Thus let $I = \{i_1 \cdots i_p\}$ and similarly for J, K,

NOTE THAT, ALL TENSORS — INCLUDING THOSE WITH POSITIVE CONTRAVARIANT RANK — ARE "GENERALLY COVARIANT" WHEN WE USE THE TERM IN THE GENERAL SENSE.

and L. In these terms

$$\overset{G}{T^I_J} \longleftrightarrow \overset{R}{T^K_L} \quad \text{is a tensor if} \quad \overset{R}{T^K_L} = \overset{G}{T^I_J} \frac{\partial x^K}{\partial \xi^I} \frac{\partial \xi^I}{\partial x^L}.$$

Here $\partial x^K/\partial \xi^I$ denotes the product of the p matrices that involve the separate indices of K and I.

We already have two familiar examples of tensors:

- The *tangents* to a curve $\dfrac{d\xi^i}{dt} \longleftrightarrow \dfrac{dx^k}{dt}$ form a contravariant vector.
- The *metric* $\gamma_{ij} \longleftrightarrow g_{kl}$ forms a covariant tensor of rank 2.

The following propositions show that many of the multi-index quantities we introduced in the study of surfaces are tensorial. There are, however, important exceptions, and we shall explore some of them, too.

Proposition 6.5 *A function $\varphi \longleftrightarrow f$ defined consistently in each coordinate frame is a tensor of type* $(0, 0)$.

PROOF: At the outset we know that f is a function of x^k and φ is a function of ξ^i. For these to be "defined consistently," $f(x^k(\xi^i)) = \varphi(\xi^i)$. This is the transformation law; it involves no matrices because f and φ have no indices. END OF PROOF

Proposition 6.6 *The gradient of a function $\varphi \longleftrightarrow f$ is a covariant vector* $\dfrac{\partial \varphi}{\partial \xi^i} \longleftrightarrow \dfrac{\partial f}{\partial x^k}$.

PROOF: We know that $f(x^k(\xi^i)) = \varphi(\xi^i)$, so differentiation with respect to ξ^i gives

$$\frac{\partial f}{\partial x^k} \frac{\partial x^k}{\partial \xi^i} = \frac{\partial \varphi}{\partial \xi^i} \quad \text{or} \quad \frac{\partial f}{\partial x^k} = \frac{\partial \varphi}{\partial \xi^i} \frac{\partial \xi^i}{\partial x^k}.$$

This is precisely the tensor property for a covariant vector.

END OF PROOF

Proposition 6.7 *The contraction of any tensor $\overset{G}{T^I_J} \longleftrightarrow \overset{R}{T^K_L}$ of type (p, q) is a new tensor of type $(p-1, q-1)$.*

PROOF: Suppose we want to contract the mth upper index with the nth lower index. Let $i = i_m$ and redefine I as $\{i_1 \cdots \widehat{i_m} \cdots i_p\}$, where the carat indicates an element that is to be removed and placed after the rest of the indices. Define J, K, L and j, k, l similarly. We can then write the transformation law of the given tensor as

$$\overset{R}{T}{}^{K,k}_{L,l} = \overset{G}{T}{}^{I,i}_{J,j} \frac{\partial x^K}{\partial \xi^I} \frac{\partial x^k}{\partial \xi^i} \frac{\partial \xi^J}{\partial x^L} \frac{\partial \xi^j}{\partial x^l}.$$

To carry out the contraction we set $k = l = \alpha$ and sum over α; then

$$\overset{R}{T}{}^{K}_{L} = \overset{R}{T}{}^{K,\alpha}_{L,\alpha} = \overset{G}{T}{}^{I,i}_{J,j} \frac{\partial x^K}{\partial \xi^I} \underbrace{\frac{\partial x^\alpha}{\partial \xi^i}} \frac{\partial \xi^J}{\partial x^L} \underbrace{\frac{\partial \xi^j}{\partial x^\alpha}}.$$

The matrices underscored by braces are inverses; their product is δ^j_i. If we replace this product by δ^j_i, the equation becomes

$$\overset{R}{T}{}^{K}_{L} = \overset{G}{T}{}^{I,i}_{J,j} \delta^j_i \frac{\partial x^K}{\partial \xi^I} \frac{\partial \xi^J}{\partial x^L} = \overset{G}{T}{}^{I,i}_{J,i} \frac{\partial x^K}{\partial \xi^I} \frac{\partial \xi^J}{\partial x^L} = \overset{G}{T}{}^{I}_{J} \frac{\partial x^K}{\partial \xi^I} \frac{\partial \xi^J}{\partial x^L},$$

proving that $\overset{R}{T}{}^{K}_{L} \longleftrightarrow \overset{G}{T}{}^{I}_{J}$ is a tensor. END OF PROOF

Proposition 6.8 *The inverse of the metric,* $\Gamma^{-1} = (\gamma^{jm}) \longleftrightarrow (g^{kl}) = G^{-1}$, *is a contravariant tensor of rank 2.*

PROOF: In the exercises you are asked to prove this directly in terms of the components γ^{ij} and g^{kl}. Here is a different approach using matrices. In matrix form, the transformation law of the metric tensor $\Gamma \longleftrightarrow G$ is

$$G = (dM_P^{-1})^t \, \Gamma \, dM_P^{-1}.$$

Therefore,

$$G^{-1} = dM_P \, \Gamma^{-1} \, dM_P^t,$$

which is sufficient to demonstrate that $\Gamma^{-1} \longleftrightarrow G^{-1}$ is a contravariant tensor of rank 2. However, it is instructive to see the matrix product in coordinate form. We have

$$\begin{pmatrix} g^{11} & \cdots & g^{14} \\ \vdots & & \vdots \\ g^{41} & \cdots & g^{44} \end{pmatrix} = \begin{pmatrix} \frac{\partial x^1}{\partial \xi^1} & \cdots & \frac{\partial x^1}{\partial \xi^4} \\ \vdots & & \vdots \\ \frac{\partial x^4}{\partial \xi^1} & \cdots & \frac{\partial x^4}{\partial \xi^4} \end{pmatrix} \begin{pmatrix} \gamma^{11} & \cdots & \gamma^{14} \\ \vdots & & \vdots \\ \gamma^{41} & \cdots & \gamma^{44} \end{pmatrix} \begin{pmatrix} \frac{\partial x^1}{\partial \xi^1} & \cdots & \frac{\partial x^4}{\partial \xi^1} \\ \vdots & & \vdots \\ \frac{\partial x^1}{\partial \xi^4} & \cdots & \frac{\partial x^4}{\partial \xi^4} \end{pmatrix},$$

and this implies $g^{kl} = \gamma^{ij} \frac{\partial x^k}{\partial \xi^i} \frac{\partial x^l}{\partial \xi^j}$. END OF PROOF

We call $G^{-1} \longleftrightarrow \Gamma^{-1}$ the **contravariant metric tensor**. One of the ways we use the two metric tensors is to raise and lower indices. The following propositions show that these processes preserve tensors; the first proposition is a simple case, and the second is the general result.

Proposition 6.9 *Suppose $\alpha^i \longleftrightarrow a^k$ is a contravariant vector. If we lower indices by $\alpha_j = \gamma_{ij}\alpha^i$ and $a_l = g_{kl}a^k$, then $\alpha_j \longleftrightarrow a_l$ is a covariant vector.*

PROOF: We are given that $a^k = \alpha^i \frac{\partial x^k}{\partial \xi^i}$. Therefore,

$$\alpha_j = \gamma_{ij}\alpha^i = g_{kl}\frac{\partial x^k}{\partial \xi^i}\frac{\partial x^l}{\partial \xi^j}\alpha^i = g_{kl}\frac{\partial x^l}{\partial \xi^j}a^k = a_l\frac{\partial x^l}{\partial \xi^j}. \quad \text{END OF PROOF}$$

Proposition 6.10 *Suppose $\overset{G}{T}{}^{I,i}_{J,j} \longleftrightarrow \overset{R}{T}{}^{K,k}_{L,l}$ is a tensor of type (p, q). Then the multi-index quantities*

$$\overset{G}{T}{}^{I,ih}_{J} = \gamma^{jh}\overset{G}{T}{}^{I,i}_{J,j} \longleftrightarrow g^{lm}\overset{R}{T}{}^{K,k}_{L,l} = \overset{R}{T}{}^{K,km}_{L},$$

$$\overset{G}{T}{}^{I}_{J,jh} = \gamma_{ih}\overset{G}{T}{}^{I,i}_{J,j} \longleftrightarrow g_{km}\overset{R}{T}{}^{K,k}_{L,l} = \overset{R}{T}{}^{K}_{L,lm}$$

that are obtained by raising or lowering an index are tensors of type $(p+1, q-1)$ and $(p-1, q+1)$, respectively.

PROOF: See the exercises.

Tensors are not preserved under differentiation

We turn now to some examples of multi-index quantities that fail to transform as tensors. Many of the tensors we use involve derivatives of other quantities; a covariant index is frequently due to differentiation with respect to a coordinate variable. It is therefore natural to ask, *Is the derivative of a tensor again a tensor?* The answer is no, and we can see this in the simple example of a contravariant vector $\alpha^i \longleftrightarrow a^k$. The transformation law of the vector is

$$a^k(x^l(\xi^j)) = \alpha^i(\xi^j)\frac{\partial x^k}{\partial \xi^i}.$$

Now differentiate with respect to ξ^j:

$$\frac{\partial a^k}{\partial x^l}\frac{\partial x^l}{\partial \xi^j} = \frac{\partial \alpha^i}{\partial \xi^j}\frac{\partial x^k}{\partial \xi^i} + \alpha^i \frac{\partial^2 x^k}{\partial \xi^j \partial \xi^i}.$$

In standard form this is

$$\frac{\partial a^k}{\partial x^l} = \underbrace{\frac{\partial \alpha^i}{\partial \xi^j}\frac{\partial x^k}{\partial \xi^i}\frac{\partial \xi^j}{\partial x^l}}_{\text{tensorial}} + \underbrace{\alpha^i \frac{\partial^2 x^k}{\partial \xi^j \partial \xi^i}\frac{\partial \xi^j}{\partial x^l}}_{\text{nontensorial}},$$

and we see that differentiation of the product $\alpha^i \cdot \partial x^k / \partial \xi^i$ has created a "nontensorial" second derivative term in the transformation law.

Of course, if the coordinate transformation $x^k = x^k(\xi^i)$ were *linear*, then all the second derivatives would be identically zero, so the nontensorial part of the transformation would vanish. Hence $\partial a^k / \partial x^l \longleftrightarrow \partial \alpha^i / \partial \xi^j$ would indeed be a tensor if we considered only linear transformations between coordinate frames. However, general relativity requires nonlinear transformations to deal with gravity, so we can say definitively that tensors are not preserved under differentiation.

Differentiating the metric tensor

In general, when we differentiate a tensor, each index spawns a separate nontensorial second derivative term in the transformation law. We illustrate this with the metric tensor $\gamma_{ij} \longleftrightarrow g_{kl}$. Since we are taking the Greek variables as fundamental, the calculations will be clearer if we express the Greek form in terms of the Roman, as follows:

$$\gamma_{ij}(\xi^k) = g_{pq}(x^r(\xi^k)) \frac{\partial x^p}{\partial \xi^i} \frac{\partial x^q}{\partial \xi^j}.$$

Then

$$\frac{\partial \gamma_{ij}}{\partial \xi^k} = \frac{\partial g_{pq}}{\partial x^r} \frac{\partial x^p}{\partial \xi^i} \frac{\partial x^q}{\partial \xi^j} \frac{\partial x^r}{\partial \xi^k} + g_{pq} \underbrace{\frac{\partial^2 x^p}{\partial \xi^k \partial \xi^i} \frac{\partial x^q}{\partial \xi^j}}_{\text{nontensorial}} + g_{pq} \underbrace{\frac{\partial^2 x^q}{\partial \xi^k \partial \xi^j} \frac{\partial x^p}{\partial \xi^i}}_{\text{nontensorial}}.$$

One consequence is that the Christoffel symbols of the first kind are not tensors, because they are sums of derivatives of the metric tensor. Since the Christoffel symbols involve three such derivatives, we might expect six nontensorial terms in their transformation law. However, there is only one, as the following proposition shows.

Christoffel symbols

Proposition 6.11 $\overset{G}{\Gamma}_{ij,k} = \overset{R}{\Gamma}_{pq,r} \frac{\partial x^p}{\partial \xi^i} \frac{\partial x^q}{\partial \xi^j} \frac{\partial x^r}{\partial \xi^k} + g_{pq} \frac{\partial^2 x^p}{\partial \xi^i \partial \xi^j} \frac{\partial x^q}{\partial \xi^k}.$

PROOF: This involves a lot of "index bookkeeping," but it is worth going through the details to see what cancellations occur in the nontensorial part.

$$2\overset{G}{\Gamma}_{ij,k} = \frac{\partial \gamma_{ik}}{\partial \xi^j} + \frac{\partial \gamma_{jk}}{\partial \xi^i} - \frac{\partial \gamma_{ij}}{\partial \xi^k}$$

$$= \frac{\partial g_{pr}}{\partial x^q} \frac{\partial x^p}{\partial \xi^i} \frac{\partial x^r}{\partial \xi^k} \frac{\partial x^q}{\partial \xi^j} + g_{pr} \underbrace{\frac{\partial^2 x^p}{\partial \xi^j \partial \xi^i} \frac{\partial x^r}{\partial \xi^k}}_{A} + g_{pr} \underbrace{\frac{\partial^2 x^r}{\partial \xi^k \partial \xi^j} \frac{\partial x^p}{\partial \xi^i}}_{B}$$

$$+ \frac{\partial g_{qr}}{\partial x^p} \frac{\partial x^q}{\partial \xi^j} \frac{\partial x^r}{\partial \xi^k} \frac{\partial x^p}{\partial \xi^i} + g_{qr} \underbrace{\frac{\partial^2 x^q}{\partial \xi^i \partial \xi^j} \frac{\partial x^r}{\partial \xi^k}}_{A} + g_{qr} \underbrace{\frac{\partial^2 x^r}{\partial \xi^i \partial \xi^k} \frac{\partial x^q}{\partial \xi^j}}_{C}$$

$$- \frac{\partial g_{pq}}{\partial x^r} \frac{\partial x^p}{\partial \xi^i} \frac{\partial x^q}{\partial \xi^j} \frac{\partial x^r}{\partial \xi^k} - g_{pq} \underbrace{\frac{\partial^2 x^p}{\partial \xi^k \partial \xi^i} \frac{\partial x^q}{\partial \xi^j}}_{C} - g_{pq} \underbrace{\frac{\partial^2 x^q}{\partial \xi^k \partial \xi^j} \frac{\partial x^p}{\partial \xi^i}}_{B}$$

$$= \left(\frac{\partial g_{pr}}{\partial x^q} + \frac{\partial g_{qr}}{\partial x^p} - \frac{\partial g_{pq}}{\partial x^r} \right) \frac{\partial x^p}{\partial \xi^i} \frac{\partial x^q}{\partial \xi^j} \frac{\partial x^r}{\partial \xi^k} + 2g_{pq} \frac{\partial^2 x^p}{\partial \xi^k \partial \xi^i} \frac{\partial x^q}{\partial \xi^j} \quad \longrightarrow \quad 2g_{pr} \frac{\partial^2 x^p}{\partial \xi^i \partial \xi^j} \frac{\partial x^q}{\partial \xi^k}$$

$$= 2\overset{R}{\Gamma}_{pq,r} \frac{\partial x^p}{\partial \xi^i} \frac{\partial x^q}{\partial \xi^j} \frac{\partial x^r}{\partial \xi^k} + 2g_{pq} \frac{\partial^2 x^p}{\partial \xi^i \partial \xi^j} \frac{\partial x^q}{\partial \xi^k}.$$

Chapter 6 Intrinsic Geometry

The two terms labeled B cancel; to see this, replace the dummy summation index r by q in the first term. The C terms also cancel: First use the symmetry of the tensor ($g_{qr} = g_{rq}$) and then make the replacement $r \mapsto p$. Similar replacements show that the two A terms are equal and that their sum can be written as in the last two lines.   END OF PROOF

Corollary 6.2 $\overset{G}{\Gamma}{}^h_{ij} = \overset{R}{\Gamma}{}^r_{pq} \dfrac{\partial x^p}{\partial \xi^i} \dfrac{\partial x^q}{\partial \xi^j} \dfrac{\partial \xi^h}{\partial x^r} + \dfrac{\partial^2 x^p}{\partial \xi^i \partial \xi^j} \dfrac{\partial \xi^h}{\partial x^p}.$

Geodesics are covariantly defined

The following theorem and corollary show that geodesics are covariantly defined—in spite of the fact that the nontensorial Christoffel symbols appear in the geodesic differential equations.

Theorem 6.5 *If $x^r(t) = x^r(\xi^h(t))$ is a curve, then*

$$\frac{d^2\xi^h}{dt^2} + \overset{G}{\Gamma}{}^h_{ij} \frac{d\xi^i}{dt} \frac{d\xi^j}{dt} \longleftrightarrow \frac{d^2 x^r}{dt^2} + \overset{R}{\Gamma}{}^r_{pq} \frac{dx^p}{dt} \frac{dx^q}{dt}$$

is a tensor (viz., a contravariant vector) defined on that curve.

PROOF: First we determine how the second derivatives transform. We have

$$\frac{dx^r}{dt} = \frac{\partial x^r}{\partial \xi^h} \frac{d\xi^h}{dt} \quad \text{and then} \quad \frac{d^2 x^r}{dt^2} = \frac{\partial x^r}{\partial \xi^h} \frac{d^2 \xi^h}{dt^2} + \frac{\partial^2 x^r}{\partial \xi^h \partial \xi^j} \frac{d\xi^h}{dt} \frac{d\xi^j}{dt}.$$

Now solve for $d^2\xi^h/dt^2$ (and make the replacement $h \mapsto i$ in the dummy summation index in the last term):

$$\frac{d^2\xi^h}{dt^2} = \frac{d^2 x^r}{dt^2} \frac{\partial \xi^h}{\partial x^r} - \frac{\partial^2 x^r}{\partial \xi^i \partial \xi^j} \frac{d\xi^i}{dt} \frac{d\xi^j}{dt} \frac{\partial \xi^h}{\partial x^r}.$$

There is a nontensorial term, and it appears with a minus sign. It will therefore cancel the corresponding term in the transformation law for the Christoffel symbols when we add:

$$\frac{d^2\xi^h}{dt^2} + \overset{G}{\Gamma}{}^h_{ij} \frac{d\xi^i}{dt} \frac{d\xi^j}{dt} = \frac{d^2 x^r}{dt^2} \frac{\partial \xi^h}{\partial x^r} - \frac{\partial^2 x^r}{\partial \xi^i \partial \xi^j} \frac{d\xi^i}{dt} \frac{d\xi^j}{dt} \frac{\partial \xi^h}{\partial x^r}$$

$$+ \overset{R}{\Gamma}{}^r_{pq} \frac{\partial x^p}{\partial \xi^i} \frac{\partial x^q}{\partial \xi^j} \frac{\partial \xi^h}{\partial x^r} \frac{d\xi^i}{dt} \frac{d\xi^j}{dt} + \frac{\partial^2 x^p}{\partial \xi^i \partial \xi^j} \frac{\partial \xi^h}{\partial x^p} \frac{d\xi^i}{dt} \frac{d\xi^j}{dt}$$

$$= \frac{d^2x^r}{dt^2}\frac{\partial \xi^h}{\partial x^r} + \overset{R}{\Gamma}{}^r_{pq}\frac{dx^p}{dt}\frac{dx^q}{dt}\frac{\partial \xi^h}{\partial x^r}$$

$$= \left(\frac{d^2x^r}{dt^2} + \overset{R}{\Gamma}{}^r_{pq}\frac{dx^p}{dt}\frac{dx^q}{dt}\right)\frac{\partial \xi^h}{\partial x^r}.$$

To simplify the term containing the Christoffel symbol in going from the second line to the third we used

$$\frac{\partial x^p}{\partial \xi^i}\frac{d\xi^i}{dt} = \frac{dx^p}{dt}, \qquad \frac{\partial x^q}{\partial \xi^j}\frac{d\xi^j}{dt} = \frac{dx^q}{dt}. \qquad \text{END OF PROOF}$$

Corollary 6.3 *If $\xi^i(t)$ is a geodesic in G, then $x^r(t) = x^r(\xi^i(t))$ is a geodesic in R.*

PROOF: If $\xi^i(t)$ is a geodesic in G, then

$$\frac{d^2\xi^h}{dt^2} + \overset{G}{\Gamma}{}^h_{ij}\frac{d\xi^i}{dt}\frac{d\xi^j}{dt} \equiv 0$$

for all t where the curve is defined. But then

$$\frac{d^2x^r}{dt^2} + \overset{R}{\Gamma}{}^r_{pq}\frac{dx^p}{dt}\frac{dx^q}{dt} = \left(\frac{d^2\xi^h}{dt^2} + \overset{G}{\Gamma}{}^h_{ij}\frac{d\xi^i}{dt}\frac{d\xi^j}{dt}\right)\frac{\partial x^r}{\partial \xi^h} \equiv 0$$

for all t as well, so $x^r(t)$ is a geodesic in R. END OF PROOF

Since we can think of the expression for geodesics as ordinary second derivatives modified by "correction terms" that restore tensoriality,

Christoffel symbols as correction terms

$$d^2\xi^h/dt^2 + \text{correction} \longleftrightarrow d^2x^r/dt^2 + \text{correction},$$

can we do the same thing for differentiation of an arbitrary tensor (i.e., by adding Christoffel symbols)? Consider first the derivative of a contravariant vector $\alpha^j \longleftrightarrow a^r$ (so $a^r = \alpha^j \cdot \partial x^r/\partial \xi^j$); it transforms as

$$\frac{\partial a^r}{\partial x^p}\frac{\partial x^p}{\partial \xi^i} = \frac{\partial \alpha^j}{\partial \xi^i}\frac{\partial x^r}{\partial \xi^j} + \alpha^j \frac{\partial^2 x^r}{\partial \xi^i \partial \xi^j}.$$

We want to solve for $\dfrac{\partial \alpha^j}{\partial \xi^i}$; first multiply by $\dfrac{\partial \xi^k}{\partial x^r}$:

$$\frac{\partial a^r}{\partial x^p}\frac{\partial x^p}{\partial \xi^i}\frac{\partial \xi^k}{\partial x^r} = \frac{\partial \alpha^j}{\partial \xi^i}\delta^k_j + \alpha^j \frac{\partial^2 x^r}{\partial \xi^i \partial \xi^j}\frac{\partial \xi^k}{\partial x^r} = \frac{\partial \alpha^k}{\partial \xi^i} + \alpha^j \frac{\partial^2 x^r}{\partial \xi^i \partial \xi^j}\frac{\partial \xi^k}{\partial x^r}.$$

Then

$$\frac{\partial \alpha^k}{\partial \xi^i} = \frac{\partial a^r}{\partial x^p}\frac{\partial x^p}{\partial \xi^i}\frac{\partial \xi^k}{\partial x^r} - \alpha^j \frac{\partial^2 x^r}{\partial \xi^i \partial \xi^j}\frac{\partial \xi^k}{\partial x^r}.$$

Since

$$\alpha^j \overset{G}{\Gamma}{}^k_{ij} = \underbrace{\alpha^j}\; \overset{R}{\Gamma}{}^r_{pq} \frac{\partial x^p}{\partial \xi^i} \underbrace{\frac{\partial x^q}{\partial \xi^j}\frac{\partial \xi^k}{\partial x^r}} + \alpha^j \frac{\partial^2 x^p}{\partial \xi^i \partial \xi^j}\frac{\partial \xi^k}{\partial x^p},$$

when we add these equations the second terms on the right cancel. Moreover, the product of the factors underscored by braces equals a^q by the transformation law for $\alpha^j \longleftrightarrow a^q$. Therefore,

$$\frac{\partial \alpha^k}{\partial \xi^i} + \alpha^j \overset{G}{\Gamma}{}^k_{ij} = \left(\frac{\partial a^r}{\partial x^p} + a^q \overset{R}{\Gamma}{}^r_{pq} \right) \frac{\partial x^p}{\partial \xi^i}\frac{\partial \xi^k}{\partial x^r}.$$

The covariant derivative

Definition 6.3 *The **covariant derivative** of the contravariant vector α^k is*

$$\alpha^k_{;i} = \frac{\partial \alpha^k}{\partial \xi^i} + \alpha^j \overset{G}{\Gamma}{}^k_{ij}.$$

We will always use a semicolon to set off the differentiation index in the way shown here. The argument preceding the definition shows that the covariant derivative $\alpha^k_{;i} \longleftrightarrow a^r_{;p}$ of $\alpha^k \longleftrightarrow a^r$ is a tensor of type (1, 1). A covariant vector also has a covariant derivative, but the correction term needs the opposite sign.

Definition 6.4 *The **covariant derivative** of the covariant vector α_i is*

$$\alpha_{i;j} = \frac{\partial \alpha_i}{\partial \xi^j} - \alpha_k \overset{G}{\Gamma}{}^k_{ij}.$$

Proposition 6.12 *The covariant derivative $\alpha_{i;j} \longleftrightarrow a_{p;q}$ of a covariant vector $\alpha_i \longleftrightarrow a_p$ is a covariant tensor of rank 2.*

PROOF: Exercise.

Definition 6.5 The **covariant derivative** of the tensor T_J^I of type (p, q) is

$$T_{J;h}^I = \frac{\partial T_J^I}{\partial \xi^h} + T_J^{mi_2\cdots i_p}\Gamma_{mh}^{i_1} + \cdots + T_J^{i_1\cdots i_{p-1}m}\Gamma_{mh}^{i_p}$$
$$- T_{kj_2\cdots j_q}^I\Gamma_{j_1h}^k - \cdots - T_{j_1\cdots j_{q-1}k}^I\Gamma_{j_qh}^k.$$

There is a correction term corresponding to each index; the correction for each contravariant index has a plus sign, while that for each covariant index has a minus sign.

Proposition 6.13 If $T_J^I \overset{G}{\longleftrightarrow} T_Q^P$ is a tensor of type (p, q), then the covariant derivative $T_{J;h}^I \overset{G}{\longleftrightarrow} T_{Q;r}^P$ is a tensor of type $(p, q+1)$.

PROOF: Exercise.

The following is a quite remarkable result. It says that from the point of view of covariant differentiation, the metric tensor is constant.

Theorem 6.6 The covariant derivative of the metric tensor is identically zero: $g_{kl;m} \equiv 0$.

PROOF: Exercise.

We state the following results for future reference; you are asked to prove them in the exercises.

Proposition 6.14 The mixed Riemann curvature tensor

$$R_{jkl}^i = \frac{\partial \Gamma_{jl}^i}{\partial x^k} - \frac{\partial \Gamma_{jk}^i}{\partial x^l} + \Gamma_{jl}^p \Gamma_{pk}^i - \Gamma_{jk}^p \Gamma_{pl}^i$$

is a tensor of type $(1, 3)$.

Corollary 6.4 The covariant Riemann curvature tensor $R_{hjkl} = g_{ih}R_{jkl}^i$ is a covariant tensor of rank 4, and the Ricci tensor $R_{ij} = R_{ihj}^h$ is a covariant tensor of rank 2.

Parallel Transport

Problems with the definition of the derivative

Here is another approach to the differentiation of tensors that helps explain why correction factors are necessary in the definition of the covariant derivative. To be concrete, let us consider the contravariant vector field $\mathbf{a}(\mathbf{x}) = a^l(x^k)$ and its partial derivative with respect to x^1 at a point $\mathbf{p} = (p^k)$. In the usual definition, the derivative is the limit of a certain quotient:

$$\frac{\partial \mathbf{a}}{\partial x^1}(\mathbf{p}) = \lim_{h \to 0} \frac{\mathbf{a}(\mathbf{p} + h\mathbf{x}_1) - \mathbf{a}(\mathbf{p})}{h}, \qquad \mathbf{x}_1 = (1, 0, 0, 0).$$

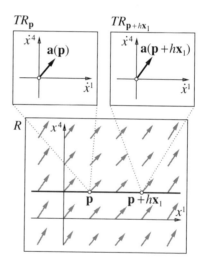

The quotient involves the difference of the vectors $\mathbf{a}(\mathbf{p} + h\mathbf{x}_1)$ and $\mathbf{a}(\mathbf{p})$. But these lie in two different tangent spaces and are therefore unrelated; in particular, *we cannot calculate their sum or difference!*

In the usual calculus of functions and vectors on $\mathbf{R}^n$ this problem does not appear because there is a natural way to identify the vectors bound at one point with those bound at another: Identify vectors that have the same coordinates. This approach doesn't work for us, though, because we must allow arbitrary coordinates. If the components of a vector agree at two different points in one coordinate frame, there is no guarantee that they will agree in

§6.5 Tensors

another coordinate frame. For suppose

$$a^l(x^1, x^2, x^3, x^4) = a^l(y^1, y^2, y^3, y^4) \qquad \text{for } l = 1, 2, 3, 4.$$

Introduce new coordinates in the usual way, by a map $M : G \to R : \xi \mapsto \mathbf{x}$, and suppose that

$$\xi^i \longleftrightarrow x^k, \qquad \alpha^j(\xi^i) \longleftrightarrow a^l(x^k),$$
$$\eta^i \longleftrightarrow y^k, \qquad \alpha^j(\eta^i) \longleftrightarrow a^l(y^k).$$

Since the Roman coordinates at the two points agree, $a^l(x^k) = a^l(y^k)$, the principle of general covariance tells us that the Greek coordinates ought to agree as well, $\alpha^i(\xi^j) = \alpha^i(\eta^j)$. Now the matrices $\partial \xi^i / \partial x^l = dM^{-1}$ translate the coordinates of vectors from Roman to Greek:

$$\alpha^i(\xi^j) = a^l(x^k) \frac{\partial \xi^i}{\partial x^l}(x^k), \qquad \alpha^i(\eta^j) = a^l(y^k) \frac{\partial \xi^i}{\partial x^l}(y^k).$$

Even though the inputs $a^l(x^k)$ and $a^l(y^k)$ are equal, we cannot expect the outputs $\alpha^i(\xi^j)$ and $\alpha^i(\eta^j)$ to be equal because the matrices

$$dM^{-1}_{(x^k)} = \frac{\partial \xi^i}{\partial x^l}(x^k) \qquad \text{and} \qquad dM^{-1}_{(y^k)} = \frac{\partial \xi^i}{\partial x^l}(y^k)$$

are generally different, because they are based at different points. Thus our scheme to identify spacetime vectors based at different points when their coordinates are equal fails to be generally covariant.

Another scheme fares better. We can illustrate the ideas in $\mathbf{R}^2$. Consider two points $\mathbf{p}$ and $\mathbf{q}$, and suppose $\mathbf{v}$ is a vector based at $\mathbf{p}$. With what vector at $\mathbf{q}$ should we identify $\mathbf{v}$? First construct a curve C that connects $\mathbf{p}$ and $\mathbf{q}$, and then slide $\mathbf{v}$ along C keeping it parallel to itself. The vector $\mathbf{w}$ at $\mathbf{q}$ that results is the one we identify with $\mathbf{v}$. We say $\mathbf{w}$ is produced by *parallel transport* of $\mathbf{v}$ along C.

Parallel transport of a vector in $\mathbf{R}^2$

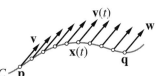

Derivatives along a curve

Suppose C is parametrized by $\mathbf{x}(t)$ and the parallel-transported vector at $\mathbf{x}(t)$ is denoted by $\mathbf{v}(t)$. By construction all the vectors $\mathbf{v}(t)$ are equal, so the derivative $\mathbf{v}'(t)$ is identically zero.

This idea, that the derivatives of a family of vectors along a curve equal zero, is one that we can carry over to spacetime, but it requires some care to get a definition of the derivative that is generally covariant. Suppose $\xi^i(t) \longleftrightarrow x^k(t)$ is a curve C and

$$\Phi^j(t) = \varphi^j(\xi^i(t)) \longleftrightarrow f^l(x^k(t)) = F^l(t)$$

is a field of contravariant vectors defined along C in the two coordinate frames G and R, respectively. (We work with coordinate descriptions in two frames to check that our results are generally covariant.) The transformation law is

$$F^l(t) = f^l(x^k(t)) = \varphi^j(\xi^i(t)) \frac{\partial x^l}{\partial \xi^j} = \Phi^j(t) \frac{\partial x^l}{\partial \xi^j}.$$

Our goal is to define the derivatives of $F^l(t)$ and $\Phi^j(t)$ so that they transform the same way—that is, so that they constitute a contravariant vector field on C. If we use the ordinary chain rule, then

$$\frac{dF^l}{dt} = \frac{\partial f^l}{\partial x^k} \frac{dx^k}{dt} = \frac{d}{dt}\left(\Phi^j(t) \frac{\partial x^l}{\partial \xi^j}\right) = \frac{d\Phi^j}{dt} \frac{\partial x^l}{\partial \xi^j} + \Phi^j \frac{\partial^2 x^l}{\partial \xi^h \partial \xi^j} \frac{d\xi^h}{dt}.$$

This is nontensorial, because the last term involves a second derivative. Let us instead use the *covariant* derivative in the chain rule and thereby define a *new* derivative of F^l, which we denote by DF^l/dt:

$$\frac{DF^l}{dt} = f^l_{;k} \frac{dx^k}{dt} = \frac{\partial f^l}{\partial x^k} \frac{dx^k}{dt} + \Gamma^l_{jk} f^j \frac{dx^k}{dt}.$$

If we use the ordinary chain rule to simplify the first term and then use the fact that $f^j = f^j(x^k(t)) = F^j(t)$ to change the second term, the resulting expression will involve only F^l, not f^l. It appears in the following definition.

Definition 6.6 *The **covariant derivative of** $F^l(t)$ **along** $x^k(t)$ is*

$$\frac{DF^l}{dt} = \frac{dF^l}{dt} + \Gamma^l_{jk} F^j \frac{dx^k}{dt}.$$

Proposition 6.15 *The covariant derivative $D\Phi^j/dt \longleftrightarrow DF^l/dt$ is a contravariant vector field on C.*

PROOF: Since $\varphi^j_{;i} \longleftrightarrow f^l_{;k}$ is a tensor of type (1, 1), we have

$$f^l_{;k} = \varphi^j_{;i} \frac{\partial x^l}{\partial \xi^j} \frac{\partial \xi^i}{\partial x^k}.$$

Therefore,

$$\frac{DF^l}{dt} = f^l_{;k} \frac{dx^k}{dt} = \varphi^j_{;i} \frac{\partial x^l}{\partial \xi^j} \frac{\partial \xi^i}{\partial x^k} \frac{dx^k}{dt} = \varphi^j_{;i} \frac{d\xi^i}{dt} \frac{\partial x^l}{\partial \xi^j} = \frac{D\Phi^j}{dt} \frac{\partial x^l}{\partial \xi^j},$$

so $D\Phi^j/dt \longleftrightarrow DF^l/dt$ transforms properly. END OF PROOF

If $F^i_j(t)$ is an arbitrary tensor defined along a curve $x^k(t)$, its covariant derivative along $x^k(t)$ can be constructed in a similar fashion. For example, the covariant derivative of the (1, 1)-tensor $F^i_j(t)$ along $x^k(t)$ is

Arbitrary tensors

$$\frac{DF^i_j}{dt} = \frac{dF^i_j}{dt} + \left(\Gamma^i_{hk} F^h_j - \Gamma^h_{jk} F^i_h \right) \frac{dx^k}{dt}.$$

The resulting object is tensorial, and you can show this by adapting the proof of the last proposition. Note that the covariant derivative *along a curve* leaves the type of a tensor unchanged, while the ordinary covariant derivative increases the covariant index by 1.

We are finally in a position to give a generally covariant definition of parallelism for vectors and tensors in spacetime.

Parallel transport in spacetime

Definition 6.7 *The tensor field $T^l_j(t)$ is **parallel along** $x^k(t)$ if its covariant derivative is identically zero along $x^k(t)$.*

If $\mathbf{v} = v^l$ is any vector in the tangent space $TR_\mathbf{p}$ and $x^k(t)$ is a curve C that starts at $\mathbf{p}$ (i.e., $x^k(0) = \mathbf{p}$), then the *parallel transport* of $\mathbf{v}$ along C is the vector field $V^l(t)$ that satisfies the initial-value problem

$$\frac{DV^l}{dt} = \frac{dV^l}{dt} + \Gamma^l_{jk} V^j \frac{dx^k}{dt} = 0, \qquad V^l(0) = v^l.$$

The parallel transport map $\tau_{\mathbf{p},\mathbf{q}}$

These are *linear* ordinary differential equations, so the solutions $V^l(t)$ exist on the entire curve $x^k(t)$ and are unique there.

Suppose $\mathbf{q}$ is a second point on C: $x^k(h) = \mathbf{q}$ for some $h \neq 0$. We can define the **parallel transport map**

$$\tau_{\mathbf{p},\mathbf{q}} : TR_{\mathbf{p}} \to TR_{\mathbf{q}} : v^l \mapsto w^l = V^l(h).$$

The existence and uniqueness of solutions to the initial-value problem implies that $\tau_{\mathbf{p},\mathbf{q}}$ is invertible and that $\tau_{\mathbf{p},\mathbf{q}}^{-1} = \tau_{\mathbf{q},\mathbf{p}}$. This relation holds for any two points $\mathbf{p}$ and $\mathbf{q}$ on C. But τ clearly depends on the curve C; there is no reason to think that parallel transport of $\mathbf{v}$ from $\mathbf{p}$ to $\mathbf{q}$ along a different curve would give the same result.

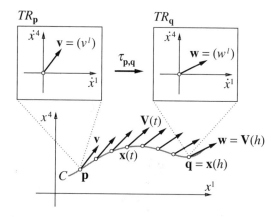

The derivative as the limit of a difference quotient

We can use the parallel transport maps to repair the difficulty we had at the outset in defining the derivative of a vector field $a^l(x^k) = \mathbf{a}(\mathbf{x})$ as the limit of a difference quotient. The computation

$$\frac{\partial \mathbf{a}}{\partial x^1}(\mathbf{p}) = \lim_{h \to 0} \frac{\mathbf{a}(\mathbf{p} + h\mathbf{x}_1) - \mathbf{a}(\mathbf{p})}{h}, \quad \mathbf{x}_1 = (1, 0, 0, 0),$$

makes no sense because the vectors $\mathbf{a}(\mathbf{p} + h\mathbf{x}_1)$ and $\mathbf{a}(\mathbf{p})$ are in different tangent spaces. But suppose we transport $\mathbf{a}(\mathbf{p} + h\mathbf{x}_1)$ back to $TR_{\mathbf{p}}$ by

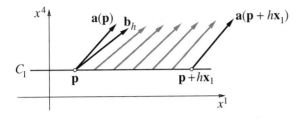

$$\mathbf{b}_h = \tau^{-1}_{\mathbf{p},\mathbf{p}+h\mathbf{x}_1} (\mathbf{a}(\mathbf{p}+h\mathbf{x}_1))$$

parallel to itself along the line $C_1 : \mathbf{x}(t) = \mathbf{p} + t\mathbf{x}_1$.

Then there is no difficulty calculating the difference quotient, so we *define* the partial derivative of $\mathbf{a}$ with respect to x^1 as

$$\frac{\partial \mathbf{a}}{\partial x^1}(\mathbf{p}) = \lim_{h \to 0} \frac{\mathbf{b}_h - \mathbf{b}_0}{h} = \lim_{h \to 0} \frac{\tau^{-1}_{\mathbf{p},\mathbf{p}+h\mathbf{x}_1}(\mathbf{a}(\mathbf{p}+h\mathbf{x}_1)) - \mathbf{a}(\mathbf{p})}{h}.$$

We make similar definitions for the partial derivative of $\mathbf{a}$ with respect to the other variables x^2, x^3, and x^4. The parallel transport is carried out along the line C_k parallel to the x^k-axis. The lines are

$$C_1 : \mathbf{y}_1(t) = (p^1 + t, p^2, p^3, p^4),$$
$$C_2 : \mathbf{y}_2(t) = (p^1, p^2 + t, p^3, p^4),$$
$$C_3 : \mathbf{y}_3(t) = (p^1, p^2, p^3 + t, p^4),$$
$$C_4 : \mathbf{y}_4(t) = (p^1, p^2, p^3, p^4 + t);$$

in components, C_k is $y_k^m(t) = p^m + \delta_k^m t$. The vector transported along C_k is

$$\mathbf{b}_h = \tau^{-1}_{\mathbf{p},\mathbf{y}_k(h)}(\mathbf{a}(\mathbf{y}_k(h))).$$

The following theorem shows that the components of the partial derivative of a contravariant vector, computed as the limit of a difference quotient, are the components of the usual covariant derivative.

Theorem 6.7 *Suppose* $\mathbf{a} = a^l$ *is a vector field on* R; *then the lth component of the partial derivative*

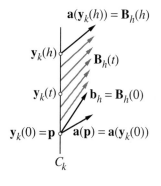

is

$$\frac{\partial \mathbf{a}}{\partial x^k}(\mathbf{p}) = \lim_{h \to 0} \frac{\tau^{-1}_{\mathbf{p}, \mathbf{y}_k(h)}(\mathbf{a}(\mathbf{y}_k(h))) - \mathbf{a}(\mathbf{y}_k(0))}{h}$$

$$a^l_{;k} = \frac{\partial a^l}{\partial x^k} + a^j \Gamma^l_{jk}.$$

PROOF: The vector field along C_k that determines

$$\tau^{-1}_{\mathbf{p}, \mathbf{y}_k(h)}(\mathbf{a}(\mathbf{y}_k(h))) = \tau_{\mathbf{y}_k(h), \mathbf{p}}(\mathbf{a}(\mathbf{y}_k(h))) = \mathbf{b}_h = \left(b^l_h\right)$$

is the solution $\mathbf{B}_h(t) = \left(B^l_h(t)\right)$ to the initial-value problem

$$\frac{dB^l_h}{dt} + \Gamma^l_{jm} B^j \frac{dy^m_k}{dt} = 0, \qquad B^l_h(h) = a^l(y^m_k(h)).$$

The initial condition is specified at the point $y^m_k(h)$ because the parallel transport map $\tau_{\mathbf{y}_k(h), \mathbf{p}}$ that defines b^l_k originates at this point. In these terms, $b^l_h = B^l_h(0)$, and the difference quotient we seek to evaluate as $h \to 0$ becomes

$$\frac{B^l_h(0) - a^l(y^m_k(0))}{h}.$$

By Taylor's theorem,

$$B^l_h(h) = B^l_h(0) + h \frac{dB^l_h}{dt}(0) + O(h^2)$$
$$= B^l_h(0) - h \Gamma^l_{jm} B^j_h(0) \frac{dy^m_k}{dt} + O(h^2),$$

Functions evaluated at different points can *be compared*

where we have used the differential equation for B^l_h to replace dB^l_h/dt. (Note: This calculation involves the values of the components of the vectors $\mathbf{B}_h$ at different points, but the components are merely functions, not vectors; they are defined on the whole domain of R, so there is no restriction to adding or subtracting them.) Now solve the previous equation for $B^l_h(0)$:

$$B^l_h(0) = B^l_h(h) + h \Gamma^l_{jm} B^j_h(0) \frac{dy^m_k}{dt} + O(h^2)$$
$$= a^l(y^m_k(h)) + h \Gamma^l_{jk} b^j_h + O(h^2).$$

We have used the initial condition $B_h^l(h) = a^l(y_k^m(h))$ and the facts that $B_h^l(0) = b_h^l$ and $dy_k^m/dt = \delta_k^m$. If we substitute the last expression for $B_h^l(0)$ into the difference quotient, we get

$$\frac{a^l(y_k^m(h)) - a^l(y_k^m(0))}{h} + \Gamma_{jk}^l b_h^j + O(h).$$

In the limit as $h \to 0$, the first term becomes

$$\frac{d}{dt} a^l(y_k^m(0)) = \frac{\partial a^l}{\partial x^m} \frac{dy_k^m}{dt} = \frac{\partial a^l}{\partial x^m} \delta_k^m = \frac{\partial a^l}{\partial x^k}.$$

Since $b_h^l \to a^l$ as well, we can finally conclude that

$$\frac{\partial \mathbf{a}}{\partial x^k}(\mathbf{p}) = \frac{\partial a^l}{\partial x^k} + \Gamma_{jk}^l a^j = a_{;k}^l. \qquad \text{END OF PROOF}$$

All this work was precipitated by the observation that the difference $\mathbf{a}(\mathbf{y}_k(h)) - \mathbf{a}(\mathbf{y}_k(0))$ is not meaningful in itself. In particular, it is *not* true that

Correction terms in parallel transport

$$\frac{\mathbf{a}(\mathbf{y}_k(h)) - \mathbf{a}(\mathbf{y}_k(0))}{h} = \frac{a^l(y_k^m(h)) - a^l(y_k^m(0))}{h}.$$

The correct way to proceed is to transport the first vector to the same tangent space as the second by using $\tau_{\mathbf{y}_k(0),\mathbf{y}_k(h)}^{-1}$. The proof just given shows that when this is done, we get the *correct* expression

$$\frac{\mathbf{a}(\mathbf{y}_k(h)) - \mathbf{a}(\mathbf{y}_k(0))}{h} = \frac{a^l(y_k^m(h)) - a^l(y_k^m(0))}{h} + \Gamma_{jk}^l b_h^j + O(h).$$

Thus the effect of parallel transport is to add correction terms involving the Christoffel symbols. If we replace the vector field by a tensor field of any type, we need simply add the appropriate correction term for each index. This leads to the following corollary; the corollary and the original theorem are both generally covariant results.

Corollary 6.5 *Suppose* $\mathbf{T} = T_J^I(x^m)$ *is a tensor field of type* (p, q) *on R. Then*

$$\lim_{h \to 0} \frac{\mathbf{T}(\mathbf{y}_k(h)) - \mathbf{T}(\mathbf{y}_k(0))}{h} = T_{J;k}^I.$$

Exercises

1. Suppose that $T^{I,i}_{J,j}$ is a tensor of type (p, q). Show that the multi-index quantities obtained by raising or lowering an index, $T^{I,ih}_J$ and $T^I_{J,jh}$, are tensors of type $(p+1, q-1)$ and $(p-1, q+1)$, respectively.

2. Prove that the mixed Riemann curvature tensor R^i_{jkl} is a tensor of type $(1, 3)$, that $R_{hjkl} = g_{hi}R^i_{jkl}$ is a covariant tensor of rank 4, and that $R_{jl} = R^h_{jhl}$ is a covariant tensor of rank 2.

3. Let $S^{J_1}_{I_1}$ and $T^{J_2}_{I_2}$ be tensors of type (p_1, q_1) and (p_2, q_2), respectively. Show that
$$U^{J_1,J_2}_{I_1,I_2} = S^{J_1}_{I_1} T^{J_2}_{I_2}$$
is a tensor of type (p_1+p_2, q_1+q_2) (called the **tensor product** of S and T).

4. Show that the determinant of the metric tensor does *not* transform like a tensor of type $(0, 0)$. Instead, $\gamma = (\det dM_P)^2 g$ when $\Gamma_P = dM^t_P \, G_{M(P)} \, dM_P$.

5. The aim of this exercise is to prove by direct calculation that the inverse of the metric tensor $\gamma_{ij} \longleftrightarrow g_{pq}$ on an ordinary 2-dimensional surface is a contravariant tensor of rank 2.

    (a) Suppose $M : G \to R$ and $dM_P : TG_P \to TR_{M(P)}$; let $m = \det dM_P$. Suppose $dM_P = (\partial x^p / \partial \xi^i)$ and $d(M^{-1})_{M(P)} = (\partial \xi^j / \partial x^q)$ in the usual way. Show that

    $$\frac{\partial \xi^1}{\partial x^1} = \frac{1}{m} \frac{\partial x^2}{\partial \xi^2}, \qquad \frac{\partial \xi^1}{\partial x^2} = -\frac{1}{m} \frac{\partial x^1}{\partial \xi^2},$$

    $$\frac{\partial \xi^2}{\partial x^1} = -\frac{1}{m} \frac{\partial x^2}{\partial \xi^1}, \qquad \frac{\partial \xi^2}{\partial x^2} = \frac{1}{m} \frac{\partial x^1}{\partial \xi^1}.$$

    Write down similar expressions for the inverses of γ_{ij} and g_{pq}, starting with $\gamma^{11} = \gamma_{22}/\gamma$.

    (b) Using the transformation law for $\gamma \longleftrightarrow g$ (see Exercise 4) and the fact that $\gamma_{ij} \longleftrightarrow g_{pq}$ is a covariant tensor of rank 2, show that

$$\gamma^{11} = \frac{\partial \xi^1}{\partial x^2}\frac{\partial \xi^1}{\partial x^2}g^{22} + \frac{\partial \xi^1}{\partial x^1}\frac{\partial \xi^1}{\partial x^2}g^{12} + \frac{\partial \xi^1}{\partial x^2}\frac{\partial \xi^1}{\partial x^1}g^{21} + \frac{\partial \xi^1}{\partial x^1}\frac{\partial \xi^1}{\partial x^1}g^{11}$$

and similarly for γ^{12}, γ^{21}, and γ^{22}. Conclude that $\gamma^{ij} \longleftrightarrow g^{pq}$ is a contravariant tensor of rank 2.

6. (a) Define the contravariant vector field $\mathbf{a}(q^1, q^2) = (a^1, a^2) = (0, 1)$ on the sphere:

$$R : \begin{cases} 0 \le q^1 \le 2\pi, \\ -\pi/2 < q^2 < \pi/2; \end{cases} \quad G = \begin{pmatrix} g_{11} & g_{12} \\ g_{21} & g_{22} \end{pmatrix} = \begin{pmatrix} \cos^2 q^2 & 0 \\ 0 & 1 \end{pmatrix}.$$

Compute the covariant derivative $a^i_{;j}$.

(b) Suppose $b^i = f^i(q^1, q^2)$, $i = 1, 2$ is an arbitrary contravariant vector field; compute $b^i_{;j}$.

7. Suppose a_p is a covariant vector field. Show that the covariant derivative

$$a_{p;q} = \frac{\partial a_p}{\partial x^q} - a_r \Gamma^r_{pq}$$

is a covariant tensor of rank 2.

8. Suppose $\varphi \longleftrightarrow f$ is a tensor of type $(0, 0)$. Show that the covariant derivative is the same as the gradient: $\varphi_{;i} = \varphi_i$.

9. Let g_{ij} be the metric tensor of any surface. Show that $g_{ij;k} = 0$ and $g^{ij}_{;k} = 0$ for all i, j, k.

10. Show that covariant differentiation commutes with raising indices; that is, $T^I_{J,j;l} = T^{I,i}_{J;l}$.

11. Prove the product rule: $(T^I_J S^K_L)_{;m} = T^I_{J;m} S^K_L + T^I_J S^K_{L;m}$.

12. (a) Let a^h be an arbitrary contravariant vector field. Show that

$$a^h_{;j;k} - a^i_{;k;j} = -R^h_{ijk} a^i.$$

Thus covariant derivatives (in different directions) do not, in general, commute. Moreover, their failure to commute can be attributed to the curvature of the underlying space.

(b) For an arbitrary covariant vector field b_i, express $b_{h;k;l} - b_{h;l;k}$ in terms of the vector field and the Riemann curvature tensor.

13. Suppose $x^k(t)$ is a geodesic on the surface $\{R, g_{ij}\}$. Let $y^k(t) = dx^k/dt$ be the tangent vector along x^k. Compute the covariant derivative Dy^k/dt along x^k to prove that y^k is parallel along x^k.

14. (a) Let $\{R, g_{ij}(q^1, q^2)\}$ be the sphere as defined in Exercise 6. Let $\mathbf{q}(t) = (t, \pi/6)$ be the parallel of latitude at 30° N and let $\mathbf{y}(t)$ be its tangent vector. Compute the covariant derivative of $\mathbf{y}$ along $\mathbf{q}$.

    (b) Determine the parallel vector field $\mathbf{z}(t)$ along $\mathbf{q}(t)$ that agrees with $\mathbf{y}(0)$ at $\mathbf{q}(0)$. Sketch the vector field $\mathbf{z}(t)$ along $\mathbf{x}(t)$ in the (q^1, q^2)-plane. How much does $\mathbf{z}$ change in one circuit; that is, what is the angle Δ between $\mathbf{z}(0)$ and $\mathbf{z}(2\pi)$?

    (c) Let $\mathbf{q}_k(t) = (t, k)$ be the parallel at an arbitrary latitude k. Determine the parallel vector field $\mathbf{z}_k(t)$ along $\mathbf{q}_k(t)$ that agrees with the tangent vector $\mathbf{q_k}'(0)$ at $\mathbf{q}_k(0)$. Show that the angle $\Delta(k)$ between $\mathbf{z}_k(0)$ and $\mathbf{z}_k(2\pi)$ is $2\pi \sin k$.

    (d) Now take the sphere embedded in $\mathbf{R}^3$ and view it from above the north pole. Take k very near $\pi/2$ and sketch $\mathbf{q}_k(t)$ as a small circle concentric with the pole, using a magnified view of the polar region. Sketch the parallel field $\mathbf{z}_k(t)$ along $\mathbf{q}(t)$, and show that it makes nearly one complete circuit around the tangent vector along $\mathbf{q}(t)$. Explain how this is related to the result that $\lim_{k \to \pi/2} \Delta(k) = 2\pi$ in part (c).

Further Reading for Chapter 6

The material in this chapter is found in the standard texts on differential geometry, including those mentioned at the end of the previous chapter ([16], [5], [6], [7], [15], [22]). For parallel transport we follow the treatment of McCleary [22] most closely.

7 General Relativity

CHAPTER

While gravitational and electric fields both have physical "sources" —masses on the one hand and electric charges on the other— there is a crucial distinction between them. In an electric field, objects accelerate differently, depending on their charge; in a gravitational field, all objects experience identical acceleration. So it is possible to "fall" along with those objects—in an orbiting spacecraft, for example—and if we do, the gravitational accelerations will seem to disappear. Conversely, in a spaceship far from other matter, where there is no perceptible gravitational field, we can create one in the spaceship simply by making it accelerate. In other words, with a coordinate change we can make a gravitational field appear or disappear—at least in a small region of space and time.

Now, the physical effects of coordinate changes are the stuff of relativity. But special relativity deals only with linear transformations of inertial frames, so it is unable to handle gravity: The transformations that create or eliminate gravitational fields are nonlinear. If it is to encompass gravity, relativity must be generalized to allow arbitrary frames and transformations. In this chapter we see how general relativity leads to a theory of gravity.

A theory of gravity based on coordinate changes

7.1 The Equations of Motion

The Newtonian equations of motion

Consider the motion of a *freely falling* test particle, which is one that moves solely under the influence of a gravitational field. In Newtonian mechanics, the field is given by a gravitational potential function $\Phi(\mathbf{x}) = \Phi(x^1, x^2, x^3)$. The gravitational force on a particle of mass m is $-m\nabla\Phi = -m\,\text{grad}\,\Phi$; cf. Section 4.3. Therefore, if the position of the particle at time t is $\mathbf{x}(t)$, then $m\mathbf{x}'' = -m\nabla\Phi$, by Newton's second law of motion. Canceling m from both sides, we get the equation for gravitational *acceleration*, $\mathbf{x}'' = -\nabla\Phi$; in terms of coordinates, this becomes the three equations

Motion is independent of mass

$$\frac{d^2 x^k}{dt^2}(t) = -\frac{\partial \Phi}{\partial x^k}(x^1(t), x^2(t), x^3(t)), \qquad k = 1, 2, 3.$$

Because these equations are independent of the mass m of the test particle, they tell us that particles of different sizes must all accelerate the same way. This demonstrates the distinctive feature of gravity that makes it amenable to a relativistic treatment.

The Gravitational Field

Motion viewed in a freely falling frame

What are the equations of motion of a freely falling test particle from the point of view of relativity? Let us first consider the motion in a frame R that falls with the particle (like an orbiting spacecraft, for example). We saw in Section 4.3 that in a sufficiently small region $\mathcal{R}$ in spacetime, the gravitational field essentially disappears. Since by assumption there are no other forces present, all objects move in straight lines with constant velocity. Indeed, R is an inertial frame, and all the laws of special relativity hold in the region $\mathcal{R}$. By an appropriate linear change to new coordinates $(t, x, y, z) = (x^0, x^1, x^2, x^3)$, we can make the metric for R become

$$g_{kl}(x^m) = \begin{pmatrix} 1 & 0 & 0 & 0 \\ 0 & -1 & 0 & 0 \\ 0 & 0 & -1 & 0 \\ 0 & 0 & 0 & -1 \end{pmatrix}, \qquad k, l = 0, 1, 2, 3,$$

for all (x^0, x^1, x^2, x^3) in $\mathcal{R}$. We shall use the indices 0, 1, 2, 3 henceforth because they help distinguish between timelike and spacelike coordinates. Roman letters i, j, etc. shall continue to refer to all four indices, but we shall often use Greek letters when we want to refer only to the three spacelike indices 1, 2, 3. Thus, for example, $g_{00} = +1$, while $g_{\alpha\alpha} = -1$.

Since the g_{kl} are constants, their derivatives are zero. Consequently, the Christoffel symbols are identically zero, so the geodesic equations in this frame take the particularly simple form

For R, worldcurves are geodesics

$$\frac{d^2 x^m}{dt^2} = 0.$$

The solutions are straight lines. But the worldcurve of every freely falling particle is a straight line; therefore, from R's point of view, *the worldcurve of a freely falling particle is a geodesic.*

Now consider an arbitrary frame $G : (\xi^0, \xi^1, \xi^2, \xi^3)$ that is related to R by a smooth map $M : G \to R$. How will G describe the same motion? The key is Einstein's **principle of general covariance** ([10], page 117):

The principle of general covariance

> The general laws of nature are to be expressed by equations which hold good for all systems of coordinates, that is, are covariant with respect to any substitutions whatever.

Because contravariant and covariant tensors transform in this fashion, Einstein was led to argue as follows ([10], page 121):

> If, therefore, a law of nature is expressed by equating all the components of a tensor to zero, it is generally covariant. By examining the laws of the formation of tensors, we acquire the means of formulating generally covariant laws.

Since the motion of a freely falling particle should certainly be described by objective physical laws, and since Theorem 6.5 assures us that the geodesic equations are tensorial, the principle of general covariance dictates that the worldcurve of a freely falling particle must be a geodesic in G's frame as well as R's. If the worldcurve of the particle is given by $\boldsymbol{\xi} = (\xi^h(t))$ in G's frame,

For G, worldcurves are geodesics, too

then its equations of motion are

$$\frac{d^2\xi^h}{dt^2} + \overset{G}{\Gamma}{}^h_{ij}\frac{d\xi^i}{dt}\frac{d\xi^j}{dt} = 0.$$

This holds in the small region $M^{-1}(\mathcal{R})$ in G's frame that corresponds to $\mathcal{R}$ in R's frame.

Freely falling particles *always* move on geodesics

The argument we just used depends on starting with a spacetime region $\mathcal{R}$ so small that we can "transform away" gravitational effects and work in an inertial frame R. But this is a serious restriction, especially if we consider the astronomical distances and time scales over which relativity is usually applied. For example, even if $\mathcal{R}$ is a ball of radius 10,000 kilometers (less than 1 second in geometric units!) concentric with the earth for a duration of a second, the gravitational field is nontrivial; it cannot be transformed away over the entire region $\mathcal{R}$. In these more general circumstances, what should the equations of motion be? Einstein's answer is to take what happens in a small region as a guide ([10], page 143):

> We now make this assumption, which readily suggests itself, that the geodesic equations also define the motion of the point in the gravitational field in the case when there is no system of reference with respect to which the special theory of relativity holds good for a finite region.

In other words, we assume henceforth that the worldcurve of a freely falling particle is a geodesic in whatever frame it is described.

The $-\Gamma^h_{ij}$ are the components of the gravitational field

The geodesic equations are second-order differential equations that determine the acceleration of a test particle, just like the Newtonian equations. If we write them side by side, we see remarkable similarities:

$$\frac{d^2\xi^h}{dt^2} = -\overset{G}{\Gamma}{}^h_{ij}\frac{d\xi^i}{dt}\frac{d\xi^j}{dt}, \qquad \frac{d^2x^h}{dt^2} = -\frac{\partial\Phi}{\partial x^h}.$$

In each case, the right-hand side gives the acceleration due to gravity; hence, $-\Gamma^h_{ij}$ *are the components of the gravitational field in the frame G*.

We think of the Newtonian field $-\partial \Phi/\partial x^h = -\operatorname{grad}\Phi$ as being derived from a single gravitational potential function Φ. In a very similar way, the Christoffel symbols Γ^h_{ij} are certain combinations of derivatives of the metric tensor, so they are a kind of "gradient" of the metric γ_{ij}, which therefore plays the role of the gravitational potential in relativity. However, while there is just a single Newtonian potential, the relativistic potential consists of 10 functions, the 10 distinct components of the symmetric tensor γ_{ij}:

The γ_{ij} form the gravitational potential

	Newton	Einstein
gravitational field	$-\nabla\Phi$	$-\Gamma^h_{ij}$
gravitational potential	Φ	γ_{ij}

Incidentally, a potential that is made up of several component functions is not without precedent: Maxwell's electromagnetic field can be described by a potential, but a 4-vector $\mathbb{A}(t, \mathbf{x}) = (A^0, A^1, A^2, A^3)$ is needed.

The metric tensor combines geometry and physics at a profound level. It tells us how spacetime is curved, and it tells us how objects move. It tells us, moreover, that objects move the way they do *because* spacetime is curved, and it even says that their worldcurves are the "straight lines" in the geometry of that curved spacetime.

The fusion of geometry and physics

A Constant Gravitational Field

Let us go back and consider in detail the motion of a test particle in a region of spacetime that is so small that the gravitational field is essentially constant. Let the frame $G : (\tau, \xi, \eta, \zeta) = (\xi^0, \xi^1, \xi^2, \xi^3)$ be stationary in the field, which we suppose has strength α and points in the direction of the negative ζ-axis. How is the gravitational field expressed in this frame, and what motions does this field then define?

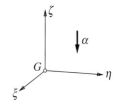

Einstein says we should first answer these questions for a second frame $R : (t, x, y, z) = (x^0, x^1, x^2, x^3)$ that is in free-fall, and then use the principle of general covariance to transfer the answers back to G. Now, R is an inertial frame (from R's point

The metric in R

of view, particles move uniformly in straight lines), and the laws of special relativity hold within it. In particular, we can take the metric in R to be

$$g_{kl}(x^0, x^1, x^2, x^3) = \begin{pmatrix} 1 & 0 & 0 & 0 \\ 0 & -1 & 0 & 0 \\ 0 & 0 & -1 & 0 \\ 0 & 0 & 0 & -1 \end{pmatrix}.$$

Let us assume that R's origin moves along the ζ-axis. Align the frames so that corresponding spatial axes in R and G are parallel, and set $t = \tau = 0$ at the instant G is at rest relative to R. Since $x = \xi$ and $y = \eta$ for all time, we ignore these variables and work in the 2-dimensional (t, z)- and (τ, ζ)-planes. We saw in Theorem 4.1 that these variables are connected by the map

$$M : \begin{cases} t = \dfrac{e^{\alpha \zeta}}{\alpha} \sinh \alpha \tau, \\ z = \dfrac{e^{\alpha \zeta}}{\alpha} \cosh \alpha \tau. \end{cases}$$

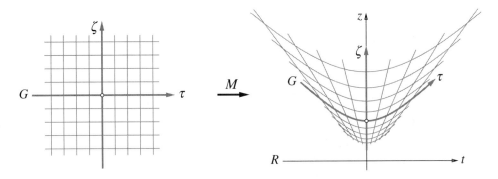

The metric in G

Because we are ignoring the x and y variables, we can write the metrics in G and R as 2×2 matrices:

$$\Gamma = \begin{pmatrix} \gamma_{00} & \gamma_{03} \\ \gamma_{30} & \gamma_{33} \end{pmatrix}, \quad G = \begin{pmatrix} g_{00} & g_{03} \\ g_{30} & g_{33} \end{pmatrix} = \begin{pmatrix} 1 & 0 \\ 0 & -1 \end{pmatrix}.$$

Since the metric tensor is covariant, we can determine γ_{ij} from g_{kl}. To do this we need the differential of M at $P = (\tau, \zeta)$:

§7.1 The Equations of Motion

$$dM_P = \begin{pmatrix} e^{\alpha\zeta}\cosh\alpha\tau & e^{\alpha\zeta}\sinh\alpha\tau \\ e^{\alpha\zeta}\sinh\alpha\tau & e^{\alpha\zeta}\cosh\alpha\tau \end{pmatrix} = e^{\alpha\zeta}H_{\alpha\tau},$$

where $H_{\alpha\tau}$ is hyperbolic rotation by the hyperbolic angle $\alpha\tau$. By the properties of hyperbolic rotations,

$$\Gamma_P = dM_P^t \, G \, dM_P = e^{2\alpha\zeta} H_{\alpha\tau} \, G \, H_{\alpha\tau} = e^{2\alpha\zeta} G,$$

or

$$\Gamma_P = \begin{pmatrix} \gamma_{00} & \gamma_{03} \\ \gamma_{30} & \gamma_{33} \end{pmatrix} = \begin{pmatrix} e^{2\alpha\zeta} & 0 \\ 0 & -e^{2\alpha\zeta} \end{pmatrix} = \begin{pmatrix} e^{2\alpha\xi^3} & 0 \\ 0 & -e^{2\alpha\xi^3} \end{pmatrix}.$$

These equations define the gravitational potential in G. We shall also need

$$\Gamma_P^{-1} = \begin{pmatrix} \gamma^{00} & \gamma^{03} \\ \gamma^{30} & \gamma^{33} \end{pmatrix} = \begin{pmatrix} e^{-2\alpha\zeta} & 0 \\ 0 & -e^{-2\alpha\zeta} \end{pmatrix} = \begin{pmatrix} e^{-2\alpha\xi^3} & 0 \\ 0 & -e^{-2\alpha\xi^3} \end{pmatrix}.$$

We are now in a position to determine the components $-\Gamma_{ij}^h$ of the relativistic gravitational field. You should verify that all the Christoffel symbols are zero except

The gravitational field

$$\Gamma_{30,0} = \Gamma_{03,0} = \alpha e^{2\alpha\xi^3}, \qquad \Gamma_{30}^0 = \Gamma_{03}^0 = \alpha,$$
$$\Gamma_{00,3} = -\alpha e^{2\alpha\xi^3}, \qquad \Gamma_{00}^3 = \alpha,$$
$$\Gamma_{33,3} = -\alpha e^{2\alpha\xi^3}, \qquad \Gamma_{33}^3 = \alpha.$$

If we assume that the geodesics are parametrized by t, then the geodesic differential equations are

$$\frac{d^2\xi^0}{dt^2} = -2\alpha \frac{d\xi^0}{dt} \frac{d\xi^3}{dt},$$

$$\frac{d^2\xi^3}{dt^2} = -\alpha\left(\frac{d\xi^0}{dt}\right)^2 - \alpha\left(\frac{d\xi^3}{dt}\right)^2.$$

Compare these with the Newtonian equations of motion,

$$\frac{d^2\xi^1}{dt^2} = 0, \qquad \frac{d^2\xi^2}{dt^2} = 0, \qquad \frac{d^2\xi^3}{dt^2} = -\alpha,$$

determined by the field $-\nabla\Phi = (0, 0, -\alpha)$.

336   Chapter 7   General Relativity

Motions in the gravitational field

Free-fall motions in the gravitational field in G are the solutions to these geodesic differential equations. As usual, we do not attempt to find analytic solutions from first principles, but rather simply verify that certain expressions do indeed satisfy the equations. In this case we expect that the images of straight worldlines in R under the map M^{-1} will be geodesics in G. (In fact, since the geodesic equations are covariant, we know in advance that this will be so; nevertheless, it is valuable to see the theory confirmed.)

The worldlines of objects at rest in R

Consider first the horizontal lines $z = b$ in R. These are the worldlines of objects that are *at rest* in R. In Section 4.2 we saw that their worldlines in G are the curves

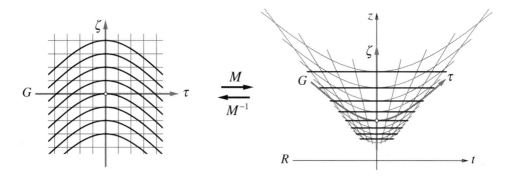

$$\xi^0 = \tau = \frac{1}{\alpha} \tanh^{-1}\left(\frac{t}{b}\right), \qquad \xi^3 = \zeta = \frac{1}{2\alpha} \ln(\alpha^2(b^2 - t^2))$$

parametrized by t. We have

$$\frac{d\xi^0}{dt} = \frac{b}{\alpha} \frac{1}{b^2 - t^2}, \qquad \frac{d\xi^3}{dt} = -\frac{1}{\alpha} \frac{t}{b^2 - t^2},$$

$$\frac{d^2\xi^0}{dt^2} = \frac{b}{\alpha} \frac{2t}{(b^2 - t^2)^2}, \qquad \frac{d^2\xi^3}{dt^2} = -\frac{1}{\alpha} \frac{b^2 + t^2}{(b^2 - t^2)^2},$$

and you can quickly check that $\xi^0(t)$ and $\xi^3(t)$ satisfy the geodesic equations.

Horizontal translates

These curves are vertical translates of one another in G; we can show that each *horizontal* translate

$$\xi^0 \mapsto \xi^0 + k, \qquad \xi^3 \mapsto \xi^3$$

is a geodesic, too. First, the coefficients of the geodesic equations are unchanged by horizontal translations (in fact, the coefficients are constants); second, derivatives are likewise unchanged. Hence the translated curves still satisfy the geodesic equations. By the principle of covariance, the image under M of one of these horizontal translates must be a geodesic—that is, a straight line—in R. What is the equation of that line, in terms of the parameters b and k?

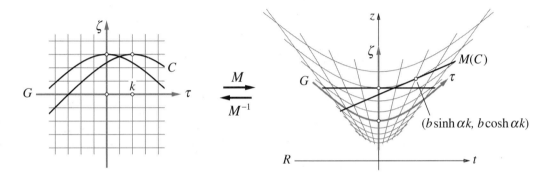

The translate C that has its apex at $\tau = k$, as in the figure above, can be described as the graph of the function

$$\zeta = \frac{1}{\alpha} \ln\left(\frac{\alpha b}{\cosh \alpha (\tau - k)}\right).$$

(We revert to the original variables τ and ζ because they are easier to write.) Now let $M : G \to R$ map C into R. By using the equation

$$\frac{e^{\alpha \zeta}}{\alpha} = \frac{b}{\cosh \alpha (\tau - k)}$$

(which follows immediately from the formula for ζ) and the shift $\sigma = \tau - k$, we can write the image $M(C)$ as

$$t = \frac{b}{\cosh \alpha (\tau - k)} \sinh \alpha \tau = \frac{b \sinh \alpha (\sigma + k)}{\cosh \alpha \sigma},$$

$$z = \frac{b}{\cosh \alpha (\tau - k)} \cosh \alpha \tau = \frac{b \cosh \alpha (\sigma + k)}{\cosh \alpha \sigma}.$$

The addition formulas for the hyperbolic functions then give

$$t = \frac{b(\sinh\alpha\sigma \cosh\alpha k + \cosh\alpha\sigma \sinh\alpha k)}{\cosh\alpha\sigma} = (b\cosh\alpha k)\tanh\alpha\sigma + b\sinh\alpha k,$$

$$z = \frac{b(\cosh\alpha\sigma \cosh\alpha k + \sinh\alpha\sigma \sinh\alpha k)}{\cosh\alpha\sigma} = (b\sinh\alpha k)\tanh\alpha\sigma + b\cosh\alpha k,$$

and thus finally,

$$z - b\cosh\alpha k = \tanh\alpha k\, (t - b\sinh\alpha k).$$

This is the straight line of slope $v = \tanh\alpha k$ that passes through the point $(t, z) = (b\sinh\alpha k, b\cosh\alpha k)$.

All possible motions in the gravitational field

Since these straight lines constitute *all* possible worldlines of freely falling particles in R, the corresponding curves in G,

$$\zeta = \frac{1}{\alpha}\ln\left(\frac{\alpha b}{\cosh\alpha(\tau - k)}\right),$$

give us the worldcurves of all possible motions under the influence of the gravitational field $-\Gamma^h_{ij}$, at least if we ignore ξ and η. If we bring these two variables back in, then the gravitational potential is

$$(\gamma_{ij}) = \begin{pmatrix} e^{2\alpha\xi^3} & 0 & 0 & 0 \\ 0 & -1 & 0 & 0 \\ 0 & 0 & -1 & 0 \\ 0 & 0 & 0 & -e^{2\alpha\xi^3} \end{pmatrix}.$$

This adds no new nonzero components to the gravitational field $-\Gamma^h_{ij}$, so the equations of motion in the full 4-dimensional spacetime are

$$\frac{d^2\xi^0}{dt^2} = -2\alpha\frac{d\xi^0}{dt}\frac{d\xi^3}{dt},$$

$$\frac{d^2\xi^1}{dt^2} = 0,$$

$$\frac{d^2\xi^2}{dt^2} = 0,$$

$$\frac{d^2\xi^3}{dt^2} = -\alpha\left(\frac{d\xi^0}{dt}\right)^2 - \alpha\left(\frac{d\xi^3}{dt}\right)^2,$$

and the solutions are

$$\xi^0(t) = k_1 + \frac{1}{\alpha}\tanh^{-1}\left(\frac{t}{k_2}\right),$$
$$\xi^1(t) = k_3 t + k_4,$$
$$\xi^2(t) = k_5 t + k_6,$$
$$\xi^3(t) = \frac{1}{2\alpha}\ln(\alpha^2(k_2^2 - t^2)).$$

This involves six arbitrary constants (and we have set $k = k_1$, $b = k_2$). But four second-order differential equations should involve *eight* constants, not six; we get the other two by making an affine change of the independent variable $t \mapsto k_7 t + k_8$.

Although we have constructed a full eight-parameter family of solutions to the geodesic equations, we do not yet have *all* solutions. First of all, our typical solution

$$\zeta = \frac{1}{\alpha}\ln\left(\frac{\alpha b}{\cosh\alpha(\tau - k)}\right) \longleftrightarrow z - b\cosh\alpha k = \tanh\alpha k\,(t - b\sinh\alpha k)$$

Worldcurves of photons

is a timelike curve that describes a motion with velocity $v = \tanh\alpha k$ in R. Since $|\tanh\alpha k| < 1$, the moving object has *positive* mass. But gravity also affects the motion of objects of zero mass—e.g., photons—that travel at the speed of light $c = 1$, so we need formulas for them, too. Those formulas will have to be additional solutions to the geodesic equations. Beyond that, there are spacelike geodesics.

To find the formulas for lightlike curves we look first in the inertial frame R. Here the worldline of a photon has the formula $z = a \pm t$, where a is arbitrary. In G the worldcurve of the same photon will be

$$\xi^0 = \frac{1}{\alpha}\tanh^{-1}\left(\frac{t}{a \pm t}\right), \qquad \xi^3 = \frac{1}{2\alpha}\ln(\alpha^2((a \pm t)^2 - t^2)).$$

We can rewrite these equations using

$$\tanh^{-1} u = \frac{1}{2}\ln\left(\frac{1+u}{1-u}\right)$$

and some algebra to get finally

$$\xi^0 = \pm \frac{1}{2\alpha} \ln(a \pm 2t) + \text{constant},$$

$$\xi^3 = \frac{1}{2\alpha} \ln(a \pm 2t) + \text{constant}.$$

Thus $\xi^3 = \pm\xi^0 + \text{constant}$, so the worldcurve is a straight line of slope ± 1 in G as well. Furthermore, these functions do indeed satisfy the geodesic differential equations, and they are clearly not members of the general eight-parameter family of solutions we just found. They are called *singular solutions*.

Singular solutions to differential equations can arise in various ways. In this case they are the common asymptotes of one-parameter families of ordinary solutions. To see this, consider a one-parameter family of solution curves chosen so that that their apexes lie on a single line of slope 1. Now, the apex of the solution

$$\zeta = \frac{1}{\alpha} \ln\left(\frac{\alpha b}{\cosh \alpha(\tau - k)} \right)$$

is at the point $(\tau, \zeta) = (k, \ln(b\alpha)/\alpha)$; to put this on the line $\zeta = \tau$, for example, we should set

$$k = \frac{\ln(b\alpha)}{\alpha}, \quad \text{or} \quad b = \frac{e^{\alpha k}}{\alpha}.$$

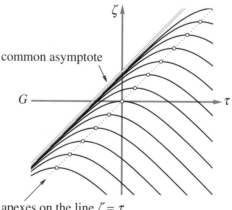

apexes on the line $\zeta = \tau$

§7.1 The Equations of Motion   341

The ordinary solutions whose apexes lie on a line of slope -1 will also have a common asymptote that is a singular solution. For geodesics that are spacelike curves, see the exercises.

Exercises

1. **The electromagnetic field potential**. The purpose of this exercise is to see how the components of the electromagnetic field $\{\mathbf{E}, \mathbf{H}\}$ (cf. Section 1.4) can arise from a 4-potential $\mathbb{A} = (A^0, A^1, A^2, A^3)$. Let $\mathbf{A}(t, x, y, z)$ be a 3-vector with spatial components $\mathbf{A} = (A^1, A^2, A^3)$; assume that (t, x, y, z) are coordinates in an inertial frame. Define the magnetic field vector $\mathbf{H} = \nabla \times \mathbf{A}$.

   (a) Show that $\mathbf{H}$ satisfies the Maxwell equation $\nabla \cdot \mathbf{H} = 0$.

   (b) Show that if the Maxwell equation $\nabla \times \mathbf{E} = -\partial \mathbf{H}/\partial t$ is to hold, then $\nabla \times (\mathbf{E} + \partial \mathbf{A}/\partial t) = \mathbf{0}$.

   (c) According to a theorem in advanced calculus, if $\nabla \times \mathbf{F} = \mathbf{0}$, then there is a scalar function ψ for which $\mathbf{F} = \nabla \psi$. Deduce that there is a function $A^0 = -\psi$ for which
   $$\mathbf{E} = -\frac{\partial \mathbf{A}}{\partial t} - \nabla A^0.$$

   (d) Show that the remaining two Maxwell equations (which involve electric charge density ρ and electric current density $\mathbf{J}$) determine the following conditions on the 4-potential $\mathbb{A} = (A^0, \mathbf{A})$:
   $$\rho = -\frac{\partial}{\partial t}(\nabla \cdot \mathbf{A}), \qquad \mathbf{J} = \nabla(\nabla \cdot \mathbf{A}) - \nabla^2 \mathbf{A} + \frac{\partial^2 \mathbf{A}}{\partial t^2} + \frac{\partial}{\partial t}(\nabla A^0).$$

   (e) Deduce the conservation of charge equation: $\dfrac{\partial \rho}{\partial t} = -\nabla \cdot \mathbf{J}$.

   The next five exercises concern the metric (γ_{ij}) of a constant gravitational field in the frame G, as described in the text.

2. Determine the Christoffel symbols $\Gamma_{ij,k}$ and Γ^h_{ij} and the geodesic equations.

3. Find the ζ-intercept β of the common asymptote $\zeta = \tau + \beta$ of the timelike geodesics

$$\zeta = \frac{1}{\alpha}\ln\left(\frac{e^{\alpha k}}{\cosh\alpha(\tau - k)}\right) = k - \frac{\ln\cosh\alpha(\tau - k)}{\alpha}.$$

4. Derive the formulas

$$\xi^0(t) = \pm\frac{1}{2\alpha}\ln(a \pm 2t) + c_1, \quad \xi^3(t) = \frac{1}{2\alpha}\ln(a \pm 2t) + c_2$$

(where c_1 and c_2 are constants) from the conditions

$$\xi^0 = \frac{1}{\alpha}\tanh^{-1}\left(\frac{t}{a \pm t}\right), \quad \xi^3 = \frac{1}{2\alpha}\ln(\alpha^2((a \pm t)^2 - t^2))$$

and prove that the curve so defined is a lightlike geodesic, for any a.

5. (a) Consider two curves $\boldsymbol{\xi}_\pm(u)$ defined by the same formulas:

$$\xi^0(u) = k_1 + \frac{1}{\alpha}\tanh^{-1}\left(\frac{k_2}{u}\right), \quad \xi^3(u) = \frac{1}{2\alpha}\ln(\alpha^2(u^2 - k_2^2)).$$

The domain of $\boldsymbol{\xi}_-$ is $u < -k_2$, and the domain of $\boldsymbol{\xi}_+$ is $k_2 < u$. Show that each is a spacelike geodesic for every k_1 and for every $k_2 > 0$.

(b) Express the curves $\boldsymbol{\xi}_\pm$ as graphs of functions $\zeta = f_\pm(\tau)$, analogous to the functions that describe the timelike geodesics in the text.

(c) Construct a full eight-parameter family of spacelike geodesics when G is a $(1 + 3)$-dimensional spacetime.

6. (a) Show that the curves $\boldsymbol{\xi}_\pm$ from the previous exercise have the vertical asymptote $\xi^0 = k_1$ and a second asymptote with slope ∓ 1.

(b) Find a one-parameter family of geodesics that has $\xi^0 = k_1$ as a common asymptote, demonstrating that the asymptote is a singular solution to the geodesic equations. Find a parametrization for the asymptote that makes it a geodesic.

The following exercises concern a frame $G : (\tau, \xi, \eta, \zeta) = (\xi^0, \xi^1, \xi^2, \xi^3)$ in uniform rotation with angular velocity ω around

§7.1 The Equations of Motion

the x-axis of an inertial frame $R : (t, x, y, z)$. If the ξ-axis coincides with the x-axis, then we can take the map $M : G \to R$ that transforms coordinates (cf. Section 4.1) as

$$M : \begin{cases} t = \tau, \\ x = \xi, \\ y = \eta \cos\omega\tau - \zeta \sin\omega\tau, \\ z = \eta \sin\omega\tau + \zeta \cos\omega\tau. \end{cases}$$

7. Determine the inverse $M^{-1} : R \to G$ and the linear maps dM_P and $(dM_P)^{-1}$.

8. Determine the metric (γ_{ij}) on G that is induced by the Minkowski metric $J_{1,3}$ on R, and show that its inverse is

$$(\gamma^{jk}) = \begin{pmatrix} 1 & 0 & \omega\zeta & -\omega\eta \\ 0 & -1 & 0 & 0 \\ \omega\zeta & 0 & -1 + \omega^2\zeta^2 & -\omega^2\eta\zeta \\ -\omega\eta & 0 & -\omega^2\eta\zeta & -1 + \omega^2\eta^2 \end{pmatrix}.$$

9. Determine the Christoffel symbols of the first kind, $\Gamma_{ij,k}$, on G. Show that the only nonzero Christoffel symbols of the second kind are

$$\Gamma^2_{00} = -\omega^2\eta, \qquad \Gamma^2_{03} = \Gamma^2_{30} = -\omega,$$
$$\Gamma^3_{00} = -\omega^2\zeta, \qquad \Gamma^3_{02} = \Gamma^3_{20} = \omega.$$

10. (a) Determine the geodesic equations on G.
    (b) Any straight line $(t, x, y, z) = (a^0 t + b^0, a^1 t + b^1, a^2 t + b^2, a^3 t + b^3)$ in R is a geodesic; why? Determine the image of this line in G, and show that the image is a geodesic in G.

11. Consider the following two curves in G:

$$\xi_1(t) = (t, 0, vt \cos\omega t, -vt \sin\omega t),$$
$$\xi_2(t) = (t, 0, (1 - vt) \cos\omega t, (vt - 1) \sin\omega t).$$

(a) Prove that ξ_1 and ξ_2 are geodesics and sketch their paths in the (η, ζ)-plane. To make the sketches take $v = 0.1$, $\omega = 1$, and $0 \le t \le \pi$.

(b) For $j = 1$ and 2, calculate $\boldsymbol{\xi}'_j(t)$ and $\boldsymbol{\xi}''_j(t)$ and sketch the images of these vectors on the curves in the (η, ζ)-plane when $t = 0$ and $t = 1$.

(c) The sketches should show that both $\boldsymbol{\xi}_1$ and $\boldsymbol{\xi}_2$ turn to the right, in the sense that the parallelogram $\boldsymbol{\xi}'_j \wedge \boldsymbol{\xi}''_j$ (cf. Section 5.1) has negative area. Prove that area $\boldsymbol{\xi}'_j(t) \wedge \boldsymbol{\xi}''_j(t) < 0$ for $j = 1$ and 2 and for all t.

12. Now consider a curve $\boldsymbol{\xi}(t)$ in G of the form

$$\tau = t,$$
$$\xi = 0,$$
$$\eta = (a^2 t + b^2) \cos \omega t + (a^3 t + b^3) \sin \omega t,$$
$$\zeta = -(a^2 t + b^2) \sin \omega t + (a^3 t + b^3) \cos \omega t.$$

Prove that this is a geodesic in the (η, ζ)-plane and show that it turns to the right, in the sense that area $\boldsymbol{\xi}'(t) \wedge \boldsymbol{\xi}''(t) < 0$ for all t.

The gravitational force that underlies this metric and that causes all geodesics to turn to the right is more commonly known as the *Coriolis force*.

7.2 The Vacuum Field Equations

The Newtonian field equation

At a point $\mathbf{x} = (x^1, x^2, x^3)$ in space where the density of matter is $\rho(\mathbf{x})$, the Newtonian gravitational field

$$-\nabla \Phi(\mathbf{x}) = -\operatorname{grad} \Phi(\mathbf{x}) = -\left(\frac{\partial \Phi}{\partial x^1}, \frac{\partial \Phi}{\partial x^2}, \frac{\partial \Phi}{\partial x^3} \right)$$

satisfies the **field equation**

$$\nabla^2 \Phi(\mathbf{x}) = \operatorname{div} \operatorname{grad} \Phi(\mathbf{x}) = \frac{\partial^2 \Phi}{(\partial x^1)^2} + \frac{\partial^2 \Phi}{(\partial x^2)^2} + \frac{\partial^2 \Phi}{(\partial x^3)^2} = 4\pi G \rho(\mathbf{x}).$$

This equation tells us how matter determines the gravitational field and, ultimately, how matter makes matter move.

A relativistic theory of gravity must answer the same question: How does matter determine the gravitational field $-\Gamma^k_{ij}$? But in relativity the question has an even deeper meaning, because the field is derived from the spacetime metric and so is a reflection of the geometric structure of spacetime. The presence of matter alters this geometry; it introduces curvature. The field equations—when we find them—must therefore tell us *how matter determines the curvature of spacetime.*

Classical Theory

The Newtonian field equation takes its simplest form at a point of empty space (the "vacuum"), where $\rho(\mathbf{x}) = 0$:

$$\nabla^2 \Phi(\mathbf{x}) = 0.$$

This is the *vacuum field equation*; we devote this section to determining its analogue in general relativity. We begin by reviewing the role it plays in classical Newtonian mechanics.

In Section 4.3, the vacuum field equation emerged in a discussion of the tidal effects of gravity. In a strictly uniform—that is, constant—gravitational field, there are no tides; the tides reveal how the field varies from point to point. Consider the gravitational force acting at each point of a large body E near a massive gravitational source S, as in the figure below. The force will vary in magnitude and direction from point to point.

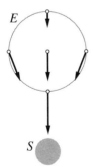

coordinate frame fixed with respect to gravitational source S

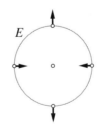

coordinate frame fixed at center of E

A family of trajectories

To focus on the *differences*, let us subtract off the force felt at the center of E. This has the effect of translating us to a coordinate frame that falls with the center of E. In this frame the net forces that appear are the tides. The tidal force draws some points together and pushes others apart.

To study these differences in a precise way, consider a collection of particles that start on a given curve $y^k(q)$ in $\mathbf{R}^3$ and fall freely in a gravitational field $-\nabla \Phi$. Let the trajectory of the particle that started at $y^k(q)$ be $x^k(t, q)$; that is, $x^k(0, q) = y^k(q)$.

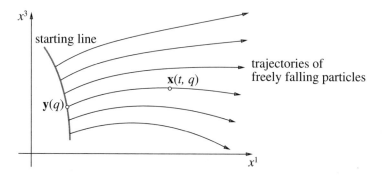

Each particle satisfies Newton's equations with acceleration provided solely by the gravitational potential Φ. For two particles that start at nearby points q and $q + \Delta q$ the equations are

$$\frac{d^2 x^k}{dt^2}(t, q) = -\frac{\partial \Phi}{\partial x^k}(x^j(t, q)),$$

$$\frac{d^2 x^k}{dt^2}(t, q + \Delta q) = -\frac{\partial \Phi}{\partial x^k}(x^j(t, q + \Delta q)).$$

(Notice that indices do not behave well in expressions in the classical theory; on one side k is a superscript, on the other it is a subscript.) Now take their difference and divide by Δq:

$$\frac{\frac{d^2 x^k}{dt^2}(t, q + \Delta q) - \frac{d^2 x^k}{dt^2}(t, q)}{\Delta q} = -\frac{\frac{\partial \Phi}{\partial x^k}(x^j(t, q + \Delta q)) - \frac{\partial \Phi}{\partial x^k}(x^j(t, q))}{\Delta q}.$$

§7.2 The Vacuum Field Equations

In the limit as $\Delta q \to 0$ we obtain

$$\frac{d^2}{dt^2}\frac{\partial x^k}{\partial q}(t, q) = -\frac{\partial^2 \Phi}{\partial q\, \partial x^k}(x^j(t, q)),$$

using d for derivatives with respect to t but ∂ for derivatives with respect to q. Applying the chain rule to the right-hand side we get the following second-order linear differential equations for $\partial x^j/\partial q$:

$$\frac{d^2}{dt^2}\frac{\partial x^k}{\partial q} = -\sum_j \frac{\partial^2 \Phi}{\partial x^j\, \partial x^k}\frac{\partial x^j}{\partial q}.$$

The equations of tidal acceleration

These are the **equations of tidal acceleration** we derived in Section 4.3, applied here to the vector $\partial \mathbf{x}/\partial q = \partial x^j/\partial q(t, q)$.

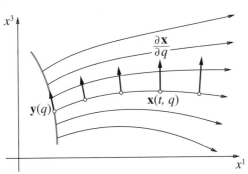

Why does the vector $\partial \mathbf{x}/\partial q = \partial x^j/\partial q$ indicate tidal effects? Follow $\partial \mathbf{x}/\partial q$ over time along the trajectory of a single particle. If $\partial \mathbf{x}/\partial q$ gets shorter, nearby particles are drawn together; if it gets longer, they drift apart. Thus $\partial \mathbf{x}/\partial q$ gives the *rate of separation of nearby trajectories*—precisely the tidal effect we want to measure.

The separation rate $\partial \mathbf{x}/\partial q$ indicates the tides

Since tidal acceleration is a linear function of the separation vector $\partial \mathbf{x}/\partial q$, we can write it as a matrix multiplication:

$$\frac{d^2}{dt^2}\frac{\partial \mathbf{x}}{\partial q} = -d^2\Phi_\mathbf{x} \cdot \frac{\partial \mathbf{x}}{\partial q}, \qquad \text{where} \qquad d^2\Phi_\mathbf{x} = \left(\frac{\partial^2 \Phi(\mathbf{x})}{\partial x^j\, \partial x^k}\right).$$

The vacuum field equation is just the statement that the trace of the matrix $d^2\Phi_\mathbf{x}$ is zero:

$$\operatorname{tr} d^2\Phi_\mathbf{x} = \sum_j \frac{\partial^2 \Phi(\mathbf{x})}{\partial x^j\, \partial x^j} = 0.$$

Vacuum field equation: $\operatorname{tr} d^2\Phi_\mathbf{x} = 0$

Character of the tidal force

Since $d^2\Phi_\mathbf{x}$ is a 3×3 symmetric matrix, it has three linearly independent eigenvectors with real eigenvalues. When $\partial\mathbf{x}/\partial q$ is one of these eigenvectors, tidal acceleration is in a parallel direction:

$$\frac{d^2}{dt^2}\frac{\partial \mathbf{x}}{\partial q} = -\lambda \frac{\partial \mathbf{x}}{\partial q}, \quad \lambda \text{ the eigenvalue for } \frac{\partial \mathbf{x}}{\partial q}.$$

When λ is positive, the acceleration points back toward the origin, so the tidal force *attracts*. By contrast, when λ is negative, acceleration points away from the origin, so the tidal force *repels*. Since the sum of the eigenvalues of any matrix equals its trace, and the trace of $d^2\Phi_\mathbf{x}$ is 0, its eigenvalues have different signs. Hence the tidal force attracts in some directions but repels in others.

Summary: tidal acceleration is a 3×3 symmetric matrix with trace zero

Here is a summary of the classical theory. If $x(t, q)$ describes the trajectory of a particle whose initial position varies with q, then the rate of separation of nearby trajectories, $\partial\mathbf{x}/\partial q$, is governed by the tidal acceleration law

$$\frac{d^2}{dt^2}\frac{\partial \mathbf{x}}{\partial q} = -d^2\Phi_\mathbf{x} \cdot \frac{\partial \mathbf{x}}{\partial q}.$$

The nature of tidal acceleration is therefore encapsulated by $d^2\Phi_\mathbf{x}$: *It is a symmetric 3×3 matrix whose trace is zero*.

Separation of Geodesics

Tides alter the separation of geodesics

How does general relativity describe the tidal acceleration experienced by freely falling bodies in a gravitational field? The short answer is this: Since free-fall occurs along geodesics, we expect that tidal effects will be manifested in the rate of separation of geodesics.

Let us review free-fall in a spacetime coordinate frame R : (x^0, x^1, x^2, x^3) endowed with a gravitational potential g_{kl}. Choose coordinates such that the x^0-axis is timelike and the other three are spacelike. Suppose the observer G is falling freely along the worldcurve $x^k = z^k(\tau)$ in R, where τ is G's proper time. Then G experiences zero acceleration by definition: The rate of change of the velocity vector $dz^k/d\tau$ is zero. Since the velocity defines a vector field along the worldcurve, to calculate its rate of change in a generally covariant way we should use the covariant derivative along G. This is

acceleration of G: $$\frac{D}{d\tau}\frac{dz^k}{d\tau} = \frac{d^2 z^k}{d\tau^2} + \Gamma^k_{ij}\frac{dz^i}{d\tau}\frac{dz^j}{d\tau}.$$

Therefore, the zero-acceleration condition is precisely the condition that the worldcurve of G be a geodesic:

$$\frac{D}{d\tau}\frac{dz^k}{d\tau} = \frac{d^2 z^k}{d\tau^2} + \Gamma^k_{ij}\frac{dz^i}{d\tau}\frac{dz^j}{d\tau} = 0.$$

To study the separation of geodesics near G, let us embed the worldcurve of G in a family of geodesics $x^k = x^k(\tau, q)$. Assume that $q = 0$ gives the curve G itself, so $x^k(\tau, 0) = z^k(\tau)$, and assume that τ is the proper-time parameter for each geodesic. The rate of separation is the vector

Embed G in a family of geodesics

$$\frac{\partial x^k}{\partial q}(\tau, q),$$

which we consider to be a function of τ for fixed q. The fundamental result about the rate of separation of geodesics is contained in the following theorem.

Theorem 7.1 *If $x^h(\tau, q)$ is a family of geodesics, then*

$$\frac{D^2}{d\tau^2}\frac{\partial x^h}{\partial q} + R^h_{ijk}\frac{dx^i}{d\tau}\frac{\partial x^j}{\partial q}\frac{dx^k}{d\tau} = 0,$$

where R^h_{ijk} is the Riemann curvature tensor defined by the metric g_{kl}.

PROOF: To underscore the different roles of τ and q, we shall use d to denote derivatives with respect to τ when possible. We must calculate the first and second covariant derivatives along $x^h(\tau, q)$, for fixed q. First,

$$\frac{D}{d\tau}\frac{\partial x^h}{\partial q} = \frac{d}{d\tau}\left(\frac{\partial x^h}{\partial q}\right) + \Gamma^h_{ij}\frac{dx^i}{d\tau}\frac{\partial x^j}{\partial q}.$$

Next,

$$\frac{D^2}{d\tau^2}\frac{\partial x^h}{\partial q} = \frac{d}{d\tau}\left(\frac{\partial^2 x^h}{\partial \tau \partial q} + \Gamma^h_{ij}\frac{dx^i}{d\tau}\frac{\partial x^j}{\partial q}\right) + \Gamma^h_{mk}\left(\frac{\partial^2 x^m}{\partial \tau \partial q} + \Gamma^m_{ij}\frac{dx^i}{d\tau}\frac{\partial x^j}{\partial q}\right)\frac{dx^k}{d\tau}$$

$$= \underbrace{\frac{\partial}{\partial q}\frac{d^2 x^h}{d\tau^2}}_{A} + \frac{\partial \Gamma^h_{ij}}{\partial x^k}\frac{dx^k}{d\tau}\frac{dx^i}{d\tau}\frac{\partial x^j}{\partial q} + \Gamma^h_{ij}\underbrace{\frac{d^2 x^i}{d\tau^2}\frac{\partial x^j}{\partial q}}_{A} + \Gamma^h_{ij}\frac{dx^i}{d\tau}\frac{\partial^2 x^j}{\partial \tau \partial q}$$

$$+ \Gamma^h_{mk}\frac{\partial^2 x^m}{\partial \tau \partial q}\frac{dx^k}{d\tau} + \Gamma^h_{mk}\Gamma^m_{ij}\frac{dx^i}{d\tau}\frac{\partial x^j}{\partial q}\frac{dx^k}{d\tau}.$$

At this point we use the fact that each x^h is a geodesic to replace

$$\frac{d^2 x^h}{d\tau^2} \quad \text{by} \quad -\Gamma^h_{ik}\frac{dx^i}{d\tau}\frac{dx^k}{d\tau}$$

at the two places marked A (and making the substitution $i \mapsto m$ in the second). Then

$$\frac{D^2}{d\tau^2}\frac{\partial x^h}{\partial q} = \frac{\partial}{\partial q}\left(-\Gamma^h_{ik}\frac{dx^i}{d\tau}\frac{dx^k}{d\tau}\right) + \frac{\partial \Gamma^h_{ij}}{\partial x^k}\frac{dx^i}{d\tau}\frac{\partial x^j}{\partial q}\frac{dx^k}{d\tau} + \Gamma^h_{mj}\left(-\Gamma^m_{ik}\frac{dx^i}{d\tau}\frac{dx^k}{d\tau}\right)\frac{\partial x^j}{\partial q}$$

$$+ \Gamma^h_{ij}\frac{dx^i}{d\tau}\frac{\partial^2 x^j}{\partial \tau \partial q} + \Gamma^h_{mk}\frac{\partial^2 x^m}{\partial \tau \partial q}\frac{dx^k}{d\tau} + \Gamma^h_{mk}\Gamma^m_{ij}\frac{dx^i}{d\tau}\frac{\partial x^j}{\partial q}\frac{dx^k}{d\tau}$$

$$= -\frac{\partial \Gamma^h_{ik}}{\partial x^j}\frac{\partial x^j}{\partial q}\frac{dx^i}{d\tau}\frac{dx^k}{d\tau} - 2\Gamma^h_{ik}\frac{\partial^2 x^i}{\partial q \partial \tau}\frac{dx^k}{d\tau} + \frac{\partial \Gamma^h_{ij}}{\partial x^k}\frac{dx^i}{d\tau}\frac{\partial x^j}{\partial q}\frac{dx^k}{d\tau}$$

$$- \Gamma^h_{mj}\Gamma^m_{ik}\frac{dx^i}{d\tau}\frac{\partial x^j}{\partial q}\frac{dx^k}{d\tau} + 2\Gamma^h_{ik}\frac{\partial^2 x^i}{\partial q \partial \tau}\frac{dx^k}{d\tau} + \Gamma^h_{mk}\Gamma^m_{ij}\frac{dx^i}{d\tau}\frac{\partial x^j}{\partial q}\frac{dx^k}{d\tau}$$

$$= \left(\frac{\partial \Gamma^h_{ij}}{\partial x^k} - \frac{\partial \Gamma^h_{ik}}{\partial x^j} + \Gamma^h_{mk}\Gamma^m_{ij} - \Gamma^h_{mj}\Gamma^m_{ik}\right)\frac{dx^i}{d\tau}\frac{\partial x^j}{\partial q}\frac{dx^k}{d\tau}$$

$$= R^h_{ikj}\frac{dx^i}{d\tau}\frac{\partial x^j}{\partial q}\frac{dx^k}{d\tau} = -R^h_{ijk}\frac{dx^i}{d\tau}\frac{\partial x^j}{\partial q}\frac{dx^k}{d\tau}. \qquad \text{END OF PROOF}$$

Corollary 7.1 (The tidal acceleration equations) Let $z^k(\tau) = x^k(\tau, 0)$ be the worldline of the observer G, embedded in the family of geodesics $x^k(\tau, q)$. Then the rate of separation of geodesics along $z^k(\tau)$ ~~is~~ satisfies the equations

$$\frac{D^2}{d\tau^2}\frac{\partial x^h}{\partial q} = -K^h_j\frac{\partial x^j}{\partial q}, \qquad \text{where} \qquad K^h_j = R^h_{ijk}\frac{dz^i}{d\tau}\frac{dz^k}{d\tau}.$$

Fermi Coordinates

The tensor K^h_j plays the same role in the relativistic tidal acceleration equations that the matrix

$$d^2\Phi_{\mathbf{x}} = \left(\frac{\partial^2 \Phi(\mathbf{x})}{\partial x^j \partial x^k}\right)$$

plays in the classical equations. Indeed, the components of K^h_j involve first derivatives of the gravitational field, just like the components of $d^2\Phi_{\mathbf{x}}$. However, there are some differences: K^h_j is a 4×4 matrix, not a 3×3, and it is not obviously symmetric, either. By recomputing K^h_j in a new coordinate system—introduced by the physicist Enrico Fermi in 1922—we shall find that it has a special form that will allow us to identify it in all respects with the classical matrix.

Fermi coordinates form a coordinate frame $G : (\xi^0, \xi^1, \xi^2, \xi^3)$ for the observer who is falling freely in the gravitational field defined by the metric g_{ij} in the frame R. From G's point of view, there *is* no field. Fermi coordinates are constructed to achieve precisely this: All the components of the gravitational field $-\Gamma^h_{ij}$ in G's frame will vanish everywhere on G's worldline. To define the Fermi coordinates we shall describe how they are connected to the coordinates for R—that is, we shall describe the map $M : G \to R$. This is best done in a number of steps.

Comparing K^h_j and $d^2\Phi_{\mathbf{x}}$

$\overset{G}{\Gamma}{}^h_{ij}$ must vanish along the worldline of G

STEP 1. Let $\xi^0 = \tau$, and make G's worldcurve the ξ^0-axis in the frame G. Since $x^k = z^k(\tau) = z^k(\xi^0)$ is G's worldcurve in R, this tells us how M maps the ξ^0-axis.

Image of ξ^0-axis

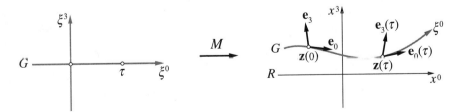

Orthonormal basis along G

STEP 2. The velocity vector $\mathbf{e}_0(\tau) = dz^k/d\tau$ is a timelike unit vector at every point on G's worldcurve in R. Choose unit spacelike vectors $\mathbf{e}_1$, $\mathbf{e}_2$, and $\mathbf{e}_3$ at the point $z^k(0)$ so that $\{\mathbf{e}_0, \mathbf{e}_1, \mathbf{e}_2, \mathbf{e}_3\}$ form an orthonormal basis for $TR_{\mathbf{z}(0)}$. This means that in terms of the metric g_{kl} defined in $TR_{\mathbf{z}(\tau)}$,

$$\mathbf{e}_0 \cdot \mathbf{e}_0 = 1, \qquad \mathbf{e}_\alpha \cdot \mathbf{e}_\alpha = -1, \ \alpha = 1, 2, 3, \qquad \mathbf{e}_k \cdot \mathbf{e}_l = 0, \ k \neq l.$$

Then define $\mathbf{e}_i(\tau)$ to be the *parallel transport* of $\mathbf{e}_i$ along the worldcurve G in R, for $i = 1, 2, 3$. Since parallel transport preserves inner products, the vectors

$$\{\mathbf{e}_0(\tau), \mathbf{e}_1(\tau), \mathbf{e}_2(\tau), \mathbf{e}_3(\tau)\}$$

form an orthonormal basis for the tangent space $TR_{\mathbf{z}(\tau)}$, for each τ.

Representing a point off the ξ^0-axis

STEP 3. Now consider a point

$$P = (\xi^0, \xi^1, \xi^2, \xi^3) = (\tau, \boldsymbol{\xi})$$

in G that lies off the ξ^0-axis; its spatial component $\boldsymbol{\xi} = (\xi^1, \xi^2, \xi^3)$ is nonzero. Think of $\boldsymbol{\xi}$ as a vector in the tangent space $TG_{(\tau,\mathbf{0})}$, and let $\mathbf{n}$ be the unit vector in the direction of $\boldsymbol{\xi}$, where we measure length with the ordinary Euclidean metric

$$\lambda = \|\boldsymbol{\xi}\| = \sqrt{(\xi^1)^2 + (\xi^2)^2 + (\xi^3)^2}.$$

Therefore, $\mathbf{n} = \boldsymbol{\xi}/\lambda = (\xi^1/\lambda, \xi^2/\lambda, \xi^3/\lambda) = (n^1, n^2, n^3)$, and we can write $P = (\tau, \lambda\mathbf{n}) = (\tau, \lambda n^1, \lambda n^2, \lambda n^3)$.

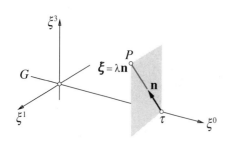
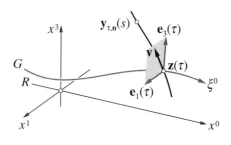

§7.2 The Vacuum Field Equations

STEP 4. Now move to R. Let $\mathbf{v}$ be the vector in the tangent space $TR_{\mathbf{z}(\tau)}$ defined by the condition

The geodesic $\mathbf{y}_{\tau,\mathbf{n}}$

$$\mathbf{v} = n^1 \mathbf{e}_1(\tau) + n^2 \mathbf{e}_2(\tau) + n^3 \mathbf{e}_3(\tau) = n^\alpha \mathbf{e}_\alpha(\tau).$$

(Recall that we use Greek letters for the spatial indices 1, 2, 3.) By definition, $\mathbf{v}$ has no component in the $\mathbf{e}_0(\tau)$ direction. In terms of the inner product g_{kl} on $TR_{\mathbf{z}(\tau)}$, $\mathbf{v}$ is a spacelike unit vector:

$$\mathbf{v} \cdot \mathbf{v} = n^\alpha n^\beta \, \mathbf{e}_\alpha(\tau) \cdot \mathbf{e}_\beta(\tau)$$
$$= -n^\alpha n^\beta \delta_{\alpha\beta} = -(n^1)^2 - (n^2)^2 - (n^3)^2 = -\|\mathbf{n}\|^2 = -1.$$

Let $\mathbf{y}_{\tau,\mathbf{n}}(s)$ be the unique geodesic in R determined by the initial conditions

$$\mathbf{y}_{\tau,\mathbf{n}}(0) = \mathbf{z}(\tau), \qquad \mathbf{y}'_{\tau,\mathbf{n}}(0) = \mathbf{v}.$$

Since $\mathbf{v}$ is a unit (spacelike) vector in R, $\mathbf{y}'_{\tau,\mathbf{n}}(s)$ is a unit speed vector at $s = 0$. Since $\mathbf{y}_{\tau,\mathbf{n}}$ is a geodesic, $\mathbf{y}'_{\tau,\mathbf{n}}(s)$ is a unit speed vector for *all* s; thus, s measures arc length along $\mathbf{y}_{\tau,\mathbf{n}}$.

STEP 5. We define the image of the point $P = (\tau, \lambda \mathbf{n})$ in G to be the point $\mathbf{y}_{\tau,\mathbf{n}}(\lambda)$ in R:

Completing the map $M: G \to R$

$$M: G \to R: (\tau, \lambda \mathbf{n}) \mapsto \mathbf{y}_{\tau,\mathbf{n}}(\lambda).$$

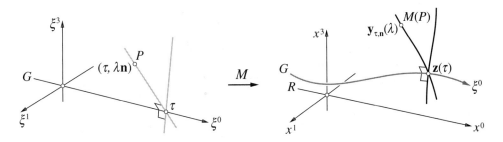

We can describe the map geometrically in the following way. Consider the collection of lines perpendicular to the ξ^0-axis at the point τ, where we mean perpendicular in the Euclidean sense. Two of these lines are shown in gray in the figure above. Each line will map to a geodesic in R that is perpendicular (in the sense of the metric in R!) to G's worldcurve at the corresponding point

$\mathbf{z}(\tau)$. Along a given gray line, points are mapped isometrically to the geodesic. In other words, a point at distance λ from the ξ^0-axis is mapped to a point at arc length λ from G's worldline along the corresponding geodesic.

Proposition 7.1 *The map $M : G \to R$ is invertible in a neighborhood of the ξ^0-axis.*

PROOF: According to the inverse function theorem, a map $M : G \to R$ will be invertible in a neighborhood of a point P in its domain if the differential $dM_P : TG_P \to TR_{M(P)}$ is invertible. Thus, to prove the proposition it suffices to show that dM_P is invertible at every point $P = (\tau, 0, 0, 0)$ on the ξ^0-axis.

To prove that a given dM_P is invertible it is sufficient to show that the image of a basis of $TG_P = TG_{(\tau,0,0,0)}$ is linearly independent in the target $TR_{M(P)} = TR_{\mathbf{z}(\tau)}$. We take the standard basis in $TG_{(\tau,0,0,0)}$:

$$\varepsilon_0 = \begin{pmatrix} 1 \\ 0 \\ 0 \\ 0 \end{pmatrix}, \quad \varepsilon_1 = \begin{pmatrix} 0 \\ 1 \\ 0 \\ 0 \end{pmatrix}, \quad \varepsilon_2 = \begin{pmatrix} 0 \\ 0 \\ 1 \\ 0 \end{pmatrix}, \quad \varepsilon_3 = \begin{pmatrix} 0 \\ 0 \\ 0 \\ 1 \end{pmatrix}.$$

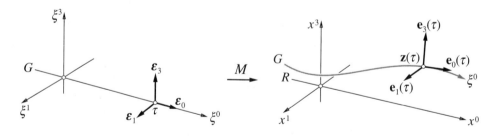

Now, M maps the ξ^0-axis itself to G's worldcurve in R; specifically,

$$M(\tau, 0, 0, 0) = \mathbf{z}(\tau).$$

Therefore, $dM_P(\varepsilon_0) = \mathbf{z}'(\tau) = \mathbf{e}_0(\tau)$. Furthermore, by definition of M, the vector $\boldsymbol{\xi} = \varepsilon_\alpha$ maps to the vector $\mathbf{v} = \mathbf{e}_\alpha(\tau)$. In other words,

$$dM_{(\tau,0,0,0)} : \{\varepsilon_0, \varepsilon_1, \varepsilon_2, \varepsilon_3\} \mapsto \{\mathbf{e}_0(\tau), \mathbf{e}_1(\tau), \mathbf{e}_2(\tau), \mathbf{e}_3(\tau)\},$$

showing that $dM_{(\tau,0,0,0)}$ is invertible.   END OF PROOF

We use the map $M : G \to R$ to "pull back" the metric from R to G in a generally covariant way. Suppose $\boldsymbol{\xi}$ and $\boldsymbol{\eta}$ are two vectors in the tangent space TG_P. Let $\mathbf{x} = dM_P(\boldsymbol{\xi})$ and $\mathbf{y} = dM_P(\boldsymbol{\eta})$ be their images in $TR_{M(P)}$. Then *by definition*,

$$\underset{\text{in } TG_P}{\boldsymbol{\xi} \cdot \boldsymbol{\eta}} = \underset{\text{in } TR_{M(P)}}{\mathbf{x} \cdot \mathbf{y}}.$$

Defining the metric in G

Let us see what this equation implies when we use it to determine the components γ_{ij} of the metric tensor in G. If

$$\xi = \frac{d\xi^i}{dt}, \qquad \mathbf{x} = \frac{dx^k}{dt} = \frac{\partial x^k}{\partial \xi^i}\frac{d\xi^i}{dt},$$

$$\eta = \frac{d\eta^j}{dt}, \qquad \mathbf{y} = \frac{dy^l}{dt} = \frac{\partial x^l}{\partial \xi^j}\frac{d\eta^j}{dt},$$

then

$$\boldsymbol{\xi} \cdot \boldsymbol{\eta} = \underbrace{\gamma_{ij}}\frac{d\xi^i}{dt}\frac{d\eta^j}{dt} \quad \text{and} \quad \mathbf{x} \cdot \mathbf{y} = g_{kl}\frac{dx^k}{dt}\frac{dy^l}{dt} = \underbrace{g_{kl}\frac{\partial x^k}{\partial \xi^i}\frac{\partial x^l}{\partial \xi^j}}\frac{d\xi^i}{dt}\frac{d\eta^j}{dt}.$$

Since we want these expressions to be equal for all possible vectors $d\xi^i/dt$ and $d\eta^j/dt$, the factors underscored by braces must be equal, giving us the familiar coordinate transformation

$$\gamma_{ij} = g_{kl}\frac{\partial x^k}{\partial \xi^i}\frac{\partial x^l}{\partial \xi^j},$$

which we can take as the definition of γ_{ij}.

Corollary 7.2 *The following lines are geodesics in the Fermi coordinate frame G:*

Special geodesics in G

- *the ξ^0-axis;*
- *the lines $\zeta(s) = (\tau, s\mathbf{n})$ orthogonal to the ξ^0-axis at every point τ on the ξ^0-axis.*

PROOF: By design, the map $M : G \to R$ carries each of these lines to a geodesic in R. By general covariance, they are therefore geodesics in G. END OF PROOF

Corollary 7.3 *At every point $P = (\tau, 0, 0, 0)$ in the Fermi coordinate frame G, the metric is the standard Minkowski metric,*

$$\gamma_{ij}(P) = \begin{pmatrix} 1 & 0 & 0 & 0 \\ 0 & -1 & 0 & 0 \\ 0 & 0 & -1 & 0 \\ 0 & 0 & 0 & -1 \end{pmatrix}.$$

PROOF: The value of $\gamma_{ij}(P)$ is the value of the inner product $\varepsilon_i \cdot \varepsilon_j$ as calculated in TG_P. By definition of the metric in G,

$$\underbrace{\varepsilon_i \cdot \varepsilon_j}_{\text{in } TG_P} = \underbrace{dM_P(\varepsilon_i) \cdot dM_P(\varepsilon_j)}_{\text{in } TR_{M(P)}} = \mathbf{e}_i(\tau) \cdot \mathbf{e}_j(\tau).$$

Since $\{\mathbf{e}_i(\tau)\}$ is an orthonormal basis in $TR_{M(P)}$, by construction, the result follows. END OF PROOF

The Fermi conditions

Theorem 7.2 (The Fermi conditions) *The gravitational field $\overset{G}{\Gamma}{}^h_{ij}(P)$ is zero at every point $P = (\tau, 0, 0, 0)$ on the worldline of a Fermi coordinate frame G.*

PROOF: Since each line $\zeta(s) = (\tau, s\mathbf{n}) = (\tau, sn^\alpha)$ is a geodesic, the components $\zeta^h(s)$ satisfy the geodesic equations

$$\frac{d^2\zeta^h}{ds^2} + \overset{G}{\Gamma}{}^h_{ij}(\tau, s\mathbf{n})\frac{d\zeta^i}{ds}\frac{d\zeta^j}{ds} = 0$$

at each point s. Since

$$\frac{d\zeta^0}{ds} = 0, \quad \frac{d\zeta^\alpha}{ds} = n^\alpha, \quad \frac{d^2\zeta^h}{ds^2} = 0$$

(where $\alpha, \beta = 1, 2, 3$ and $h = 0, 1, 2, 3$ as usual), the geodesic equations reduce to

$$\overset{G}{\Gamma}{}^h_{\alpha\beta}(\tau, sn^1, sn^2, sn^3)n^\alpha n^\beta = 0.$$

These hold for all s and all unit 3-vectors $\mathbf{n} = n^\alpha$. Now set $s = 0$; then

$$\overset{G}{\Gamma}{}^h_{\alpha\beta}(\tau, 0, 0, 0) n^\alpha n^\beta = 0.$$

Since the Christoffel symbols are now being evaluated on the ξ^0-axis and, in particular, are independent of $\mathbf{n}$, we can finally conclude that

$$\overset{G}{\Gamma}{}^h_{\alpha\beta}(\tau, 0, 0, 0) = 0$$

for every τ and for $h = 0, 1, 2, 3$, and $\alpha, \beta = 1, 2, 3$.

To deal with the remaining symbols, $\overset{G}{\Gamma}{}^h_{k0} = \overset{G}{\Gamma}{}^h_{0k}$, we use the fact that the vectors $\mathbf{e}_k(\tau)$ in R are defined by parallel transport along G's worldline. Consequently, their preimages ε_k under M are "parallel-transported" along the ξ^0-axis in G. In terms of components,

$$\varepsilon_k = \left(\delta^h_k\right),$$

so the parallel transport equations for ε_k in G are

$$\frac{d}{d\tau}\delta^h_k + \overset{G}{\Gamma}{}^h_{ij}(\tau, 0, 0, 0)\delta^i_k \frac{d\zeta^j}{d\tau} = 0.$$

We use $\zeta(\tau) = (\tau, 0, 0, 0)$ to parametrize the ξ^0-axis; hence $d\zeta^0/d\tau = 1$ and $d\zeta^\alpha/d\tau = 0$. Therefore,

$$\overset{G}{\Gamma}{}^h_{i0}(\tau, 0, 0, 0)\delta^i_k = \overset{G}{\Gamma}{}^h_{k0}(\tau, 0, 0, 0) = 0. \qquad \text{END OF PROOF}$$

Corollary 7.4 *At every point $P = (\tau, 0, 0, 0)$ on the ξ^0-axis,*

$$\overset{G}{\Gamma}_{ij,k}(P) = 0, \qquad \frac{\partial \gamma_{ij}}{\partial \xi^k}(P) = 0, \qquad \frac{\partial^2 \gamma_{ij}}{\partial \xi^0 \partial \xi^k}(P) = 0.$$

PROOF: Every Christoffel symbol of the first kind is a linear combination of Christoffel symbols of the second kind:

$$\overset{G}{\Gamma}_{ij,k}(P) = \gamma_{kh}(P)\overset{G}{\Gamma}{}^h_{ij}(P) = 0.$$

Similarly, $\dfrac{\partial \gamma_{ij}}{\partial \xi^k}(P) = \overset{G}{\Gamma}_{ik,j}(P) + \overset{G}{\Gamma}_{jk,i}(P) = 0$. Finally,

$$\frac{\partial^2 \gamma_{ij}}{\partial \xi^0 \partial \xi^k}(P) = \lim_{\theta \to 0} \frac{\frac{\partial \gamma_{ij}}{\partial \xi^k}(\tau + \theta, 0, 0, 0) - \frac{\partial \gamma_{ij}}{\partial \xi^k}(\tau, 0, 0, 0)}{\theta}$$

$$= \lim_{\theta \to 0} \frac{0 - 0}{\theta} = 0. \qquad \text{END OF PROOF}$$

The Relativistic Vacuum Field Equations

K_j^h in G's frame

Let us return to the tidal acceleration equations, but expressed now in terms of the Fermi coordinate frame of a freely falling observer G whose worldcurve has the simple form

$$\zeta^h(\tau) = (\tau, 0, 0, 0).$$

We assume that this curve is embedded in a family of geodesics $\xi^h(\tau, q)$, with $\xi^h(\tau, 0) = \zeta^h(\tau)$. The tidal acceleration equations for G are

$$\frac{D^2}{d\tau^2} \frac{\partial \xi^h}{\partial q} = -\overset{G}{K}{}_j^h \frac{\partial \xi^j}{\partial q},$$

but because $\zeta^h(\tau)$ is so simple, $\overset{G}{K}{}_j^h$ is just a single term:

$$\overset{G}{K}{}_j^h = \overset{G}{R}{}_{ijk}^h \frac{d\zeta^i}{d\tau} \frac{d\zeta^k}{d\tau} = \overset{G}{R}{}_{0j0}^h.$$

Let us determine $\overset{G}{K}{}_j^h(P)$ explicitly at each point $P = (\tau, 0, 0, 0)$ along the ξ^0-axis, in terms of the metric γ_{ij}. Since all calculation will be done in G's frame, we will stop including the super-symbol G in the various multi-index quantities. First of all, $R_{0j0}^h(P)$ has only two terms,

$$R_{0j0}^h(P) = \frac{\partial \Gamma_{00}^h}{\partial \xi^j}(P) - \frac{\partial \Gamma_{0j}^h}{\partial \xi^0}(P),$$

because the Christoffel symbols are all zero along the ξ^0-axis. Next, since

$$\Gamma_{0j}^h(\tau, 0, 0, 0) = 0 \quad \text{for all } \tau,$$

we have

$$\frac{\partial \Gamma^h_{0j}}{\partial \xi^0}(P) = \frac{\partial \Gamma^h_{0j}}{\partial \tau}(\tau, 0, 0, 0) = 0.$$

Therefore,

$$K^h_j(P) = R^h_{0j0}(P) = \frac{\partial \Gamma^h_{00}}{\partial \xi^j}(P).$$

Theorem 7.3 $\dfrac{\partial \Gamma^h_{00}}{\partial \xi^j}(P) = \pm \dfrac{1}{2} \dfrac{\partial^2 \gamma_{00}}{\partial \xi^j \partial \xi^h}(P).$

PROOF: At an arbitrary point near the ξ^0-axis,

$$\Gamma^h_{00} = \gamma^{hm} \Gamma_{00,m} = \frac{\gamma^{hm}}{2}\left(2\frac{\partial \gamma_{0m}}{\partial \xi^0} - \frac{\partial \gamma_{00}}{\partial \xi^m}\right).$$

Therefore,

$$\frac{\partial \Gamma^h_{00}}{\partial \xi^j} = \frac{1}{2}\frac{\partial \gamma^{hm}}{\partial \xi^j}\underbrace{\left(2\frac{\partial \gamma_{0m}}{\partial \xi^0} - \frac{\partial \gamma_{00}}{\partial \xi^m}\right)} + \frac{\gamma^{hm}}{2}\left(2\underbrace{\frac{\partial^2 \gamma_{0m}}{\partial \xi^0 \partial \xi^j}} - \frac{\partial^2 \gamma_{00}}{\partial \xi^j \partial \xi^m}\right).$$

The terms underscored by braces vanish at the point $P = (\tau, 0, 0, 0)$ on the ξ^0-axis, according to Corollary 7.4. Hence

$$\frac{\partial \Gamma^h_{00}}{\partial \xi^j}(P) = -\frac{\gamma^{hm}(P)}{2}\frac{\partial^2 \gamma_{00}}{\partial \xi^j \partial \xi^m}(P).$$

Now, $\gamma^{00}(P) = 1$, $\gamma^{\alpha\alpha}(P) = -1$, and $\gamma^{hm}(P) = 0$ if $m \neq h$; hence

$$\frac{\partial \Gamma^h_{00}}{\partial \xi^j}(P) = -\frac{\pm 1}{2}\frac{\partial^2 \gamma_{00}}{\partial \xi^j \partial \xi^h}(P). \qquad \text{END OF PROOF}$$

Corollary 7.5 $K^h_0(P) = K^0_j(P) = 0$ and $K^\beta_\alpha(P) = K^\alpha_\beta(P)$, $\alpha, \beta = 1, 2, 3$.

PROOF: By Corollary 7.4,

$$K^h_0(P) = \frac{\partial \Gamma^h_{00}}{\partial \xi^0}(P) = \pm\frac{1}{2}\frac{\partial^2 \gamma_{00}}{\partial \xi^0 \partial \xi^h}(P) = 0.$$

In the same way we obtain $K_j^0(P) = 0$. To prove the last claim, just note that the ± sign is always a plus sign in the index range 1, 2, 3:

$$K_\alpha^\beta(P) = \frac{1}{2} \frac{\partial^2 \gamma_{00}}{\partial \xi^\alpha \partial \xi^\beta}(P) = \frac{1}{2} \frac{\partial^2 \gamma_{00}}{\partial \xi^\beta \partial \xi^\alpha}(P) = K_\beta^\alpha(P). \quad \text{END OF PROOF}$$

K is a matrix of second partial derivatives

Thus in G's special Fermi coordinate frame the tidal acceleration equations are expressed in terms of the matrix defined along G's worldcurve:

$$\overset{G}{K} = \frac{1}{2} \begin{pmatrix} 0 & 0 & 0 & 0 \\ 0 & \frac{\partial^2 \gamma_{00}}{\partial \xi^1 \partial \xi^1} & \frac{\partial^2 \gamma_{00}}{\partial \xi^2 \partial \xi^1} & \frac{\partial^2 \gamma_{00}}{\partial \xi^3 \partial \xi^1} \\ 0 & \frac{\partial^2 \gamma_{00}}{\partial \xi^1 \partial \xi^2} & \frac{\partial^2 \gamma_{00}}{\partial \xi^2 \partial \xi^2} & \frac{\partial^2 \gamma_{00}}{\partial \xi^3 \partial \xi^2} \\ 0 & \frac{\partial^2 \gamma_{00}}{\partial \xi^1 \partial \xi^3} & \frac{\partial^2 \gamma_{00}}{\partial \xi^2 \partial \xi^3} & \frac{\partial^2 \gamma_{00}}{\partial \xi^3 \partial \xi^3} \end{pmatrix}.$$

Therefore, in both relativity and classical mechanics, information about tidal acceleration is found in a symmetric 3×3 matrix of second partial derivatives. The corresponding functions are

$$\frac{1}{2} \gamma_{00}(\xi^0, \xi^1, \xi^2, \xi^3) \longleftrightarrow \Phi(x^1, x^2, x^3).$$

In other words, among the 10 components of the gravitational potential in G's frame, $\gamma_{00}/2$ corresponds most closely to the scalar potential Φ of Newtonian theory. We cannot yet say they are equal, though; we shall look more closely at the correspondence in Sections 7.3 and 8.1.

The vacuum field equations: $R_{ij} = 0$

In the classical theory the trace of the tidal acceleration matrix is zero; *by analogy we impose the same requirement on the relativistic matrix*:

$$\overset{G}{K}{}_h^h = 0.$$

The super-symbol G emphasizes that our analogy is based, at the outset, on the special character of K in G's frame. However, by general covariance, what is true in G's frame must be true in any

§7.2 The Vacuum Field Equations

other frame R:

$$\overset{R}{K}{}^h_h = \overset{R}{R}{}^h_{ihk} \frac{dz^i}{d\tau}\frac{dz^k}{d\tau} = 0.$$

Here $z^i(\tau)$ is the worldcurve of G in R. This is a geodesic, and since G could be moving on any geodesic worldcurve, the derivatives $dz^i/d\tau$ and $dz^k/d\tau$ are arbitrary (timelike) vectors. This implies

$$\overset{R}{R}{}^h_{ihk} = 0.$$

This particular contraction of the Riemann curvature tensor is clearly important; it is the Ricci tensor.

Ricci tensor: $R_{ik} = R^h_{ihk}.$

The relativistic vacuum field equations are expressed in terms of the Ricci tensor:

The vacuum field equations: $R_{ik} = 0.$

Like the classical equations, these are second-order partial differential equations in the gravitational potentials. However, they also have geometric significance; they say that certain combinations of the components of the curvature tensor must vanish. This puts constraints on the curvature of spacetime at a point in the vacuum, but does not say that the curvature is necessarily zero.

Albert Einstein. Photo: THE GRANGER COLLECTION, New York. Reproduced with permission.

"The happiest thought of my life"

To formulate the vacuum equations we made essential use of a coordinate frame in which the gravitational field disappears—at least along a single worldcurve. In 1907, early on in his thinking about general relativity, Einstein described his own delight in realizing how important it would be to take the point of view of a freely falling observer in order to eliminate the gravitational field (quoted by Pais [26], page 178):

> Then there occurred to me the happiest thought of my life, in the following form. The gravitational field has only a relative existence in a way similar to the electric field generated by magnetoelectric induction. *Because for an observer falling freely from the room of a house there exists—at least in his immediate surroundings—no gravitational field.* Indeed, if the observer drops some bodies, then these remain relative to him in a state of rest or of uniform motion, independent of their particular chemical or physical nature (in consideration of which the air resistance is, of course, ignored). The observer has the right to interpret his state as "at rest."

Exercises

The first three exercises concern the sphere of radius r with its usual metric,

$$(g_{ij}) = \begin{pmatrix} r^2 \cos^2 q^2 & 0 \\ 0 & r^2 \end{pmatrix}, \quad 0 \leq q^1 \leq 2\pi, \quad -\pi/2 < q^2 < \pi/2.$$

1. Consider the one-parameter family of geodesics $x^1(t, q) = q$, $x^2(t, q) = t$; verify directly that $x^h(t, q)$ satisfy the geodesic separation equations

$$\frac{D^2}{dt^2} \frac{\partial x^h}{\partial q} + R^h_{ijk} \frac{dx^i}{dt} \frac{dx^k}{dt} \frac{\partial x^j}{\partial q} = 0.$$

2. Are (q^1, q^2) Fermi coordinates for the equator $(q^1, q^2) = (t, 0)$? Are they Fermi coordinates for the prime meridian $(q^1, q^2) = (0, t)$?

§7.2 The Vacuum Field Equations

3. (a) Determine the geodesic separation matrix $K^h_j = R^h_{ijk}\dfrac{dz^i}{ds}\dfrac{dz^k}{ds}$ for the unit-speed parametrization of the equator: $z^1(s) = s/r$, $z^2(s) = 0$.

   (b) Show that the general solution to $\dfrac{D^2 w^h}{ds^2} = -K^h_j w^j$ is
   $$w^1 = As + B, \qquad w^2 = C\sin\left(\dfrac{s}{r}\right) + D\cos\left(\dfrac{s}{r}\right).$$

4. Consider the one-parameter family of geodesics $x^1(\tau, q) = \tau$, $x^2(\tau, q) = q$ in the $(1+1)$-dimensional de Sitter spacetime given by the metric
   $$(g_{ij}) = \begin{pmatrix} 1 & 0 \\ 0 & -\cosh^2(q^1) \end{pmatrix}, \qquad -\infty < q^1 \leq \infty, \quad 0 < q^2 < 2\pi.$$

   (a) Determine the geodesic separation matrix $K^h_j = R^h_{ijk}\dfrac{dx^i}{d\tau}\dfrac{dx^k}{d\tau}$.

   (b) Verify that $x^h(\tau, q)$ satisfy the geodesic separation equations
   $$\dfrac{D^2}{d\tau^2}\dfrac{\partial x^h}{\partial q} + K^h_j \dfrac{\partial x^j}{\partial q} = 0.$$

5. Consider the two-parameter family of geodesics $\mathbf{q}(s, c, r)$,
   $$q^1(s, c, r) = c + r\tanh s, \qquad q^2(s, c, r) = r\,\mathrm{sech}\, s,$$
   in the hyperbolic plane H, which is defined on the upper half-plane $0 < q^2$ by the metric (cf. Section 6.3)
   $$g_{11} = g_{22} = \dfrac{1}{(q^2)^2}, \qquad g_{12} = g_{21} = 0.$$

   (a) Show that the second covariant derivative of the contravariant vector field w^h along $\mathbf{q}$ is
   $$\dfrac{D^2 w^1}{ds^2} = \dfrac{d^2 w^1}{ds^2} + 2\dfrac{dw^1}{ds}\tanh s - 2\dfrac{dw^2}{ds}\,\mathrm{sech}\, s + w^1\tanh^2 s - w^2 \tanh s\,\mathrm{sech}\, s,$$
   $$\dfrac{D^2 w^2}{ds^2} = \dfrac{d^2 w^2}{ds^2} + 2\dfrac{dw^1}{ds}\,\mathrm{sech}\, s + 2\dfrac{dw^2}{ds}\tanh s + w^1\tanh s\,\mathrm{sech}\, s + w^2\tanh^2 s.$$

(b) Determine the geodesic separation matrix $K_j^h = R_{ijk}^h \dfrac{dq^i}{ds} \dfrac{dq^k}{ds}$.

(c) Fix r and let $x^k(s,c)$ be the one-parameter family of geodesics defined by $x^k(s,c) = q^k(s,c,r)$. Verify the geodesic separation equations

$$\frac{D^2 w^h}{ds^2} + K_j^h w^j = 0$$

for $w^h = \partial x^h / \partial c$.

(d) Now fix c and let $x^k(s,r) = q^k(s,c,r)$. Verify the geodesic separation equations for $w^h = \partial x^h / \partial r$.

6. Suppose $A_{ij} v^i w^j = B_{ij} v^i w^j$ for all vectors v^i and w^j. Prove that $A_{ij} = B_{ij}$. This argument is frequently used to show that certain tensors are equal. In this section it shows that when γ_{ij} is defined as the "pullback" of g_{kl} by the map $M : G \to R : \xi^i = \xi^i(x^k)$, then

$$\gamma_{ij} = g_{kl} \frac{\partial x^k}{\partial \xi^i} \frac{\partial x^l}{\partial \xi^j}.$$

7. (a) Compute the Ricci tensor R_{ik} for the sphere of radius r and for the hyperbolic plane H. Does the fact that the Ricci tensor in each case is nonzero contradict the vacuum field equations?

(b) Show that the scalar curvature $R = R_i^i$ of the sphere is $2/r^2$, while that of the hyperbolic plane is -2.

8. Compute the Ricci tensor R_{ik} and the scalar curvature $R = R_i^i$ of the $(1+1)$-dimensional de Sitter spacetime.

9. (a) Let $G : (\tau, \xi, \eta, \zeta) = (\xi^0, \xi^1, \xi^2, \xi^3)$ be the frame in uniform rotation with angular velocity ω around the x-axis of an inertial frame R, as in Exercises 7–12 of Section 7.1. Determine the Ricci tensor R_{ik} on G, assuming that the metric (γ_{ij}) on G is induced from the Minkowski metric on R.

(b) Show that the τ-axis is a geodesic in G and determine whether G's coordinates are Fermi coordinates along this geodesic.

10. The purpose of this exercise is to construct Fermi coordinates for an observer C falling freely in the field given by the metric

$$\begin{pmatrix} \gamma_{00} & \gamma_{03} \\ \gamma_{30} & \gamma_{33} \end{pmatrix} = \begin{pmatrix} e^{2\alpha\zeta} & 0 \\ 0 & -e^{2\alpha\zeta} \end{pmatrix} = e^{2\alpha\zeta}\begin{pmatrix} 1 & 0 \\ 0 & -1 \end{pmatrix}$$

in the frame $G : (\tau, \zeta) = (\xi^0, \xi^3)$. In a small neighborhood of the origin this represents a constant gravitational acceleration α in the negative ζ-direction. Let the worldcurve of C be

$$\zeta(T) = (\tau(T), \zeta(T)) = \left(\frac{1}{\alpha}\tanh^{-1}(\alpha T), \frac{1}{2\alpha}\ln(1 - \alpha^2 T^2)\right).$$

(a) Show that ζ is a geodesic and T is proper time for C. Sketch $\zeta(T)$ for $\alpha = 1$.

(b) Let $\mathbf{e}_0 = (1, 0)$ and $\mathbf{e}_3 = (0, 1)$ in the tangent space $TG_{\zeta(0)} = TG_{(0,0)}$. Show that they are unit vectors and determine their parallel translates $\mathbf{e}_0(T)$ and $\mathbf{e}_3(T)$ in the tangent space $TG_{\zeta(T)}$.

(c) Fix T and consider the vector $\mathbf{v} = \mathbf{e}_3(T)$. Show that the curve

$$\eta_{T,\mathbf{n}}(s) = \left(\frac{1}{\alpha}\tanh^{-1}\left(\frac{T}{s+1/\alpha}\right), \frac{1}{2\alpha}\ln(\alpha^2((s+1/\alpha)^2 - T^2))\right)$$

is a geodesic in G and satisfies the initial conditions

$$\eta_{T,\mathbf{n}}(0) = \zeta(T), \qquad \eta'_{T,\mathbf{n}}(0) = \mathbf{v}.$$

(At least verify that the curve is a geodesic; solve the geodesic differential equations *ab initio* if you can.)

(d) Conclude that (T, S) are Fermi coordinates for C. Show that the curves $S =$ constant are vertical translates of the graph $\zeta = -\ln(\cosh\alpha\tau)/\alpha$ in the frame G, while the curves $T -$ constant are the graphs

$$\tau = \frac{1}{\alpha}\sinh^{-1}\left(\frac{\alpha T}{e^{\alpha\zeta}}\right).$$

Sketch the (T, S)-coordinate grid as it appears in G.

7.3 The Matter Field Equations

Aspects of the relativistic equation

In the presence of matter whose density is ρ at (x^1, x^2, x^3), the Newtonian gravitational field satisfies the equation

$$\frac{\partial^2 \Phi}{(\partial x^1)^2} + \frac{\partial^2 \Phi}{(\partial x^2)^2} + \frac{\partial^2 \Phi}{(\partial x^3)^2} = 4\pi G \rho.$$

We saw in the last section that Φ is analogous to $\gamma_{00}/2$ in a Fermi coordinate frame, suggesting that the relativistic field equation might be

$$\frac{\partial^2 \gamma_{00}}{(\partial \xi^1)^2} + \frac{\partial^2 \gamma_{00}}{(\partial \xi^2)^2} + \frac{\partial^2 \gamma_{00}}{(\partial \xi^3)^2} = \kappa \rho,$$

for a suitable universal constant κ.

Even though our proposed equation has only a heuristic basis, it still faces several immediate difficulties. For a start, the left-hand side needs to be restated in a generally covariant form. This brings in the Ricci tensor. But the modification $R_{ij} = \kappa \rho$ is not adequate, because mass-density ρ is not generally covariant, either. We need to replace mass-density by a suitable relativistic expression that is, like R_{ij}, a rank-2 covariant tensor so that the equation makes sense. Finally, we need to determine the constant κ.

The Energy–Momentum Tensor

Mass and volume both vary in relativity

In relativity, the mass-density of an object increases on two counts when the object is moving. If μ is its rest mass (in an inertial frame G, say), then we know that its mass is $m = \mu/\sqrt{1-v^2}$ in an inertial frame R in which it moves with velocity v. It also undergoes Fitzgerald contraction by the factor $\sqrt{1-v^2}$ in the direction it moves—but not in perpendicular directions. Therefore, if its "rest volume" is Σ in G, its volume in R will be $S = \Sigma\sqrt{1-v^2}$. The values of mass-density in the two frames are thus

$$\rho_R = \frac{m}{S} = \frac{1}{1-v^2}\frac{\mu}{\Sigma} = \frac{1}{1-v^2}\rho_G > \rho_G.$$

§7.3 The Matter Field Equations

We saw in Section 3.1 that the proper covariant way to deal with the relativistic mass m of an object is to make it a component of the 4-momentum

$$\mathbb{P}_R = (m, \mathbf{p}) = (m, m\mathbf{v}) = \mu \left(\frac{1}{\sqrt{1-v^2}}, \frac{\mathbf{v}}{\sqrt{1-v^2}} \right) = \mu \mathbb{U}_R.$$

4-momentum

Recall that $\mathbb{U}_R$ is the proper 4-velocity of the object in R's frame; that is, $\mathbb{U}_R = \mathbb{X}'_R$ when $\mathbb{X}_R(\tau)$ is the object's worldcurve in R parametrized by its proper time τ. If $\mathbb{P}_C$ is the 4-momentum of the same object in another inertial frame C that is connected to R by the Lorentz transformation $L: R \to C$, then $\mathbb{P}_C = L(\mathbb{P}_R)$. In particular, the components of $\mathbb{P}_C$ are linear combinations of the components of $\mathbb{P}_R$; the object's mass in C's frame is related to its mass *and* 3-momentum in R's frame — not to the mass alone.

While we can incorporate relativistic mass into a generally covariant 4-vector, we cannot readily do the same with volume. However, by rethinking the whole question in the context of a swarm of particles we can get a good covariant treatment of mass-density.

Mass-density of a swarm of particles

Consider a swarm of identical noninteracting particles that are at rest in an inertial frame G. Assume that they are uniformly distributed over space, with ν particles per unit 3-volume in G, and have individual rest mass μ. The product $\mu\nu$ of the individual mass by the particle density gives us the mass-density of the swarm in G.

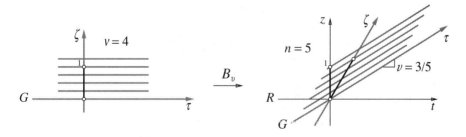

Let us calculate the mass-density in a second inertial frame R in which G and the swarm move with velocity $\mathbf{v}$. It will again be the product mn of the individual mass $m = \mu/\sqrt{1-v^2}$ by the particle density n. To determine n in the new frame, note that a region containing ν particles occupies the smaller volume $\sqrt{1-v^2}$ according to R; this is Fitzgerald contraction. (Note carefully the difference between density ν and speed v.) Hence there are

$$n = \frac{\nu}{\sqrt{1-v^2}}$$

particles per unit volume in R, and

$$\rho_R = mn = \frac{\mu\nu}{1-v^2} = \frac{1}{1-v^2}\rho_G,$$

in agreement with what we found above. In the figure above, where $v = \frac{3}{5}$, we can see that particle density increases by the factor $\frac{5}{4}$. Since individual mass also increases by $\frac{5}{4}$, mass density increases altogether by $\frac{25}{16} = \frac{5}{4} \times \frac{5}{4}$.

The particle density 4-vector

Notice that particle density transforms exactly like relativistic mass. This suggests that we construct a 4-vector $\mathbb{N}$ to carry the "rest density" ν the same way 4-momentum carries the rest mass μ:

$$\mathbb{N}_R = (n, n\mathbf{v}) = \left(\frac{\nu}{\sqrt{1-v^2}}, \frac{\nu\mathbf{v}}{\sqrt{1-v^2}}\right) = \nu\mathbb{U}_R.$$

Then, in any frame, mass-density is the product of the first components of $\mathbb{P}$ and $\mathbb{N}$. We call $\mathbb{N}$ the **particle density 4-vector**.

The tensor product of $\mathbb{P}$ and $\mathbb{N}$

Of course, the product of those first components is not itself a covariant quantity. However, the 4-vectors $\mathbb{P} = (p^0, p^1, p^2, p^3)$ and $\mathbb{N} = (n^0, n^1, n^2, n^3)$ *are* covariant; they are tensors of type $(1, 0)$. Therefore, if we take all possible products of a component of $\mathbb{P}$ and a component of $\mathbb{N}$,

IN THE GENERAL SENSE (CF. PAGE 308)

$$T^{ij} = p^i n^j,$$

the result is also covariant. It is a tensor of type $(2, 0)$ called the **tensor product** of $\mathbb{P}$ and $\mathbb{N}$; cf. Exercise 3 in Section 6.5. So mass-density is now embedded as the T^{00} component of a covariant object. What does the whole tensor T^{ij} represent?

The physical meaning of T^{ij}

Since this is a question of physics, not geometry, we will get the clearest answer by using dimensionally homogeneous coor-

§7.3 The Matter Field Equations

dinates $(x^0, x^1, x^2, x^3) = (ct, x, y, z)$. Then $\mathbb{U} = (c, \mathbf{v})$ and

$$m = \frac{\mu}{\sqrt{1 - (v/c)^2}}, \qquad n = \frac{\nu}{\sqrt{1 - (v/c)^2}},$$

$$\mathbb{P} = (mc, \mathbf{p}) = (E/c, p^1, p^2, p^3), \qquad \mathbb{N} = (nc, n\mathbf{v}) = (nc, nv^1, nv^2, nv^3),$$

where $E = mc^2$ is the relativistic energy of a particle in the swarm. Therefore,

$$T^{ij} = \begin{pmatrix} En & Env^1/c & Env^2/c & Env^3/c \\ cp^1n & p^1nv^1 & p^1nv^2 & p^1nv^3 \\ cp^2n & p^2nv^1 & p^2nv^2 & p^2nv^3 \\ cp^3n & p^3nv^1 & p^3nv^2 & p^3nv^3 \end{pmatrix}.$$

Since multiplication by n gives the spatial density of a quantity, we see that $T^{00} = En = mnc^2 = \rho c^2$ is actually the density of the relativistic *energy* of the swarm.

Energy density

We can interpret the other components of the tensor T^{ij} by analogy with the simpler example of fluid flow that we considered in Section 4.3. We represented the flow through space of an incompressible fluid of density ρ by a 3-vector $\mathbf{F} = \rho \mathbf{V}$, where $\mathbf{V} = (V_x, V_y, V_z)$ was its 3-velocity. Then the component $F_x = \rho V_x$ gives the rate at which fluid mass flows across a plane perpendicular to the x-direction, in kilograms per second per square meter. If we abbreviate this as x-flow of mass, then we have

	ρV_x	ρV_y	ρV_z
$\mathbf{F}$:	x-flow of mass	y-flow of mass	z-flow of mass

While an element of fluid is characterized by a single quantity, its mass, particles in the swarm need four: their energy E and their momentum components p^i. Each of these has a density and a component of flow in the three spatial directions—accounting for the 16 quantities we find in T^{ij}. For example, p^1nv^2 is the 2-flow of 1-momentum, while Env^1 (without the factor of $1/c$) is the 1-flow of energy. Here is the full list:

The energy–momentum tensor

T^{00} density of energy	T^{01} 1-flow of energy $\div c$	T^{02} 2-flow of energy $\div c$	T^{03} 3-flow of energy $\div c$
T^{10} $c \times$ density of 1-momentum	T^{11} 1-flow of 1-momentum	T^{12} 2-flow of 1-momentum	T^{13} 3-flow of 1-momentum
T^{20} $c \times$ density of 2-momentum	T^{21} 1-flow of 2-momentum	T^{22} 2-flow of 2-momentum	T^{23} 3-flow of 2-momentum
T^{30} $c \times$ density of 3-momentum	T^{31} 1-flow of 3-momentum	T^{32} 2-flow of 3-momentum	T^{33} 3-flow of 3-momentum

Elaborating T^{ij}

Because of this interpretation we call T^{ij} the **energy–momentum tensor**. In dimensionally homogeneous coordinates the components of T^{ij} all have the same dimensions: energy per 3-volume. This requires the compensating factors that we see in $T^{\alpha 0}$ and $T^{0\beta}$. It is also true that the energy–momentum tensor is symmetric: $T^{ji} = T^{ij}$. You are asked to verify these facts in the exercises.

Matter and energy usually take more complicated forms than a swarm of identical particles moving in parallel without interacting. For example, the particles may have different masses or velocities; they may collide; they may form a fluid that exerts internal pressure; and they may form a solid that has internal stresses. Even if there is no matter present, a region of spacetime may have energy in the form of an electromagnetic field. Each such manifestation of matter and energy gives rise to a further contribution to the energy–momentum tensor that can be calculated using the appropriate laws of physics. However, the resulting object continues to be a rank-2 symmetric tensor T^{kl} that represents in a generally covariant way the matter and energy at any point in spacetime.

Conservation of Energy–Momentum

Let T_{ij} be the covariant version of the energy–momentum tensor, obtained in the usual way by lowering indices:

$$T_{ij} = g_{ik}g_{jl}T^{kl}.$$

T_{ij} replaces ρ in the field equations

This is the natural candidate to replace mass-density ρ in the field equations in the presence of matter–energy: $R_{ij} = \kappa T_{ij}$. Alas, this equation is still not correct, but the difficulty has moved to the left-hand side. Energy–momentum is a conserved quantity; as we shall see, this implies that the energy–momentum tensor has divergence zero. The same is not true of the Ricci tensor, so the two cannot be proportional.

We shall now analyze the energy–momentum tensor the same way we analyzed the fluid flow vector **F** in Section 4.3—and we shall reach the same conclusion. Consider the flow of energy–momentum through a small 4-dimensional box in an inertial frame R. Suppose the box is centered at $(ct, x, y, z) = (x^0, x^1, x^2, x^3)$ and has sides whose dimensions are Δx^0, Δx^1, Δx^2, and Δx^3. While there may be collisions and other events that transform matter and energy inside the box, the conservation of energy–momentum means that the net flow is zero: What goes in must come out.

Energy–momentum is conserved

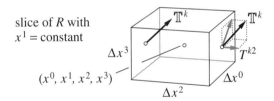

Let $\mathbb{T}^k$ denote the kth row of the energy–momentum tensor $\mathbb{T} = T^{kl}$. The net flow of $\mathbb{T}^k$ out of the box is the algebraic sum of the outflows across its eight faces. On each face, we can break down $\mathbb{T}^k$ into a parallel component and a normal one; only the normal component contributes to the sum. The figure illustrates the flow in the 2-direction; the normal contribution is made by the 2-flow component T^{k2}. (Because the figure is a slice of R with

Flow across a single face

x^1 = constant, we see only a 3-dimensional slice of the box and slices of only six of its faces, not eight.)

If the dimensions of the box are small enough, T^{k2} is essentially constant on each face. The outflow on the right face is therefore approximately

$$T^{k2}\left(x^0, x^1, x^2 + \frac{\Delta x^2}{2}, x^3\right) \Delta x^0 \Delta x^1 \Delta x^3.$$

In the same way, the flow from left to right on the *left* face is approximately

$$T^{k2}\left(x^0, x^1, x^2 - \frac{\Delta x^2}{2}, x^3\right) \Delta x^0 \Delta x^1 \Delta x^3.$$

But this is an *inflow*, not an outflow. The combined outflow across these two faces is approximately equal to the difference

$$\left[T^{k2}\left(x^0, x^1, x^2 + \frac{\Delta x^2}{2}, x^3\right) - T^{k2}\left(x^0, x^1, x^2 - \frac{\Delta x^2}{2}, x^3\right)\right] \Delta x^0 \Delta x^1 \Delta x^3$$

$$= \frac{T^{k2}\left(x^0, x^1, x^2 + \frac{\Delta x^2}{2}, x^3\right) - T^{k2}\left(x^0, x^1, x^2 - \frac{\Delta x^2}{2}, x^3\right)}{\Delta x^2} \Delta \Sigma,$$

where $\Delta \Sigma = \Delta x^0 \Delta x^1 \Delta x^2 \Delta x^3$; $\Delta \Sigma$ is the negative of the 4-volume of the box in the Minkowski metric.

Total outflow is the divergence

The expressions above involve vectors based at different points of R, so we cannot calculate their differences unless we use parallel transport in the way described in Section 6.5. In that case the difference quotient above is approximately equal to the covariant derivative of T^{k2} with respect to x^2. Therefore, the net outflow across the pair of x^2-faces is approximately

$$T^{k2}_{;2} \Delta \Sigma.$$

In a similar way we can approximate the net outflow across the other three pairs of faces by $T^{k0}_{;0} \Delta\Sigma$, $T^{k1}_{;1} \Delta\Sigma$, and $T^{k3}_{;3} \Delta\Sigma$. Therefore, using the summation convention we have this approximate expression:

$$\text{total net flow out of the box} = T^{kl}_{;l} \Delta\Sigma.$$

§7.3 The Matter Field Equations

This is proportional to the 4-volume of the box (ignoring the sign); the proportionality factor gives the outflow *per unit volume*:

$$T^{kl}_{;l}.$$

Since this sum of derivatives is like the classical divergence of a vector field, we give it the same name:

Definition 7.1 *Let $S^{l,k}_j$ be a tensor of type $(p, q \neq 1)$* (p, q); *its **divergence** is the tensor* div $S = S^{l,k}_{j;k}$ *of type* (p, q). (p, q+1)

Divergence of a tensor

Theorem 7.4 *The divergence of the energy-momentum tensor is identically zero:* $T^{kl}_{;l} = 0$. *Moreover,* $T^l_{i;l} = 0$.

PROOF: According to our calculations, the net flow of energy-momentum through a box of 4-volume $-\Delta\Sigma$ is approximately $T^{kl}_{;l}\Delta\Sigma$, and the approximation becomes exact as the dimensions $\Delta x^j \to 0$. Since energy-momentum is conserved, the limiting net flow is zero, so $T^{kl}_{;l} = 0$. The second statement follows from the product rule and the fact that $g_{ik;l} = 0$:

$$T^l_{i;l} = \left(g_{ik}T^{kl}\right)_{;l} = g_{ik;l}T^{kl} + g_{ik}T^{kl}_{;l} = 0. \qquad \text{END OF PROOF}$$

Divergence of the Ricci Tensor

We turn now to determining the divergence of the Ricci tensor. The first step is the following result, called the **Bianchi identity**.

Theorem 7.5 $R^h_{ijk;l} + R^h_{ikl;j} + R^h_{ilj;k} = 0.$

Bianchi identity

PROOF: We prove the result at the origin of a Fermi coordinate frame; by general covariance, it will then be true at any point in any coordinate frame. All the Christoffel symbols vanish at the origin, so both the Riemann tensor and its covariant derivative have simple forms:

$$R^h_{ijk} = \frac{\partial \Gamma^h_{ik}}{\partial x^j} - \frac{\partial \Gamma^h_{ij}}{\partial x^k}, \qquad R^h_{ijk;l} = \frac{\partial^2 \Gamma^h_{ik}}{\partial x^l \partial x^j} - \frac{\partial^2 \Gamma^h_{ij}}{\partial x^l \partial x^k},$$

$$R^h_{ikl} = \frac{\partial \Gamma^h_{il}}{\partial x^k} - \frac{\partial \Gamma^h_{ik}}{\partial x^l}, \qquad R^h_{ikl;j} = \frac{\partial^2 \Gamma^h_{il}}{\partial x^j \partial x^k} - \frac{\partial^2 \Gamma^h_{ik}}{\partial x^j \partial x^l},$$

$$R^h_{ilj} = \frac{\partial \Gamma^h_{ij}}{\partial x^l} - \frac{\partial \Gamma^h_{il}}{\partial x^j}, \qquad R^h_{ilj;k} = \frac{\partial^2 \Gamma^h_{ij}}{\partial x^k \partial x^l} - \frac{\partial^2 \Gamma^h_{il}}{\partial x^k \partial x^j}.$$

The six terms on the right sum to zero. END OF PROOF

Scalar curvature

The following theorem on the divergence of the Ricci tensor is necessarily expressed in terms of the *mixed* form $R^h_j = g^{ih} R_{ij}$. It also involves the **scalar curvature** $R = R^h_h$, the contraction of the mixed Ricci tensor. For the scalar curvature, as for any function (tensor of rank 0), the covariant derivative is just the ordinary partial derivative.

Theorem 7.6 $R^h_{k;h} = \frac{1}{2} \delta^h_k R_{;h} = \frac{1}{2} \delta^h_k \frac{\partial R}{\partial x^h}$.

PROOF: In the Bianchi identity, contract on $j = h$:

$$0 = R^h_{ihk;l} + R^h_{ikl;h} + R^h_{ilh;k} = R_{ik;l} + R^h_{ikl;h} - R_{il;k}.$$

The third term is a result of the fact that the Riemann tensor is antisymmetric in its second pair of indices: $R^h_{ilh;k} = -R^h_{ihl;k} = -R_{il;k}$. Since the covariant derivative of g^{im} is zero, we can rewrite our equation as

$$0 = \left(g^{im} R_{ik}\right)_{;l} + \left(g^{im} R^h_{ikl}\right)_{;h} - \left(g^{im} R_{il}\right)_{;k}$$
$$= R^m_{k;l} + \left(g^{im} R^h_{ikl}\right)_{;h} - R^m_{l;k}.$$

Now contract on $m = l$:

$$0 = R^l_{k;l} + \left(g^{il} R^h_{ikl}\right)_{;h} - R^l_{l;k} = R^l_{k;l} + \left(g^{il} R^h_{ikl}\right)_{;h} - R_{;k}.$$

Notice that scalar curvature appears in the last term. We can simplify the second term by first noting

$$\begin{aligned}
g^{il}R^h_{ikl} &= g^{il}g^{hm}R_{mikl} \\
&= g^{il}g^{hm}R_{imlk} \quad \text{(symmetry of } R_{mikl}\text{)} \\
&= g^{hm}g^{il}R_{imlk} \\
&= g^{hm}R^l_{mlk} = g^{hm}R_{mk} = R^h_k.
\end{aligned}$$

Therefore, $(g^{il}R^h_{ikl})_{;h} = R^h_{k;h}$. If we change the dummy summation index $l \mapsto h$ in the first term of the equation, we get

$$2R^h_{k;h} = R_{;k} = \delta^h_k R_{;h}. \qquad \text{END OF PROOF}$$

Since the scalar curvature is not generally constant, the Ricci tensor has nonzero divergence.

$\operatorname{div} R^h_k \neq 0$

The Relativistic Field Equations

If we write the last theorem in the form

A divergence-free curvature tensor

$$\left(R^h_k - \tfrac{1}{2}\delta^h_k R\right)_{;h} = 0,$$

we find a tensor, namely $R^h_k - \tfrac{1}{2}\delta^h_k R$, that *is* divergence-free and completely determined by curvature—that is, by the gravitational potentials and their first and second derivatives. The covariant form of this tensor is $R_{ij} - \tfrac{1}{2}g_{ij}R$; using it we can finally write the correct relativistic equations for the gravitational field:

$$R_{ij} - \tfrac{1}{2}g_{ij}R = \kappa T_{ij}.$$

We still need to determine the constant κ. This is easier to do if we convert the equations to an equivalent form that Einstein used to express them. First rewrite the equations with mixed tensors,

Einstein's form

$$R^h_k - \tfrac{1}{2}\delta^h_k R = \kappa T^h_k,$$

and then contract on the two indices:

$$R^h_h - \tfrac{1}{2}\delta^h_h R = R - 2R = \kappa T^h_h = \kappa T.$$

The contraction $T = T^h_h$ is called *Laue's scalar*. The left-hand side reduces the way it does because δ^h_k is the 4×4 identity matrix, so

Determine κ in conventional units

its trace δ_h^h is 4. Thus $R = -\kappa T$; use this result in the covariant form of the field equations to solve for R_{ij}:

$$R_{ij} = \kappa T_{ij} + \tfrac{1}{2} g_{ij} R = \kappa (T_{ij} - \tfrac{1}{2} g_{ij} T).$$

We turn now to the task of finding κ; for clarity, we will use dimensionally homogeneous coordinates (ct, x, y, z). Our strategy is to solve the single equation

$$R_{00} = \kappa (T_{00} - \tfrac{1}{2} g_{00} T)$$

in a simple physical situation where we can determine R_{00}, T_{00}, and T.

The simplest possibility is a swarm of noninteracting particles moving together. In a Fermi coordinate system $(\xi^0, \xi^1, \xi^2, \xi^3)$ in which the worldline of one of the particles is the ξ^0-axis, the proper 4-velocity for the swarm is just $\mathbb{U} = (c, 0, 0, 0)$, and the energy–momentum tensor has only the single term

$$T_{00} = T_0^0 = T^{00} = \rho c^2,$$

the energy-density of the swarm. Since T is also equal to ρc^2 and since $\gamma_{00} = 1$ on the ξ^0-axis,

$$T_{00} - \tfrac{1}{2} \gamma_{00} T = \tfrac{1}{2} \rho c^2.$$

In Section 7.2 we showed that along the ξ^0-axis in a Fermi coordinate system,

$$R_{00} = K_h^h = \frac{1}{2} \sum_{\alpha=1}^{3} \frac{\partial^2 \gamma_{00}}{(\partial \xi^\alpha)^2}.$$

Furthermore, this corresponds to the Laplacian of the Newtonian potential,

$$\nabla^2 \Phi = \sum_{\alpha=1}^{3} \frac{\partial^2 \Phi}{(\partial \xi^\alpha)^2}.$$

However, R_{00} and $\nabla^2 \Phi$ have different dimensions: The units for R_{00} are $1/m^2$, while the units for $\nabla^2 \Phi$ are $1/sec^2$ (exercise). To equate the two expressions we must factor in an appropriate power of c to get the dimensions to balance; thus

$$\frac{1}{c^2}\nabla^2\Phi = R_{00}.$$

(Although we got this equation only by heuristic reasoning, we shall justify it by other means in the next chapter.) The classical field equation tells us that $\nabla^2\Phi = 4\pi G\rho$, so $R_{00} = \kappa(T_{00} - \frac{1}{2}g_{00}T)$ becomes

$$\frac{1}{c^2}4\pi G\rho = \kappa \frac{1}{2}\rho c^2; \quad \text{hence} \quad \kappa = \frac{8\pi G}{c^4} \approx 2 \times 10^{-43}\,\frac{\text{sec}^2}{\text{kg-m}}.$$

Here, then, are the relativistic field equations in two equivalent forms. The culmination of Einstein's theory of gravity, they describe how matter and energy determine the curvature of spacetime.

How matter–energy determines curvature

Relativistic equations of the gravitational field:

$$R_{ij} = \frac{8\pi G}{c^4}\left(T_{ij} - \tfrac{1}{2}g_{ij}T\right), \quad \text{or} \quad T_{ij} = \frac{c^4}{8\pi G}\left(R_{ij} - \tfrac{1}{2}g_{ij}R\right).$$

These are second-order partial differential equations for the 10 components of the gravitational potential; their solutions determine the gravitational field, which, in turn, determines the geodesic paths that both matter and massless particles must follow.

The Cosmological Constant

In 1917, Einstein considered an extension of the gravitational field equations under the assumption that the spatial universe is finite and has the form of a 3-dimensional sphere S^3; cf. [8]. One way to describe the 3-sphere of radius r is as the set of points in $\mathbf{R}^4 : (x, y, z, u)$ that satisfy the equation

A finite universe

$$S^3(r): \quad x^2 + y^2 + z^2 + u^2 = r^2.$$

Compare this to the 2-sphere in $\mathbf{R}^3$. Like the 2-sphere, the 3-sphere can be parametrized by angle variables, but three are needed now, instead of two:

$$\mathbf{X}: \begin{cases} x = r\cos\theta\cos\varphi\cos\omega, \\ y = r\sin\theta\cos\varphi\cos\omega, \\ z = r\sin\varphi\cos\omega, \\ u = r\sin\omega, \end{cases} \quad \begin{array}{l} 0 \le \theta \le 2\pi, \\ -\pi/2 < \varphi < \pi/2, \\ -\pi/2 < \omega < \pi/2. \end{array}$$

In the exercises you are asked to verify this claim and all the others to follow. Notice that we already have a new question about the relation between physics and geometry to consider: What is the radius r of the universe?

Modifying the Newtonian field equation

Einstein also assumed that this universe lasts forever in essentially the same form. Thus spacetime is the product $E = \mathbf{R} \times S^3(r)$, and it can be described by the $(1+3)$-dimensional frame R provided with dimensionally homogeneous coordinates $(ct, \theta, \varphi, \omega)$ and the Minkowski metric, as adapted to the 3-sphere. Because the spatial universe itself is finite (its volume is $2\pi^2 r^3$), the total amount of matter—and thus its average density ρ_0—is *also* finite. Therefore, Einstein observed, if we replace the Newtonian field equation $\nabla^2 \Phi = 4\pi G \rho$ by

$$\nabla^2 \Phi - \Lambda \Phi = 4\pi G \rho,$$

then the constant potential

$$\Phi_0 = -\frac{4\pi G}{\Lambda} \rho_0$$

is a solution to the new equation (if we set $\rho = \rho_0$ on the right-hand side of the equation). Of course, matter is not distributed uniformly throughout space, so ρ generally differs from ρ_0. To deal with this, let Φ_1 solve the equation

$$\nabla^2 \Phi - \Lambda \Phi = 4\pi G (\rho - \rho_0).$$

Then $\Phi = \Phi_0 + \Phi_1$ solves the field equation $\nabla^2 \Phi = 4\pi G \rho$. If, by contrast, we assume that space is an ordinary Euclidean 3-dimensional space, then it can be shown that the Newtonian theory requires the average density of matter to be zero and $\Phi \to 0$ far from gravitating bodies.

The cosmological constant

How should the modification be carried over from the Newtonian field equations to the relativistic? We know that the "geometric" side (the one involving the Ricci tensor) must have zero divergence for the equations to be valid. Einstein therefore proposed

$$R_{ij} - \tfrac{1}{2} g_{ij} R - g_{ij} \Lambda = \frac{8\pi G}{c^4} T_{ij},$$

because the new term $g_{ij} \Lambda$ has zero divergence and g_{ij} is analogous to Φ. Under the assumption that the only contribution to T_{ij} comes from matter of uniform spatial density ρ, Einstein solved these modified field equations (see Exercise 12) and showed that

$$\Lambda = \frac{4\pi G \rho}{c^2} = \frac{1}{r^2}.$$

The total amount of matter is therefore

$$\rho \times \text{vol } S^3(r) = \rho \cdot 2\pi^2 r^3 = \frac{2\rho \pi^2 c^2 r}{4\pi G \rho} = \frac{\pi c^2}{2G\sqrt{\Lambda}}.$$

Since Λ determines two fundamental data of the "cosmos," its radius and its total mass, it has come to be called the **cosmological constant**.

Exercises

The aim of the first three exercises is to show that the divergence of a contravariant vector field a^h can be expressed in terms of the determinant g of the metric tensor—and without the Christoffel symbols—as

$$\text{div } a^h = a^h_{;h} = \frac{1}{\sqrt{-g}} \frac{\partial \left(a^h \sqrt{-g} \right)}{\partial x^h}.$$

1. Show that $g^{ij} \dfrac{\partial g_{ij}}{\partial x^h} = 2 \Gamma^k_{kh}$. Suggestion: First show that

   $$\frac{\partial g_{ij}}{\partial x^h} = \Gamma_{ih,j} + \Gamma_{jh,i};$$

   then contract with g^{ij} and use the symmetries of the Christoffel symbols.

2. Suppose $G = (g_{ij})$ is a 2×2 symmetric matrix, so $g = \det G = g_{11} g_{22} - g_{12} g_{21}$. Suppose also that $g < 0$.

(a) Show by direct computation that $g^{ij} = \dfrac{1}{g}\dfrac{\partial g}{\partial g_{ij}}$, $i,j = 1,2$.

(b) Justify these equalities:
$$\Gamma^k_{kh} = \frac{1}{2}\frac{1}{g}\frac{\partial g}{\partial x^h} = \frac{1}{2}\frac{\partial \ln|g|}{\partial x^h} = \frac{\partial \ln\sqrt{-g}}{\partial x^h} = \frac{1}{\sqrt{-g}}\frac{\partial \sqrt{-g}}{\partial x^h}.$$

(c) Deduce that $a^h_{;h} = \dfrac{\partial a^h}{\partial x^h} + \dfrac{a^h}{\sqrt{-g}}\dfrac{\partial \sqrt{-g}}{\partial x^h} = \dfrac{1}{\sqrt{-g}}\dfrac{\partial(a^h\sqrt{-g})}{\partial x^h}$.

3. Now consider the general case in relativity: $G = (g_{ij})$ is a 4×4 symmetric matrix with $g = \det G < 0$. Most of the following can be deduced from standard facts about determinants and inverses that you can find in a text on linear algebra.

(a) Let G^{ij} be the 3×3 minor obtained by deleting the ith row and jth column from G, and let $\Delta^{ij} = (-1)^{i+j}\det G^{ij}$ be the corresponding cofactor. Show that
$$g = g_{i1}\Delta^{i1} + g_{i2}\Delta^{i2} + g_{i3}\Delta^{i3} + g_{i4}\Delta^{i4} \quad \text{(no sum on } i\text{)},$$
for any $i = 1, 2, 3, 4$.

(b) Explain why none of Δ^{i1}, Δ^{i2}, Δ^{i3}, Δ^{i4} depend on g_{ij} and use this fact to conclude that $\partial g/\partial g_{ij} = \Delta^{ij}$.

(c) Show that $g^{ij} = \Delta^{ji}/g$, where g^{ij} is, as usual, the element in the ith row and jth column of the inverse of G. Note the order of the indices, which is needed for an arbitrary matrix but is irrelevant here because G is symmetric.

(d) Deduce that $g^{ij} = \dfrac{1}{g}\dfrac{\partial g}{\partial g_{ij}}$, then argue as in Exercise 2 to show that
$$a^h_{;h} = \frac{1}{\sqrt{-g}}\frac{\partial(a^h\sqrt{-g})}{\partial x^h}.$$

4. (a) Show that the Ricci tensor can be written in the form
$$R_{ik} = \frac{\partial \Gamma^h_{ik}}{\partial x^h} - \frac{\partial^2 \ln\sqrt{-g}}{\partial x^i \partial x^k} + \Gamma^p_{ik}\frac{\partial \ln\sqrt{-g}}{\partial x^p} - \Gamma^p_{ih}\Gamma^h_{pk}.$$

(b) Deduce from the equation in part (a) that R_{ik} is symmetric.

(c) To simplify calculations, Einstein frequently assumed a coordinate frame in which $g \equiv -1$. What simple form does R_{ik} take then?

5. Suppose R is an inertial frame in which there is a swarm of noninteracting particles with particle density 4-vector $\mathbb{N}_R$. Assume that R has traditional coordinates (t, x, y, z), so $\mathbb{N}_R = (n, n\mathbf{v})$. Suppose C is a Galilean observer with proper 4-velocity $\mathbb{U}_R$ in R. Show that the numerical density of the swarm in C's frame is $\mathbb{N}_R \cdot \mathbb{U}_R$. (This is the ordinary Minkowski inner product in R.) Suggestion: First calculate the corresponding quantity $\mathbb{N}_G \cdot \mathbb{U}_G$ in the inertial frame G in which the swarm is at rest.

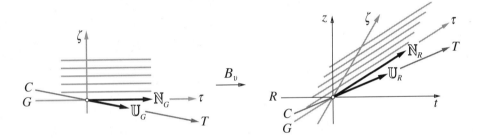

6. (a) Consider a swarm of noninteracting particles of individual rest mass μ and particle density ν in a rest frame. Suppose the swarm has proper 4-velocity $\mathbb{U} = (u^0, u^1, u^2, u^3)$ in a frame R. Show that the energy–momentum tensor of the swarm in R can be written as $T^{ij} = \mu\nu u^i u^j$. Does this result depend on whether traditional or dimensionally homogeneous coordinates are used in R?

(b) Prove that T^{ij} is symmetric: $T^{ji} = T^{ij}$.

7. (a) Consider the energy–momentum tensor T^{ij} in a frame R provided with dimensionally homogeneous coordinates $(ct, x, y, z) = (x^0, x^1, x^2, x^3)$. Show that every component of T^{ij} has the dimensions energy/length^3.

(b) Show that this implies that $T^{\alpha 0}/c$, rather than $T^{\alpha 0}$, can be interpreted as density of momentum in the α-direction.

Similarly, show that $cT^{0\beta}$, but not $T^{0\beta}$, can be interpreted as the flow of energy in the α-direction.

8. Show that the components of the Ricci tensor all have dimensions meters^{-2} in a frame R with dimensionally homogeneous coordinates (ct, x, y, z). Determine the dimensions of the Newtonian potential Φ and conclude that its Laplacian $\nabla^2\Phi$ has the dimensions seconds^{-2}.

9. Consider the relativistic equations of the gravitational field with the cosmological modification:

$$R_{ij} - \tfrac{1}{2}g_{ij}R - g_{ij}\Lambda = \frac{8\pi G}{c^4}T_{ij}.$$

Assume, in this exercise, an *arbitrary* spacetime frame with dimensionally homogeneous coordinates. In particular, do not assume that the spatial universe is a 3-sphere.

(a) Determine the dimensions of g_{ij}, R_{ij}, R, and T_{ij} and show that Λ must have the dimensions of length^{-2}.

(b) Show that the left-hand side has zero divergence.

(c) Assume $T_{ij} \equiv 0$ and show that $R = -4\Lambda$ where R is the scalar curvature function $R = R^i_i$. Then show that $R_{ij} = -\Lambda g_{ij}$.

(d) The original matter field equations connect the curvature of spacetime to the presence of matter–energy; in particular, an empty universe ($T_{ij} \equiv 0$) will be flat. Show that with the "cosmological" term added, the empty universe can no longer be flat if $\Lambda \neq 0$. (This lends support to Einstein's proposal to consider a spatial universe of the form $S^3(r)$.)

(e) Show that the modified equations can also be written in the form

$$R_{ij} + g_{ij}\Lambda = \frac{8\pi G}{c^4}\left(T_{ij} - \tfrac{1}{2}g_{ij}T\right).$$

Note the change in the sign of $g_{ij}\Lambda$.

10. (a) Let r be a fixed positive number; consider the map $\mathbf{x}: \Omega \to \mathbf{R}^4$ defined by

$$\mathbf{x}: \begin{cases} x = r\cos\theta \cos\varphi \cos\omega, \\ y = r\sin\theta \cos\varphi \cos\omega, \\ z = r\sin\varphi \cos\omega, \\ u = r\sin\omega; \end{cases} \quad \Omega: \begin{cases} 0 \le \theta \le 2\pi, \\ -\pi/2 \le \varphi \le \pi/2, \\ -\pi/2 \le \omega \le \pi/2. \end{cases}$$

Show that the image of **x** is $S^3(r)$, the 3-sphere of radius r.

(b) Using the ordinary Euclidean dot product in $\mathbf{R}^4$, determine the metric $a_{ij} = \mathbf{x}_i \cdot \mathbf{x}_j$ that **x** induces on Ω.

(c) By analogy with the area of a surface (Section 5.1), show that the volume of the 3-sphere of radius r is

$$V = \iiint_\Omega \sqrt{a}\, d\theta\, d\varphi\, d\omega,$$

where $a = \det(a_{ij})$. Show that $V = 2\pi^2 r^3$ by computing the integral.

11. (a) Let $E = \mathbf{R} \times S^3(r)$ be Einstein's quasi-static spherical universe, described by the frame R with dimensionally homogeneous coordinates $(ct, \theta, \varphi, \omega) = (x^0, x^1, x^2, x^3)$. Show that the metric on E is

$$(g_{ij}) = \begin{pmatrix} 1 & & & \\ & -r^2 \cos^2\varphi \cos^2\omega & & \\ & & -r^2 \cos^2\omega & \\ & & & -r^2 \end{pmatrix}.$$

(This is a diagonal matrix; the entries not printed are all zero.)

(b) Compute the inverse matrix g^{ij} and the Christoffel symbols of the first kind on E. Show that the only nonzero Christoffel symbols of the second kind are

$$\Gamma^2_{11} = \sin\varphi \cos\varphi, \qquad \Gamma^1_{12} = \Gamma^1_{21} = -\tan\varphi,$$
$$\Gamma^3_{11} = \cos^2\varphi \sin\omega \cos\omega, \qquad \Gamma^1_{13} = \Gamma^1_{31} = -\tan\omega,$$
$$\Gamma^3_{22} = \sin\omega \cos\omega, \qquad \Gamma^2_{23} = \Gamma^2_{32} = -\tan\omega.$$

(c) Compute the determinant g and $L = \ln\sqrt{-g}$. Show that the components of the Ricci tensor are

$$R_{00} = 0, \quad \text{while} \quad R_{ij} = -\frac{2}{r^2}g_{ij} \quad \text{if } (i,j) \neq (0,0).$$

(Suggestion: Use the formula for the Ricci tensor in Exercise 4.)

12. (a) Assume that all matter–energy in the Einstein universe $E = \mathbf{R} \times S^3(r)$ appears only as a swarm of noninteracting particles of uniform density ρ at rest in the frame R. Thus $T^{00} = \rho c^2$, and all other components of the matter tensor T^{ij} are zero. Determine the mixed and covariant tensors T^i_j and T_{ij} and determine Laue's scalar $T = T^i_i$.

(b) Show that the 10 matter field equations

$$R_{ij} + \Lambda g_{ij} = \frac{8\pi G}{c^4}\left(T_{ij} - \tfrac{1}{2}g_{ij}T\right)$$

reduce to the following pair of independent equations,

$$\Lambda = \frac{4\pi G\rho}{c^2}, \qquad -\frac{2}{r^2} + \Lambda = -\frac{4\pi G\rho}{c^2},$$

which together imply $r = \dfrac{1}{\sqrt{\Lambda}}$ and $\rho = \dfrac{c^2 \Lambda}{4\pi G}$.

Further Reading for Chapter 7

Einstein's own writing on general relativity is incomparable. The fundamental paper is [10]; he considers the cosmological constant in a closed universe in [8]. Of the many substantial texts, the following provide a variety of approaches to general relativity: Adler, Bazin, and Schiffer [1]; Dubrovnin, Fomenko, and Novikov [7]; Frankel [14]; Landau and Lifshitz [17]; Lawden [18]; Misner, Thorne, and Wheeler [24]. Separation of geodesics plays a central role in Misner et al. [24] and is also discussed from a differential geometric point of view in [5] and [15]. Fermi coordinates along a geodesic are considered in detail in the paper by Manasse and Misner [21]. The physical basis of the energy–momentum (or stress–energy) tensor is carefully laid out in volume 2 of *The Feynman Lectures* [13] and developed in the context of general relativity in [18], [24], and [25].

8 Consequences

CHAPTER

Testable predictions

In the last part of his revolutionary 1916 paper on general relativity [10], Einstein draws several conclusions from the new theory that lead to testable predictions. Two of the most famous are that a massive body will bend light rays that pass near it and that the point on the orbit of Mercury that is closest to the sun (the *perihelion* point) will drift by an additional amount that the Newtonian theory cannot explain. Indeed, the perihelion drift was already known; Einstein's prediction resolved a long-standing puzzle. And in 1919, measurements taken during an eclipse of the sun confirmed that light from stars that passed near the sun was deflected in just the way Einstein had deduced.

To draw these conclusions, Einstein first demonstrated that Newton's theory is a "first approximation" to relativity. Then, from the Newtonian potential of the sun he derives a relativistic gravitational field that he uses to calculate the deflection of light and the perihelion drift. In this chapter we shall see how the Newtonian approximation leads to the predictions that established general relativity.

8.1 The Newtonian Approximation

How Einstein's theory reduces to Newton's

The gravitational theories of Newton and Einstein arise out of fundamentally different basic conceptions. Nevertheless, for objects that move at low speeds in weak gravitational fields, the two theories give results that are in close agreement. This agreement occurs because general relativity *reduces to* Newtonian mechanics—in a sense we shall make precise—when

- the gravitational field is weak;
- objects move slowly in relation to the speed of light.

Briefly stated, the reduction involves neglecting small quantities in the relativistic theory. In this section we shall say what makes a quantity "small," why that means it can be neglected, and how the parts that are left lead to a good approximation of the Newtonian theory.

Use conventional units

The standard by which we shall measure the size of a quantity is the speed of light, so it is instructive to use conventional units, in which $c \approx 3 \times 10^8$ m/sec and velocities have the dimensions of meters per second, rather than geometric units, in which $c = 1$ and velocities are dimensionless.

Small Quantities

How a Lorentz boost reduces to a Galilean shear

Lorentz and Galilean transformations provide a good example of the way reduction works. When one observer in an inertial frame G moves with velocity v with respect to a second observer, R, then classical mechanics relates the two frames by a Galilean shear $S_v : G \to R$, while special relativity uses a Lorentz velocity boost $B_v : G \to R$: (cf. Exercise 5, §3.2)

$$S_v = \begin{pmatrix} 1 & 0 \\ v & 1 \end{pmatrix}, \qquad B_v = \frac{1}{\sqrt{1-(v/c)^2}} \begin{pmatrix} 1 & \dfrac{v}{c^2} \\ v & 1 \end{pmatrix}.$$

The matrices take these forms when we use conventional units together with traditional coordinates (t, z). If v is small in relation to c, then we can write

$$\frac{1}{\sqrt{1-(v/c)^2}} = 1 + O\left(\frac{1}{c^2}\right) \quad \text{and} \quad \frac{v}{c^2} = O\left(\frac{1}{c^2}\right),$$

giving us a strikingly simple relation between the two matrices:

$$B_v = S_v + O\left(\frac{1}{c^2}\right).$$

Therefore, B_v reduces to S_v when we neglect terms that are of the same order of magnitude as $1/c^2$. In ordinary terrestrial physics, the speed of light has an extremely large value, and its reciprocal is therefore extremely small; in those circumstances the equation tells us that there is no practical way to distinguish between a boost and a shear.

However, if v is *not* small in relation to c, the outcome is entirely different. For example, if $v = 0.9c$, then

$$B_v = 2.3 \begin{pmatrix} 1 & 0.9/c \\ 0.9c & 1 \end{pmatrix}, \quad \text{while} \quad S_v = \begin{pmatrix} 1 & 0 \\ 0.9c & 1 \end{pmatrix}.$$

The most significant difference is the factor

$$2.3 = \frac{1}{\sqrt{1-(0.9c/c)^2}}$$

that multiplies B_v. To see why this differs so much from 1, note that $v = 0.9c$ implies that we should write $v = O(c)$, so

$$\frac{v}{c} = O(1) \quad \text{and} \quad \frac{1}{\sqrt{1-(v/c)^2}} = 1 + O(1) \neq 1 + O\left(\frac{1}{c^2}\right).$$

However, if $v = O(1)$, then we do indeed have

$$\frac{1}{\sqrt{1-(v/c)^2}} = 1 + O\left(\frac{1}{c^2}\right),$$

so our original assumption—that v is small in relation to c—means that $v = O(1)$ and *not* $v = O(c)$.

Criterion for neglecting small quantities

From the point of view of our example, we expect that *any* quantity Q will have the same order of magnitude as one of those in the sequence

$$\cdots \ll \frac{1}{c^2} \ll \frac{1}{c} \ll 1 \ll c \ll c^2 \ll \cdots.$$

We will then ignore Q if it is of the same order of magnitude as a sufficiently high power of $1/c$. In practice, we will write each relativistic expression (e.g., metric tensor, components of the gravitational field, component of a worldcurve) as a Taylor series in powers of $1/c$ (including small positive powers of c, as necessary), do our computations with these series, and then truncate the result at a point that is appropriate for the context.

Weak Gravitational Fields

Dimensionally homogeneous coordinates

To explore the Newtonian approximation, we shall find it helpful to work with dimensionally homogeneous coordinates, so

$$\mathbb{X} = (x^0, x^1, x^2, x^3) = (ct, x, y, z) \quad \text{meters},$$
$$\|\mathbb{X}\|^2 = (x^0)^2 - (x^1)^2 - (x^2)^2 - (x^3)^2 \quad \text{meters}^2$$

in the frame R. Let J denote the matrix that defines the Minkowski norm; its components J_{kl} are all dimensionless:

$$\|\mathbb{X}\|^2 = \mathbb{X}^t J \mathbb{X}, \quad \text{where} \quad J = (J_{kl}) = \begin{pmatrix} 1 & 0 & 0 & 0 \\ 0 & -1 & 0 & 0 \\ 0 & 0 & -1 & 0 \\ 0 & 0 & 0 & -1 \end{pmatrix}.$$

Expand g_{kl} in powers of $1/c$

We can now say what we mean by a weak gravitational field for R. The basic idea is simple. If we had $g_{kl} = J_{kl}$, then R would be an inertial frame, and there would be no gravity. Adding small terms to g_{kl} should therefore introduce weak gravitational effects to R. Since "small" means "of order of magnitude $1/c$," we can do this systematically by expanding the gravitational potential functions g_{kl} in powers of $1/c$:

$$g_{kl} = J_{kl} + \frac{1}{c} g_{kl}^{(1)} + \frac{1}{c^2} g_{kl}^{(2)} + \frac{1}{c^3} g_{kl}^{(3)} + O\left(\frac{1}{c^4}\right).$$

Since the gravitational field $-\Gamma^h_{kl}$ involves partial derivatives of the potential functions, we shall also want those partial derivatives to be small in the same way. Our goal is to have the classical equations of motion and field equations emerge from their relativistic counterparts:

$$\frac{d^2 x^h}{d\tau^2} = -\Gamma^h_{kl} \frac{dx^k}{d\tau} \frac{dx^l}{d\tau} \quad \longrightarrow \quad \frac{d^2 x^\alpha}{dt^2} = -\frac{\partial \Phi}{\partial x^\alpha},$$

$$R_{ij} = \frac{8\pi G}{c^4}\left(T_{ij} - \tfrac{1}{2} g_{ij} T\right) \quad \longrightarrow \quad \nabla^2 \Phi = 4\pi \rho.$$

In fact, the exercises show that this will not happen unless the terms of order $1/c$ in the metric vanish: $g^{(1)}_{kl} = 0$. Here, then, is our definition of a weak field.

Definition 8.1 *A **weak gravitational field** is one where the gravitational potential is the standard Minkowski metric plus "correction" terms of order no greater than $1/c^2$:*

$$g_{kl} = J_{kl} + \frac{1}{c^2} g^{(2)}_{kl} + \frac{1}{c^3} g^{(3)}_{kl} + O\!\left(\frac{1}{c^4}\right).$$

The individual components must be small and must furthermore vary slowly in space and time:

$$g^{(m)}_{kl} = O(1), \quad \frac{\partial g^{(m)}_{kl}}{\partial t} = O(1), \quad \frac{\partial g^{(m)}_{kl}}{\partial x^\alpha} = O(1).$$

Note that

$$\frac{\partial g^{(m)}_{kl}}{\partial x^0} = \frac{\partial g^{(m)}_{kl}}{\partial t} \frac{dt}{dx^0} = O(1) \cdot \frac{1}{c} = O\!\left(\frac{1}{c}\right).$$

Proposition 8.1 *The matrix J is its own inverse: $J^{ij} = J_{ij}$. If g_{kl} is a weak gravitational field, then*

$$g^{ij} = J^{ij} + O\!\left(\frac{1}{c^2}\right), \quad g = -1 + O\!\left(\frac{1}{c^2}\right).$$

PROOF: Exercise.

Equations of Motion

World time and proper time

To transform the relativistic equations of motion into the classical ones, we must be able to pass between proper time—which is not part of the Newtonian theory—and ordinary "world time." If $\mathbf{x}(t) = (x^1(t), x^2(t), x^3(t))$ is the trajectory of a classical particle, then $\mathbb{X}(t) = (ct, \mathbf{x}(t))$ is its worldcurve. To parametrize $\mathbb{X}$ by proper time, we can use the integral

$$L(t) = \int_a^t \|\mathbb{V}(t)\| dt.$$

But because the 4-velocity $\mathbb{V} = (c, \dot{\mathbf{x}})$ is measured in meters per second, L is a *length*, not a *time*. However, we get proper time by correcting dimensions in the usual way: $\tau(t) = L(t)/c$ seconds.

The Newtonian approximation applies only to particles whose velocities are small, so we make the following definition.

Definition 8.2 *We say that a particle with trajectory* $\mathbf{x}(t) = (x^1(t), x^2(t), x^3(t))$ *is* **slow** *if* $\dot{x}^\alpha = O(1)$, $\alpha = 1, 2, 3$.

Theorem 8.1 *In a weak gravitational field the four relativistic equations of motion of a slow particle reduce to the* three *classical equations*

$$\frac{d^2 x^i}{d\tau^2} = -\Gamma^i_{kl} \frac{dx^k}{d\tau} \frac{dx^l}{d\tau} \quad \longrightarrow \quad \ddot{x}^\alpha = -\frac{\partial \Phi}{\partial x^\alpha},$$

and the **Newtonian potential** Φ *is* $\frac{1}{2} g^{(2)}_{00}$.

PROOF: The reduction has several stages. First, proper time τ has to be replaced by "world time" t; second, one of the four relativistic equations has to vanish or at least become irrelevant; third, the right-hand sides of the three remaining equations have to reduce to the gradient of a single function. We do all this in a sequence of steps.

$\dfrac{dt}{d\tau} = 1 + O\left(\dfrac{1}{c^2}\right)$

STEP 1. In the simplest situation of uniform motion in an inertial frame (t, z) with Minkowski metric $\|(t, z)\|^2 = t^2 - z^2/c^2$,

$$\Delta \tau = \sqrt{\Delta t^2 - \frac{1}{c^2} \Delta z^2} = \sqrt{1 - \frac{1}{c^2}\left(\frac{\Delta z}{\Delta t}\right)^2} \Delta t = \left[1 + O\left(\frac{1}{c^2}\right)\right] \Delta t,$$

§8.1 The Newtonian Approximation

at least when $v = \Delta z / \Delta t = O(1)$. We show that the same is true for a slow particle in a weak gravitational field. Along the worldcurve $\mathbb{X}(t)$, the proper time function $\tau(t)$ satisfies

$$d\tau = \frac{\|\mathbb{V}\|}{c} dt = \sqrt{\frac{g_{kl} \dot{x}^k \dot{x}^l}{c^2}} \, dt.$$

Because the field is weak and $\dot{x}^0 = c$, $\dot{x}^\alpha = O(1)$, we have

$$\frac{g_{kl} \dot{x}^k \dot{x}^l}{c^2} = g_{00} + 2 g_{0\alpha} \frac{\dot{x}^\alpha}{c} + g_{\alpha\beta} \frac{\dot{x}^\alpha \dot{x}^\beta}{c^2} = 1 + O\left(\frac{1}{c^2}\right).$$

Therefore,

$$d\tau = \sqrt{\frac{g_{kl} \dot{x}^k \dot{x}^l}{c^2}} \, dt = \sqrt{1 + O\left(\frac{1}{c^2}\right)} \, dt = \left[1 + O\left(\frac{1}{c^2}\right)\right] dt.$$

You should show that in addition,

$$\frac{dt}{d\tau} = 1 + O\left(\frac{1}{c^2}\right) \quad \text{and} \quad \frac{d\tau}{dt} = 1 + O\left(\frac{1}{c^2}\right).$$

STEP 2. We now turn to the equations themselves. If we reparametrize $\mathbb{X}(t)$ using the proper time τ, we can differentiate the components of $\mathbb{X}$ with respect to τ using the chain rule. As usual, the first component is different from the rest:

Comparing the time derivatives

$$\frac{dx^0}{d\tau} = \frac{dx^0}{dt} \frac{dt}{d\tau} = c \left[1 + O\left(\frac{1}{c^2}\right)\right] = c + O\left(\frac{1}{c}\right),$$

$$\frac{d^2 x^0}{d\tau^2} = \frac{d}{dt}\left[c + O\left(\frac{1}{c}\right)\right] \frac{dt}{d\tau} = O\left(\frac{1}{c}\right)\left[1 + O\left(\frac{1}{c^2}\right)\right] = O\left(\frac{1}{c}\right),$$

so the left-hand side of the first relativistic equation of motion is negligible. (In the exercises you are asked to show that this would not be true if we abandoned the requirement $g_{kl}^{(1)} = 0$ for a weak gravitational field.) For $\alpha = 1, 2, 3$,

$$\frac{dx^\alpha}{d\tau} = \frac{dx^\alpha}{dt} \frac{dt}{d\tau} = \dot{x}^\alpha \left[1 + O\left(\frac{1}{c^2}\right)\right] = \dot{x}^\alpha + O\left(\frac{1}{c^2}\right),$$

$$\frac{d^2 x^\alpha}{d\tau^2} = \frac{d}{dt}\left[\dot{x}^\alpha + O\left(\frac{1}{c^2}\right)\right] \frac{dt}{d\tau} = \ddot{x}^\alpha + O\left(\frac{1}{c^2}\right).$$

Thus, there is a negligible difference between the proper time derivatives and the world time derivatives of the spatial variables x^1, x^2, x^3. (If the particle is not slow, then we cannot draw this conclusion; see the exercises.)

Order of magnitude of the Christoffel symbols

STEP 3. We must now determine what form the right-hand sides of the relativistic equations take. The Christoffel symbols involve partial derivatives of g_{kl} with respect to the coordinates, and again we must treat $x^0 = ct$ differently from x^α. Since $g_{kl} = J_{kl} + O(1/c^2)$,

$$\frac{\partial g_{kl}}{\partial x^\alpha} = O\left(\frac{1}{c^2}\right) \quad \text{but} \quad \frac{\partial g_{kl}}{\partial x^0} = \frac{\partial g_{kl}}{\partial t}\frac{dt}{dx^0} = \frac{\partial g_{kl}}{\partial t}\frac{1}{c} = O\left(\frac{1}{c^3}\right).$$

Therefore, $\Gamma_{k0,k} = O(1/c^3)$, while $\Gamma_{kl,i} = O(1/c^2)$ in general. For the Christoffel symbols of the second kind we note in particular that

$$\Gamma^\alpha_{00} = g^{\alpha m}\Gamma_{00,m} = g^{\alpha\alpha}\Gamma_{00,\alpha} + g^{\alpha 0}\Gamma_{00,0} + g^{\alpha\beta}\Gamma_{00,\beta} \quad (\beta \neq \alpha)$$

$$= \left[-1 + O\left(\frac{1}{c^2}\right)\right]\left[\frac{1}{2}\left(2\frac{\partial g_{0\alpha}}{\partial x^0} - \frac{\partial g_{00}}{\partial x^\alpha}\right)\right]$$

$$+ O\left(\frac{1}{c^2}\right)\frac{1}{2}\frac{\partial g_{00}}{\partial x^0} + O\left(\frac{1}{c^2}\right)\left[\frac{1}{2}\left(2\frac{\partial g_{0\beta}}{\partial x^0} - \frac{\partial g_{00}}{\partial x^\beta}\right)\right]$$

$$= \frac{1}{2}\frac{\partial g_{00}}{\partial x^\alpha} + O\left(\frac{1}{c^3}\right) = \frac{1}{2c^2}\frac{\partial g^{(2)}_{00}}{\partial x^\alpha} + O\left(\frac{1}{c^3}\right).$$

For the remaining Christoffel symbols you should prove the following:

$$\Gamma^h_{h0} = O\left(\frac{1}{c^3}\right), \quad \Gamma^h_{kl} = O\left(\frac{1}{c^2}\right), \quad (k, l) \neq (0, 0).$$

Reducing the geodesic equation for x^0

STEP 4. We are now in a position to consider how the geodesic equations of relativity reduce to the equations of motion of the Newtonian theory. As always, x^0 is a special case. The equation for x^0 is

$$\frac{d^2 x^0}{d\tau^2} = -\Gamma^0_{kl}\frac{dx^k}{d\tau}\frac{dx^l}{d\tau} = -\Gamma^0_{00}\left(\frac{dx^0}{d\tau}\right)^2 - 2\Gamma^0_{0\alpha}\frac{dx^0}{d\tau}\frac{dx^\alpha}{d\tau} - \Gamma^0_{\alpha\beta}\frac{dx^\alpha}{d\tau}\frac{dx^\beta}{d\tau},$$

and we have already determined that the left-hand side is $O(1/c)$. For the three terms on the right-hand side we have

$$\Gamma^0_{00}\left(\frac{dx^0}{d\tau}\right)^2 = O\left(\frac{1}{c^3}\right)\left(c + O\left(\frac{1}{c}\right)\right)^2 = O\left(\frac{1}{c^3}\right)(c^2 + O(1)) = O\left(\frac{1}{c}\right),$$

$$\Gamma^0_{0\alpha}\frac{dx^0}{d\tau}\frac{dx^\alpha}{d\tau} = O\left(\frac{1}{c^2}\right)\left(c + O\left(\frac{1}{c}\right)\right)\left(\dot{x}^\alpha + O\left(\frac{1}{c^2}\right)\right) = O\left(\frac{1}{c}\right),$$

$$\Gamma^0_{\alpha\beta}\frac{dx^\alpha}{d\tau}\frac{dx^\beta}{d\tau} = O\left(\frac{1}{c^2}\right)\left(\dot{x}^\alpha + O\left(\frac{1}{c^2}\right)\right)\left(\dot{x}^\beta + O\left(\frac{1}{c^2}\right)\right) = O\left(\frac{1}{c^2}\right).$$

In other words, the geodesic equation for x^0 has order $1/c$; in the Newtonian approximation it is entirely negligible. Note that we used the condition $\dot{x}^\alpha = O(1)$ to guarantee that the second and third terms are negligible.

STEP 5. We turn now to the geodesic equation for x^α,

$$\frac{d^2 x^\alpha}{d\tau^2} = -\Gamma^\alpha_{00}\left(\frac{dx^0}{d\tau}\right)^2 - 2\Gamma^\alpha_{0\beta}\frac{dx^0}{d\tau}\frac{dx^\beta}{d\tau} - \Gamma^\alpha_{\beta\gamma}\frac{dx^\beta}{d\tau}\frac{dx^\gamma}{d\tau}.$$

Reducing the geodesic equation for x^α

The left-hand side reduces very simply to

$$\frac{d^2 x^\alpha}{d\tau^2} = \ddot{x}^\alpha + O\left(\frac{1}{c^2}\right).$$

On the right-hand side, the second and third terms are negligible as in the case of x^0:

$$\Gamma^\alpha_{0\beta}\frac{dx^0}{d\tau}\frac{dx^\beta}{d\tau} = O\left(\frac{1}{c^2}\right)\left(c + O\left(\frac{1}{c}\right)\right)\left(\dot{x}^\beta + O\left(\frac{1}{c^2}\right)\right) = O\left(\frac{1}{c}\right),$$

$$\Gamma^\alpha_{\beta\gamma}\frac{dx^\beta}{d\tau}\frac{dx^\gamma}{d\tau} = O\left(\frac{1}{c^2}\right)\left(\dot{x}^\beta + O\left(\frac{1}{c^2}\right)\right)\left(\dot{x}^\gamma + O\left(\frac{1}{c^2}\right)\right) = O\left(\frac{1}{c^2}\right).$$

But the first term provides a nonnegligible contribution,

$$\Gamma^\alpha_{00}\left(\frac{dx^0}{d\tau}\right)^2 = \left[\frac{1}{2c^2}\frac{\partial g^{(2)}_{00}}{\partial x^\alpha} + O\left(\frac{1}{c^3}\right)\right](c^2 + O(1)) = \frac{1}{2}\frac{\partial g^{(2)}_{00}}{\partial x^\alpha} + O\left(\frac{1}{c}\right).$$

We can therefore rewrite the geodesic equation in the form

$$\ddot{x}^\alpha = -\frac{1}{2}\frac{\partial g^{(2)}_{00}}{\partial x^\alpha} + O\left(\frac{1}{c}\right).$$

END OF PROOF

The Field Equations

Theorem 8.2 *The relativistic equations of a weak gravitational field imply the classical field equation:*

$$R_{ij} = \frac{8\pi G}{c^4}(T_{ij} - \tfrac{1}{2}g_{ij}T) \quad \longrightarrow \quad \sum_{\alpha=1}^{3} \frac{\partial^2 \Phi}{(\partial x^\alpha)^2} = 4\pi G\rho.$$

PROOF: We begin with the right-hand side of the relativistic equations and make them as simple as possible by assuming that matter–energy consists of a swarm of identical noninteracting particles at rest in the frame $R : (x^0 = ct, x^1, x^2, x^3)$. Suppose the swarm has rest density ρ. Its proper 4-velocity is $\mathbb{U} = (c, 0, 0, 0)$, and its energy–momentum tensor has the single nonzero component $T_{00} = \rho c^2$. Therefore, $T^0_0 = T = \rho c^2$, so the right-hand side of the relativistic field equation with $i = j = 0$ is

$$\frac{8\pi G}{c^4}(T_{00} - \tfrac{1}{2}g_{00}T) = \frac{8\pi G}{c^4}\left[\rho c^2 - \left(1 + O\left(\frac{1}{c^2}\right)\right)\frac{\rho c^2}{2}\right]$$

$$= \frac{8\pi G}{c^4}\left[\frac{\rho c^2}{2} + O(1)\right] = \frac{1}{c^2}4\pi G\rho + O\left(\frac{1}{c^4}\right).$$

On the left-hand side of the same equation we have

$$R_{00} = R^h_{0h0} = R^\alpha_{0\alpha0} = \frac{\partial \Gamma^\alpha_{00}}{\partial x^\alpha} - \frac{\partial \Gamma^\alpha_{0\alpha}}{\partial x^0} + \Gamma^p_{00}\Gamma^\alpha_{p\alpha} - \Gamma^p_{0\alpha}\Gamma^\alpha_{p0}.$$

(Note: You should check that $R^0_{000} = 0$, so we need only consider $\alpha = 1, 2, 3$.) In a weak gravitational field, all Christoffel symbols are at least $O(1/c^2)$, so the products in the last two terms are $O(1/c^4)$. Furthermore,

$$\Gamma^\alpha_{0\alpha} = O\left(\frac{1}{c^3}\right), \qquad \frac{\partial \Gamma^\alpha_{0\alpha}}{\partial x^0} = O\left(\frac{1}{c^4}\right),$$

and

$$\Gamma^\alpha_{00} = \frac{1}{c^2}\frac{\partial \Phi}{\partial x^\alpha} + O\left(\frac{1}{c^3}\right), \qquad \frac{\partial \Gamma^\alpha_{00}}{\partial x^\alpha} = \frac{1}{c^2}\frac{\partial^2 \Phi}{(\partial x^\alpha)^2} + O\left(\frac{1}{c^3}\right).$$

Thus

$$R_{00} = \frac{1}{c^2} \sum_{\alpha=1}^{3} \frac{\partial^2 \Phi}{(\partial x^\alpha)^2} + O\left(\frac{1}{c^3}\right),$$

which gives us the classical field equation modulo terms of order $1/c$:

$$\sum_{\alpha=1}^{3} \frac{\partial^2 \Phi}{(\partial x^\alpha)^2} = 4\pi G \rho + O\left(\frac{1}{c}\right). \qquad \text{END OF PROOF}$$

Exercises

1. Assume that g_{ij} is a weak gravitational field; show that

$$-g = 1 + \frac{g_{00}^{(2)} - g_{11}^{(2)} - g_{22}^{(2)} - g_{33}^{(2)}}{c^2} + O\left(\frac{1}{c^4}\right),$$

$$\ln\sqrt{-g} = \frac{g_{00}^{(2)} - g_{11}^{(2)} - g_{22}^{(2)} - g_{33}^{(2)}}{2c^2} + O\left(\frac{1}{c^4}\right),$$

$$g^{ij} = J^{ij} + +O\left(\frac{1}{c^2}\right).$$

2. (a) Show that $\dfrac{1}{1+O(a)} = 1 + O(a)$ and $\sqrt{1+O(a)} = 1 + O(a)$.

   (b) Show that $d\tau = \sqrt{1 + O\left(\dfrac{1}{c^2}\right)} \, dt$ implies

   $$\frac{dt}{d\tau} = 1 + O\left(\frac{1}{c^2}\right) \quad \text{and} \quad \frac{d\tau}{dt} = 1 + O\left(\frac{1}{c^2}\right).$$

3. Show that $\Gamma_{h0}^{h} = O\left(\dfrac{1}{c^3}\right)$ and $\Gamma_{kl}^{h} = O\left(\dfrac{1}{c^2}\right)$ $((k, l) \neq (0, 0))$ in a weak gravitational field.

4. What is the relation between proper time and world time for a "fast" particle (i.e., one for which $\dot{x}^\alpha = O(c)$)?

5. Suppose we were to allow the potentials of a weak gravitational field to contain terms of order $1/c$:

$$g_{kl} = J_{kl} + \frac{1}{c}g_{kl}^{(1)} + O\left(\frac{1}{c^2}\right).$$

   (a) Show that $d\tau = \left(1 + A/c + O(1/c^2)\right) dt$ for a slow particle; express A in terms of the $g_{kl}^{(m)}$.

   (b) Suppose the worldcurve $(x^0(t), x^\alpha(t)) = (ct, x^\alpha(t))$ of a slow particle is reparametrized using proper time τ. Assuming that τ and t are related as in part (a), show that $d^2x^0/d\tau^2 = O(1/c)$ is no longer true. What implication does this have for the Newtonian approximation?

   (c) Find the terms in Γ^α_{00} and $\Gamma^\alpha_{0\beta}$ that are of order $1/c$. What effect do these terms have on an attempt to reduce the relativistic geodesic equations to the classical equations of motion?

6. (a) Suppose T^{ij} is the energy–momentum tensor of a swarm of noninteracting particles moving slowly in the inertial frame R with dimensionally homogeneous coordinates (ct, x, y, z). Show that $T^{00} = O(c^2)$ and determine the order of magnitude of all the other components. Determine the order of magnitude of $T = T^i_i$.

   (b) Determine the order of magnitude of each component of the Ricci tensor R_{ij} when g_{ij} is a weak gravitational field.

8.2 Spherically Symmetric Fields

Solving the field equations

This section is about finding solutions to the gravitational field equations. These are a system of partial differential equations for the components of the gravitational potential; they are generally intractable. We shall restrict ourselves to cases that are simple enough to solve.

The Weak Field of a Point Source

The simplest gravitational field is produced by a single massive body that is concentrated in a relatively small region of space — a "point source." From Section 4.3 we know that the Newtonian potential of a single point source of mass M is

$$\Phi(\mathbf{x}) = -\frac{GM}{r}, \qquad r = \|\mathbf{x}\| = \sqrt{x^2 + y^2 + z^2}.$$

The mass sits motionless at the origin $\mathbf{x} = \mathbf{0}$. The Newtonian field it generates is spherically symmetric about this center, it is constant over time, and it approaches zero as $r \to \infty$.

Let us construct a relativistic field for this source that has the same properties: spherically symmetric, independent of time, and tending to the Minkowski metric as $r \to \infty$. We shall assume that it is a *weak* field in the sense of the last section. Thus we take the Minkowski metric and add correction terms, starting with

Solve the vacuum field equations to determine the field

$$g_{00} = 1 + \frac{2\Phi}{c^2} + O\!\left(\frac{1}{c^3}\right) = 1 - \frac{2GM}{rc^2} + O\!\left(\frac{1}{c^3}\right).$$

There are other correction terms, of course, but we don't know them at the outset. In fact, to determine them we shall use the vacuum field equations $R_{ij} = 0$, which we know will hold away from the gravitational source. These equations are partial differential equations for the components of the gravitational metric. In this case we shall see that they reduce to a single *ordinary* differential equation—one that we shall be able to solve readily.

To construct the metric, we assume that the worldcurve of the mass is the t-axis in the frame R with dimensionally homogeneous coordinates $(x^0 = ct, x, y, z)$. Furthermore, we assume, as in the classical case, that all the components g_{kl} of the metric are independent of time and depend only on the function r of the spatial coordinates. The spherical symmetry of the metric suggests that we should convert (x, y, z) to spherical coordinates:

Exploit the spherical symmetry of the field

$$x = r \sin\varphi \cos\theta,$$
$$y = r \sin\varphi \sin\theta,$$
$$z = r \cos\varphi.$$

This defines a map $M: G \to R$ where $G: (x^0, r, \varphi, \theta)$, and we can determine the new metric γ_{ij} in G from the original Minkowski metric g_{kl} in R (before we add the correction terms) in the standard way described in Section 6.4:

$$(\gamma_{ij}) = dM^t (g_{kl}) dM.$$

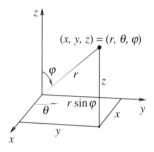

Write the metric as a differential

But there is another way. Since the norm is used to determine proper time by an integral, we can write it as a differential:

$$\tau = \frac{1}{c} \int \sqrt{g_{kl} \frac{dx^k}{dt} \frac{dx^l}{dt}} \, dt = \frac{1}{c} \int \sqrt{g_{kl} \, dx^k dx^l} = \frac{1}{c} \int ds;$$

thus $ds^2 = g_{kl} \, dx^k dx^l$. For the original Minkowski norm,

$$ds^2 = (dx^0)^2 - (dx^2 + dy^2 + dz^2).$$

Therefore, to determine the metric ds^2 in the new coordinates, we should express dx^2, dy^2, and dz^2 in terms of r, φ, θ, and their differentials. (Note: Compute ordinary products here—not the *exterior* products defined in Exercise 6 of Section 4.4—because we are merely representing products of derivatives, e.g., dx^2 for $(dx/dt)^2$.) Thus,

$$dx = \sin \varphi \cos \theta \, dr + r \cos \varphi \cos \theta \, d\varphi - r \sin \varphi \sin \theta \, d\theta,$$
$$dy = \sin \varphi \sin \theta \, dr + r \cos \varphi \sin \theta \, d\varphi + r \sin \varphi \cos \theta \, d\theta,$$
$$dz = \cos \varphi \, dr - r \sin \varphi \, d\varphi.$$

Now compute the squares and show that

$$dx^2 + dy^2 + dz^2 = dr^2 + r^2 \, d\varphi^2 + r^2 \sin^2 \varphi \, d\theta^2.$$

If we let $(ct, r, \varphi, \theta) = (\xi^0, \xi^1, \xi^2, \xi^3)$ in G, then we can construct indexed quantities like γ_{ij} and Γ^k_{ij} while still retaining the original variables for visual clarity in our formulas. We note for a start that γ_{ij} is the *diagonal* matrix

$$(\gamma_{ij}) = \begin{pmatrix} 1 & & & \\ & -1 & & \\ & & -r^2 & \\ & & & -r^2 \sin^2 \varphi \end{pmatrix}.$$

To add the correction terms that account for the point mass at the spatial origin, let us first write $\Psi(r) = \Phi(x, y, z)$ to express the Newtonian potential explicitly in terms of the distance r; thus

The correction terms

$$\gamma_{00} = 1 + \frac{2\Psi(r)}{c^2} + O\left(\frac{1}{c^3}\right).$$

Our assumption that the perturbed field should be spherically symmetric means that we do not alter any terms involving $d\varphi$ or $d\theta$, and that the correction to dr^2 should depend only on r:

$$\gamma_{11} = -1 + \frac{P(r)}{c^2} + O\left(\frac{1}{c^3}\right).$$

Thus, to determine the weak field approximation—at least to order $1/c^2$—we need only find the function $P(r)$.

Theorem 8.3 $P(r) = 2\Psi(r) + O\left(\dfrac{1}{c}\right)$.

PROOF: The vacuum field equations $R_{ij} = 0$ hold outside the mass M, which we assume to be concentrated at the spatial origin $r = 0$. Since there is only one unknown function, it turns out to be sufficient to consider the single equation

$$R_{11} = R^h_{1h1} = \frac{\partial \Gamma^h_{11}}{\partial \xi^h} - \frac{\partial \Gamma^h_{1h}}{\partial \xi^1} + \Gamma^p_{11}\Gamma^h_{ph} - \Gamma^p_{1h}\Gamma^h_{p1} = 0,$$

where $h, p = 0, 1, 2, 3$. We make the following general observations:

- Since γ_{ij} is diagonal, so is its inverse γ^{kl}; thus $\gamma^{kl} = 0$ if $k \neq l$ and $\gamma^{kk} = 1/\gamma_{kk}$.

- $\Gamma^k_{ij} = \gamma^{kl}\Gamma_{ij,l}$ (summing on l) reduces to the single term $\gamma^{kk}\Gamma_{ij,k}$.
- If i, j, k are all different, then $\Gamma^k_{ij} = 0$. This follows from the fact that

$$\Gamma_{ij,k} = \tfrac{1}{2}\left(\frac{\partial \gamma_{ik}}{\partial \xi^j} + \frac{\partial \gamma_{jk}}{\partial \xi^i} - \frac{\partial \gamma_{ij}}{\partial \xi^k}\right) = 0$$

because $\gamma_{ik} = \gamma_{jk} = \gamma_{ij} = 0$ when the subscripts are all different.

- All derivatives with respect to $\xi^0 = ct$ are zero.

By direct calculation we find that most terms that appear in R_{11} reduce to zero; the only Christoffel symbols we need are

$$\Gamma^0_{01} = \frac{\Psi'}{c^2} + O\!\left(\frac{1}{c^3}\right), \quad \Gamma^1_{11} = -\frac{P'}{2c^2} + O\!\left(\frac{1}{c^3}\right), \quad \Gamma^2_{12} = \Gamma^3_{13} = \frac{1}{r}.$$

Therefore,

$$\frac{\partial \Gamma^h_{11}}{\partial \xi^h} = \frac{\partial \Gamma^1_{11}}{\partial r} = -\frac{P''}{2c^2} + O\!\left(\frac{1}{c^3}\right),$$

$$\frac{\partial \Gamma^h_{1h}}{\partial \xi^1} = \frac{\partial \Gamma^0_{10}}{\partial r} + \frac{\partial \Gamma^1_{11}}{\partial r} + \frac{\partial \Gamma^2_{12}}{\partial r} + \frac{\partial \Gamma^3_{13}}{\partial r} = \frac{\Psi''}{c^2} - \frac{P''}{2c^2} - \frac{2}{r^2} + O\!\left(\frac{1}{c^3}\right),$$

$$\Gamma^p_{11}\Gamma^h_{ph} = \Gamma^1_{11}\Gamma^h_{1h} = \Gamma^1_{11}(\Gamma^0_{10} + \Gamma^1_{11} + \Gamma^2_{12} + \Gamma^3_{13}) = -\frac{P'}{2c^2}\cdot\frac{2}{r} + O\!\left(\frac{1}{c^3}\right),$$

$$\Gamma^p_{1h}\Gamma^h_{p1} = (\Gamma^2_{12})^2 + (\Gamma^3_{13})^2 + O\!\left(\frac{1}{c^4}\right) = \frac{2}{r^2} + O\!\left(\frac{1}{c^4}\right),$$

and $R_{11} = 0$ reduces to

$$-\frac{\Psi''}{c^2} - \frac{P'}{rc^2} + O\!\left(\frac{1}{c^3}\right) = 0,$$

or

$$P' = -r\Psi'' + O\!\left(\frac{1}{c}\right) = \frac{2GM}{r^2} + O\!\left(\frac{1}{c}\right).$$

Therefore, if we ignore terms of order $O(1/c)$,

$$P(r) = -\frac{2GM}{r} + A = 2\Psi(r) + A,$$

§8.2 Spherically Symmetric fields

where A is a constant of integration. But since we require $P(r) \to 0$ as $r \to \infty$, A must be zero.

END OF PROOF

If we write the metric as a differential and neglect terms of order $1/c^3$ in the coefficients, it has this vivid form

The metric in Cartesian coordinates

$$ds^2 = c^2\left(1 + \frac{2\Phi}{c^2}\right)dt^2 - \left(1 - \frac{2\Phi}{c^2}\right)dr^2 - r^2(d\varphi^2 + \sin^2\varphi\, d\theta^2)$$

$$= c^2\left(1 + \frac{2\Phi}{c^2}\right)dt^2 - dr^2 - r^2(d\varphi^2 + \sin^2\varphi\, d\theta^2) + \frac{2\Phi}{c^2}dr^2.$$

To write this in the Cartesian frame $R : (ct, x, y, z) = (x^0, x^1, x^2, x^3)$, we need to transform the last term. Since $r^2 = x^2 + y^2 + z^2$,

$$dr = \frac{x\,dx + y\,dy + z\,dz}{r} = \sum_\alpha \frac{x^\alpha}{r}dx^\alpha,$$

$$dr^2 = \left(\sum_\alpha \frac{x^\alpha}{r}dx^\alpha\right)\left(\sum_\beta \frac{x^\beta}{r}dx^\beta\right) = \sum_{\alpha,\beta} \frac{x^\alpha x^\beta}{r^2}dx^\alpha dx^\beta,$$

so the metric in the frame R is

$$ds^2 = c^2\left(1 + \frac{2\Phi}{c^2}\right)dt^2 - \sum_\alpha (dx^\alpha)^2 + \frac{2\Phi}{c^2}\sum_{\alpha,\beta}\frac{x^\alpha x^\beta}{r^2}dx^\alpha dx^\beta$$

$$= (dx^0)^2 - \sum_\alpha (dx^\alpha)^2 - \frac{2GM}{rc^2}\left((dx^0)^2 + \sum_{\alpha,\beta}\frac{x^\alpha x^\beta}{r^2}dx^\alpha dx^\beta\right).$$

cf. p. 398

In terms of components, this is

$$g_{00} = 1 - \frac{2GM}{c^2}\frac{1}{r}, \qquad g_{0\alpha} = 0, \qquad g_{\alpha\beta} = -\delta_{\alpha\beta} - \frac{2GM}{c^2}\frac{x^\alpha x^\beta}{r^3}.$$

The Schwarzschild Metric

In 1916 the physicist Karl Schwarzschild determined the relativistic field outside an arbitrary spherically symmetric distribution of masses. A hollow spherical shell is an example. He made no assumption that the field should be weak—as we did in Theorem 8.3—but only that it should reduce to the Minkowski metric

An arbitrary spherically symmetric gravitational source

far from the masses. His solution even allowed the field to vary over time—while retaining the spherical symmetry—but we shall look only at the simpler time-invariant case.

The form of the metric

We assume that the worldline of the center of symmetry is the t-axis in the frame $G : (ct, r, \varphi, \theta)$, and we use spherical coordinates (r, φ, θ) to take best advantage of the spherical symmetry. The Schwarzschild metric is a perturbation of the Minkowski metric that takes the following form:

$$ds^2 = e^{T(r)} c^2\, dt^2 - e^{Q(r)}\, dr^2 - r^2(d\varphi^2 + \sin^2\varphi\, d\theta^2).$$

The perturbations depend only on r and do not affect $d\varphi$ or $d\theta$ at all. Furthermore, since ds^2 must reduce to the plain Minkowski metric for r large, we require $T \to 0$ and $Q \to 0$ as $r \to \infty$. The main reason the coefficients are written as exponentials is to simplify later calculations, but the Taylor expansions of the exponentials also give us the useful form

$$ds^2 = (1 + T + O(T^2))c^2\, dt^2 - (1 + Q + O(Q^2))\, dr^2 \\ - r^2(d\varphi^2 + \sin^2\varphi\, d\theta^2).$$

This shows that T and Q themselves are, in some sense, the main components of the perturbations.

The vacuum field equations

To find $T(r)$ and $Q(r)$ we use the vacuum field equations, which hold outside the region containing the gravity-producing matter. The calculations are long and messy—patience and careful writing are as important as a knowledge of the rules of differentiation—but we shall need only the two equations $R_{00} = 0$, $R_{11} = 0$.

Again let $(\xi^0, \xi^1, \xi^2, \xi^3) = (ct, r, \varphi, \theta)$ so we can construct the usual indexed quantities with the more informative spherical coordinates. We start with the metric tensor and its inverse:

$$\gamma_{00} = e^T, \quad \gamma_{11} = -e^Q, \quad \gamma_{22} = -r^2, \quad \gamma_{33} = -r^2 \sin^2\varphi,$$
$$\gamma^{00} = e^{-T}, \quad \gamma^{11} = -e^{-Q}, \quad \gamma^{22} = -r^{-2}, \quad \gamma^{33} = -r^{-2} \sin^{-2}\varphi.$$

All off-diagonal terms are zero. This implies that the Christoffel symbols satisfy the same conditions we noted in the proof of Theorem 8.3:

- $\Gamma_{ij}^k = 0$ if i, j, k are all different;
- Γ_{ij}^k is the single term $\gamma^{kk}\Gamma_{ij,k}$.

Theorem 8.4 $e^T = 1 + \dfrac{a}{r}$, $e^Q = e^{-T} = \dfrac{1}{1 + (a/r)}$, where a is an arbitrary constant.

PROOF: Direct calculation shows that the only nonzero Christoffel symbols are the following nine and the additional four obtained by transposing the two subscripts:

$$\Gamma_{00}^1 = \frac{T'}{2} e^{T-Q}, \quad \Gamma_{11}^1 = \frac{Q'}{2}, \quad \Gamma_{22}^1 = -re^{-Q}, \quad \Gamma_{33}^1 = -re^{-Q}\sin^2\varphi,$$

$$\Gamma_{01}^0 = \frac{T'}{2}, \quad \Gamma_{12}^2 = \frac{1}{r}, \quad \Gamma_{23}^3 = \cot\varphi, \quad \Gamma_{33}^2 = -\sin\varphi\cos\varphi.$$

$$\Gamma_{13}^3 = \frac{1}{r},$$

The nonzero components of the covariant Ricci tensor are

$$R_{00} = e^{T-Q}\left(\frac{T''}{2} + \frac{(T')^2}{4} - \frac{T'Q'}{4} + \frac{T'}{r}\right),$$

$$R_{11} = -\frac{T''}{2} - \frac{(T')^2}{4} + \frac{T'Q'}{4} + \frac{Q'}{r},$$

$$R_{22} = 1 - e^{-Q} + re^{-Q}\left(\frac{Q'}{2} - \frac{T'}{2}\right),$$

$$R_{33} = R_{22}\sin^2\varphi.$$

Now consider the two vacuum field equations $R_{00} = 0$ and $R_{11} = 0$. Dividing $R_{00} = 0$ by e^{T-Q} and adding the result to $R_{11} = 0$ gives

$$\frac{T'}{r} + \frac{Q'}{r} = 0,$$

implying $T + Q = k$ for some constant k. Thus $e^T = e^{-Q}e^k$, so the metric is

$$ds^2 = e^{-Q}e^k c^2\, dt^2 - e^Q\, dr^2 - r^2(d\varphi^2 + \sin^2\varphi\, d\theta^2).$$

If we let $t = e^{k/2}u$, then $dt^2 = e^k du^2$ and
$$ds^2 = e^{-Q}c^2\, du^2 - e^Q\, dr^2 - r^2(d\varphi^2 + \sin^2\varphi\, d\theta^2).$$

In other words, by a simple rescaling of the time coordinate we can put the metric into a form in which $k = 0$. Therefore, we can assume $T = -Q$ without loss of generality.

It remains to determine T. With $Q = -T$ we can rewrite $R_{11} = 0$ as
$$-\frac{T''}{2} - \frac{(T')^2}{4} - \frac{(T')^2}{4} - \frac{T'}{r} = -\frac{1}{2r}(rT'' + r(T')^2 + 2T') = 0.$$

Since $(re^T)'' = e^T(rT'' + r(T')^2 + 2T')$, $R_{11} = 0$ implies $(re^T)'' = 0$. Therefore, $(re^T)' = b$ and then $re^T = br + a$, where b and a are constants of integration. We can write this as
$$e^T = b + \frac{a}{r},$$

and since we want $e^T \to 1$ as $r \to \infty$, it follows that $b = 1$. Therefore,
$$e^T = 1 + \frac{a}{r}, \qquad e^Q = \frac{1}{1 + \dfrac{a}{r}}. \qquad \text{END OF PROOF}$$

Schwarzschild metric

Thus, away from the field-producing matter, the Schwarzschild metric has the form
$$ds^2 = c^2\left(1 + \frac{a}{r}\right) dt^2 - \frac{1}{1 + (a/r)} dr^2 - r^2(d\varphi^2 + \sin^2\varphi\, d\theta^2).$$

As $r \to \infty$, ds^2 approaches the ordinary Minkowski metric as we required. This means that when r is large the Schwarzschild metric is a weak metric, implying
$$1 + \frac{a}{r} = \gamma_{00} \approx 1 + \frac{2\Phi}{c^2} = 1 - \frac{2GM}{rc^2}$$

and thus
$$-a \approx \frac{2GM}{c^2} = r_M.$$

Gravitational radius

This has the dimensions of a length (as it must for the term a/r in γ_{00} to be dimensionless). It is called the *gravitational radius*, or the *Schwarzschild radius*, of the mass M. If we write the metric in terms of r_M,

$$ds^2 = c^2\left(1 - \frac{r_M}{r}\right)dt^2 - \frac{1}{1-(r_M/r)}dr^2 - r^2(d\varphi^2 + \sin^2\varphi\, d\theta^2),$$

we can make the following observations:

- The metric depends only on the *total mass* M, not on the details of its distribution in space, just as in the Newtonian theory.
- When $r < r_M$, γ_{00} is negative, so the metric has a fundamentally different character.

Asymptotic Behavior

The Einstein weak-field metric and the Schwarzschild metric are both asymptotically inertial, in the sense that they approach the standard Minkowski metric as $r \to \infty$. This implies that their frames are nonrotating with respect to an inertial frame, so there are no Coriolis forces present; see Exercises 7–12 in Section 7.1. More important, when we measure the bending of light and perihelion drift in these frames—as we do in the following sections—the measurements are in reference to the background of the "fixed stars."

Gravitational Red Shift

Gravitational red shift is the decrease in characteristic frequency of light emitted by atoms that sit in a gravitational field. It is a consequence of time dilation—gravity causes all clocks, including atomic clocks, to slow down. In Section 4.4 we explained time dilation in terms of the equivalence principle, but we can also detect it by direct analysis of the gravitational metric. Specifically, consider the Schwarzschild metric in the frame of an observer R who is far from the source, and suppose a second observer G is motionless with respect to R but close to the source. Then $dr = d\varphi = d\theta = 0$ for both observers, so their proper times τ and

Time dilation in the Schwarzschild metric

t satisfy the equation

$$c^2 d\tau^2 = ds^2 = c^2\left(1 - \frac{r_M}{r}\right) dt^2, \quad \text{or} \quad d\tau = \sqrt{1 - \frac{r_M}{r}}\, dt < dt.$$

In other words, a time interval of $\Delta\tau$ seconds on G's clock will appear to be

$$\Delta t = \frac{1}{\sqrt{1 - r_M/r}}\, \Delta\tau > \Delta\tau \text{ seconds}$$

on R's clock. If we think of G as representing observers at various distances r, then we can display the *time dilation factor* $1/\sqrt{1 - r_M/r}$ for each observer as a grid in the (t, r) coordinate frame.

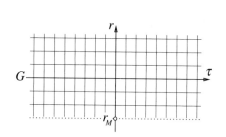
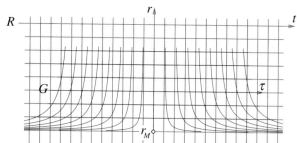

Notice that from R's point of view, the ticks on G's clock become infinitely far apart as G approaches the gravitational radius r_M. You should compare this grid with those we drew for accelerated frames and simple gravitational fields in Chapter 4.

Isotropic Coordinates

Complications in Cartesian coordinates

In our formulas for both Einstein's weak-field metric and the Schwarzschild metric, the spatial component is *not* proportional to the Euclidean metric

$$dx^2 + dy^2 + dz^2 = dr^2 + r^2\, d\varphi^2 + r^2 \sin^2\varphi\, d\theta^2,$$

because the radial term dr^2 has a factor not shared by the others. As a consequence, the Cartesian form of the metric is more complicated than the spherical. For example, when we converted the weak-field metric from spherical to Cartesian coordinates, many off-diagonal terms appeared:

$$g_{\alpha\beta} = -\delta_{\alpha\beta} - \frac{2GM}{c^2} \frac{x^\alpha x^\beta}{r^3}.$$

However, it turns out to be possible to introduce a new radial coordinate ρ in place of r so that the Schwarzschild metric takes the form

$$ds^2 = A(\rho)c^2 dt^2 - B(\rho)\left(d\rho^2 + \rho^2\, d\varphi^2 + \rho^2 \sin^2 \varphi\, d\theta^2\right).$$

Therefore, in terms of the Cartesian coordinates (ξ, η, ζ) associated with the spherical coordinates (ρ, φ, θ), the metric becomes

$$ds^2 = A(\rho)c^2 dt^2 - B(\rho)\left(d\xi^2 + d\eta^2 + d\zeta^2\right).$$

Since the metric now involves the same factor $-B(\rho)$ in every spatial direction, we say that the spatial coordinates (ρ, φ, θ) and (ξ, η, ζ) are **isotropic** for the metric. (The Greek roots are *isos*, "equal," and *trope*, "turn"; in effect, the metric "looks the same no matter which way we turn.")

Isotropic coordinates

To find the isotropic coordinates we must determine the function $r = f(\rho)$ that relates the two radial variables. By assumption,

$$\left(1 - \frac{r_M}{r}\right)c^2 dt^2 - \frac{1}{1 - r_M/r} dr^2 - r^2\, d\varphi^2 - r^2 \sin^2 \varphi\, d\theta^2$$
$$= A(\rho)c^2 dt^2 - B(\rho)d\rho^2 - B(\rho)\rho^2\, d\varphi^2 - B(\rho)\rho^2 \sin^2 \varphi\, d\theta^2,$$

so we must have

$$\frac{dr^2}{1 - r_M/r} = B(\rho)\, d\rho^2 \quad \text{and} \quad r^2 = B(\rho)\rho^2.$$

If we substitute $B = r^2/\rho^2$ into the first equation and separate the variables, we get

$$\frac{dr^2}{r^2 - r_M r} = \frac{d\rho^2}{\rho^2}, \quad \text{or} \quad \int \frac{dr}{\sqrt{r^2 - r_M r}} = \pm \int \frac{d\rho}{\rho} = \pm \ln k\rho,$$

where k is a constant of integration. To determine the integral involving r, first complete the square and let $r_M = 2\alpha$ for ease of calculation:

$$\int \frac{dr}{\sqrt{r^2 - 2\alpha r}} = \int \frac{dr}{\sqrt{(r-\alpha)^2 - \alpha^2}} = \int \frac{du}{\sqrt{u^2 - \alpha^2}},$$

where $u = r - \alpha$. With the substitution $u = \alpha \cosh v$ we get

$$\int \frac{du}{\sqrt{u^2 - \alpha^2}} = \int dv = v = \cosh^{-1}\left(\frac{r - \alpha}{\alpha}\right).$$

Therefore,

$$\cosh^{-1}\left(\frac{r - \alpha}{\alpha}\right) = \pm \ln k\rho,$$

or

$$\frac{r - \alpha}{\alpha} = \cosh(\pm \ln k\rho) = \frac{e^{\pm \ln k\rho} + e^{\mp \ln k\rho}}{2} = \frac{k\rho}{2} + \frac{1}{2k\rho}.$$

To specify the value of k we are free to impose a condition on r and ρ. It is reasonable to have r and ρ "asymptotically equal"—that is, to have $r/\rho \to 1$ as $\rho \to \infty$. This implies $k = 2/\alpha$, so solving for r gives

$$r = \rho + \alpha + \frac{\alpha^2}{4\rho} = \rho + \frac{r_M}{2} + \frac{r_M^2}{16\rho} = \rho\left(1 + \frac{r_M}{4\rho}\right)^2 = f(\rho).$$

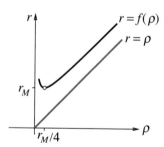

As the graph suggests, $f : (r_M/4, \infty) \to (r_M, \infty) : \rho \mapsto r$ is a valid coordinate change. The restriction $r_M < r$ is is not a serious limitation because we have already noted that the metric changes its character when $r \leq r_M$.

Using $r = f(\rho)$ we find that

Finding $A(\rho)$ and $B(\rho)$

$$B(\rho) = \frac{r^2}{\rho^2} = \left(1 + \frac{r_M}{4\rho}\right)^4$$

and

$$A(\rho) = \frac{r - r_M}{r} = \frac{\rho - \frac{r_M}{2} + \frac{r_M^2}{16\rho}}{\rho + \frac{r_M}{2} + \frac{r_M^2}{16\rho}} = \frac{\rho\left(1 - \frac{r_M}{4\rho}\right)^2}{\rho\left(1 + \frac{r_M}{4\rho}\right)^2} = \left(\frac{1 - \frac{r_M}{4\rho}}{1 + \frac{r_M}{4\rho}}\right)^2.$$

The foregoing arguments have therefore proven the following theorem.

Theorem 8.5 *In isotropic Cartesian coordinates* (t, ξ, η, ζ), *the Schwarzschild metric is*

$$ds^2 = \left(\frac{1 - r_M/4\rho}{1 + r_M/4\rho}\right)^2 c^2 dt^2 - \left(1 + \frac{r_M}{4\rho}\right)^4 \left(d\xi^2 + d\eta^2 + d\zeta^2\right).$$

Here $\rho^2 = \xi^2 + \eta^2 + \zeta^2$.

Corollary 8.1 *When ρ is large, the Schwarzschild metric is approximately*

$$ds^2 = \left(1 - \frac{r_M}{\rho}\right) c^2 dt^2 - \left(1 + \frac{r_M}{\rho}\right) \left(d\xi^2 + d\eta^2 + d\zeta^2\right).$$

PROOF: See the exercises.

In these statements we have transformed isotropic spherical to Cartesian coordinates,

$$\xi = \rho \cos\theta \sin\varphi,$$
$$\eta = \rho \sin\theta \sin\varphi,$$
$$\zeta = \rho \cos\varphi.$$

Since $\xi = (\rho/r)x$, and similarly for η and ζ, the new Cartesian coordinates are identical nonlinear rescalings of the original Cartesian coordinates x, y, and z, respectively.

Exercises

1. Compute dx^2, dy^2, and dz^2 in the transformation from spherical coordinates, and confirm that $dx^2 + dy^2 + dz^2 = dr^2 + r^2\,d\varphi^2 + r^2 \sin^2\varphi\,d\theta^2$.

2. Verify that Γ^0_{01}, Γ^1_{11}, Γ^2_{12}, and Γ^3_{13} have the values given in the proof of Theorem 8.3, and show that these are the only Christoffel symbols that appear in nonzero terms in R_{11}.

3. (a) Let γ be the determinant of the spherical weak metric of Theorem 8.3, in terms of the spherical coordinates $(\xi^0, \xi^1, \xi^2, \xi^3) = (ct, r, \varphi, \theta)$. Show that

$$\ln\sqrt{-\gamma} = 2\ln r + \ln\sin\varphi - \frac{P}{2c^2} + \frac{\Psi}{c^2} + O\left(\frac{1}{c^3}\right).$$

(b) Use the formula

$$R_{ij} = \frac{\partial \Gamma^h_{ij}}{\partial \xi^h} - \frac{\partial^2 \ln\sqrt{-\gamma}}{\partial \xi^i \partial \xi^j} + \Gamma^p_{ij} \frac{\partial \ln\sqrt{-\gamma}}{\partial \xi^p} - \Gamma^p_{ih}\Gamma^h_{pj}$$

(Exercise 4, Section 7.3) to recalculate R_{11} for the spherical weak metric and confirm the value given in the proof of Theorem 8.3.

4. Determine Christoffel symbols Γ^h_{ij} for the weak spherical metric in spherical coordinates. (Note: The nonzero ones are the same as for the Schwarzschild metric.)

5. Write the geodesic equations for the weak spherical metric up to terms of order $1/c^2$.

6. In this question, a *geodesic* is a solution to the geodesic equations of the weak spherical metric up to terms of order $1/c^2$.

   (a) Show that a radial line is a geodesic path; that is, for any given constants $\bar{\varphi}$ and $\bar{\theta}$, show that

   $$\mathbb{X}(\tau) = (t(\tau), r(\tau), \bar{\varphi}, \bar{\theta})$$

   is a solution to the geodesic equations (up to order $1/c^2$), for appropriate functions $t(\tau)$ and $r(\tau)$. (It is not necessary to solve the equations for $t(\tau)$ and $r(\tau)$.)

(b) Show that a geodesic that starts in the equatorial plane $\varphi = \pi/2$ remains there for all time. In other words, show that if $\mathbb{X}(\tau) = (t(\tau), r(\tau), \varphi(\tau), \theta(\tau))$ is a geodesic for which

$$\varphi(0) = \pi/2 \quad \text{and} \quad \frac{d\varphi}{d\tau}(0) = 0,$$

then $\varphi(\tau) = \pi/2$ for all τ. Show that this is not true if the initial value is different, that is, if $\varphi(0) \neq \pi/2$.

7. The purpose of this exercise is to show that the geodesic equations up to order $1/c^2$ for the weak spherical metric admit circular solutions in the equatorial plane.

   (a) Assume a solution of the form $r = \bar{r}$, $\varphi = \pi/2$, where $\bar{r}$ is a fixed positive number. Show that the equations for t and θ then imply that there are constants k and ω for which

   $$\frac{dt}{d\tau} = k, \quad \frac{d\theta}{d\tau} = \omega.$$

   (b) From part (a) deduce that the period T of the circular orbit of radius $\bar{r}$ is $T = 2\pi k/\omega$ in terms of the world time t.

   (c) Show that the equation for r requires

   $$\bar{r}^3 = \frac{GMk^2}{\omega^2} = \text{constant} \times T^2.$$

   This is Kepler's third law of planetary motion: The square of the period of an orbit is proportional to the cube of its radius.

8. Verify the expressions given for the nine Christoffel symbols of the Schwarzschild metric that appear in the proof of Theorem 8.4; show that all other Christoffel symbols vanish.

9. (a) Show that $\ln\sqrt{-\gamma} = 2\ln r + \ln\sin\varphi + \dfrac{T+Q}{2}$ when γ is the determinant of the Schwarzschild metric.

   (b) Using the formula for R_{ij} in Exercise 3, verify the formulas for R_{00}, R_{11}, R_{22}, and R_{33} given in the proof of Theorem 8.4, and show that all other components of the Ricci tensor are zero.

10. In the proof of Theorem 8.4 we saw that $R_{33} = R_{22} \sin^2 \varphi$. Therefore, the conditions $R_{22} = 0$ and $R_{33} = 0$ are not independent; the second follows from the first. Show that the equation $R_{22} = 0$ follows from $R_{00} = R_{11} = 0$, thus implying that at most two of the four vacuum field equations are independent.

11. (a) Show that the components of the mixed Ricci tensor R_i^j for the Schwarzschild metric are

$$R_0^0 = e^{-Q}\left(\frac{T''}{2} + \frac{(T')^2}{4} - \frac{T'Q'}{4} + \frac{T'}{r}\right),$$

$$R_1^1 = e^{-Q}\left(\frac{T''}{2} + \frac{(T')^2}{4} - \frac{T'Q'}{4} - \frac{Q'}{r}\right),$$

$$R_2^2 = -\frac{1}{r^2} + \frac{e^{-Q}}{r^2} - \frac{e^{-Q}}{r}\left(\frac{Q'}{2} - \frac{T'}{2}\right),$$

$$R_3^3 = R_2^2,$$

and hence that the scalar curvature R is

$$R = R_i^i = -\frac{2}{r^2} + e^{-Q}\left(T'' + \frac{(T')^2}{2} - \frac{T'Q'}{2} + \frac{2T'}{r} - \frac{2Q'}{r} + \frac{2}{r^2}\right).$$

(b) Show that the vacuum field equations in the form $R_i^j - \frac{1}{2}\delta_i^j R = 0$ also lead to the conclusions $e^T = 1 + a/r$, $e^Q = e^{-T}$ for the Schwarzschild metric.

12. (a) Obtain the geodesic equations for the Schwarzschild metric in terms of the spherical coordinates $(\xi^0, \xi^1, \xi^2, \xi^3) = (ct, r, \varphi, \theta)$.

(b) Show that a radial line is a geodesic path (as in Exercise 6).

(c) Show that a geodesic that starts in the equatorial plane $\varphi = \pi/2$ remains there for all time.

13. Explore the possibility of circular orbits for the Schwarzschild metric (as in Exercise 7). Is there a Kepler's law?

14. Let $r = f(\rho) = \rho\left(1 + \frac{r_M}{4\rho}\right)^2$.

(a) Show that $f'(r_M/4) = 0$ and $f(r_M/4) = r_M$.

(b) Show that $f'(\rho) > 0$ for $\rho > r_M/4$.

(c) Show that $f(\rho) \to \infty$ as $\rho \to \infty$ and thus conclude that the map $f : [r_M/4, \infty) \to [r_M, \infty)$ is a valid coordinate transformation—that is, it is one-to-one and onto.

15. Show that

$$\left(\frac{1 - \frac{r_M}{4\rho}}{1 + \frac{r_M}{4\rho}}\right)^2 = 1 - \frac{r_M}{\rho} + O\left(\frac{1}{\rho^2}\right), \quad \left(1 + \frac{r_M}{4\rho}\right)^4 = 1 + \frac{r_M}{\rho} + O\left(\frac{1}{\rho^2}\right)$$

and explain why this proves Corollary 8.1.

8.3 The Bending of Light

As early as 1907, Einstein observed that gravity alters the velocity of light and that this will cause light rays to bend. However, the effect would be too small to be detected in any terrestrial experiment that he contemplated at the time. He let the matter drop, but revived it in 1911 (in [12]) when he realized that the deflection of starlight grazing the sun would be detectable, and suggested measuring it during a solar eclipse.

From 1907 to 1916

Einstein's 1911 calculations predate the general theory of relativity; they use only the familiar physical concepts and the relativistic properties of gravity we considered in Section 4.4. Einstein calculates that the deflection should be 4×10^{-6} radians, which is an angle of 0.83 seconds. In his 1916 paper he recomputes the deflection using the perspective of general relativity and gets a value that is twice as large. What changes between 1911 and 1916 is only the way he reckons the effect of gravity; the calculation is the same. To analyze Einstein's results we look first at why the variation in the speed of light from place to place causes light rays to bend. This is called refraction, and it is explained by classical physics and geometry, not relativity.

How Deflection Depends on Speed

Light bends when it slows down

The speed of light in a vacuum, $c = 2.998 \times 10^8$ meters per second, is a theoretical maximum. The speed is less in air and in glass, and even varies from warm air to cool air, and from one type of glass to another. According to an argument based on Huygens' principle—which we shall go through—it is precisely the variation in speed that causes light rays to bend. Therefore, when we establish that a gravitational field *also* causes the speed of light to vary from place to place, we can use the same argument to decide how much the field bends light.

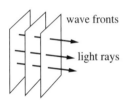

If we think of light as a wave, then we can picture it as a series of surfaces, or *wave fronts*, that represent the position of the "crest" of the wave at different moments. If we think of light as a collection of photons, then we can picture the photon paths as *light rays*. In the simplest case, the wave fronts are parallel planes, and the light rays are straight lines orthogonal to the fronts. In general, though, the fronts are curved, not flat; but when that happens, the rays curve in such a way as to remain orthogonal to the fronts.

Huygens' principle

Therefore, to determine how the rays bend we can study how the wave fronts propagate through space. Suppose we know the position of a wave front at a certain moment $t = t_0$. Where will it be a moment later, $t = t_0 + \Delta t$? If the front is a plane and light has a constant speed γ, then the new front is a parallel plane at distance $\gamma \Delta t$ from the first. But if the front is curved and the speed of light varies from point to point, we use **Huygens' principle**: Treat each point $P = (x, y, z)$ on the current front as a light source. If the speed of light at that point is $\gamma(P)$, then in the time Δt a spherical front of radius $\gamma(P)\Delta t$ will emanate from

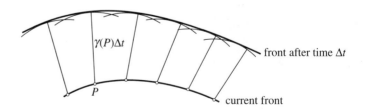

§8.3 The Bending of Light

that source. The wave front after time Δt is the *envelope* of the spherical fronts from the individual sources.

Now consider a light ray that passes through a region of space where the speed of light $\gamma(P)$ varies smoothly with P. We want to determine how the ray changes direction. Draw the ray in the plane of the paper along with a small portion of a wave front that is orthogonal to it. Construct the n-axis normal to the ray, and suppose n and $n + \Delta n$ are the coordinates of two nearby points on the wave front. The wave front after a small time increment Δt will be tangent to the spheres of radius $\gamma(n) \Delta t$ and $\gamma(n + \Delta n) \Delta t$ centered at these two points.

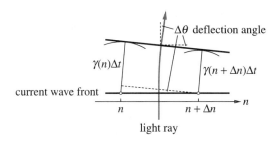

The deflection angle $\Delta \theta$ that the ray experiences during the time Δt is the same as the change in the orientation of the wave front. This angle is extremely small, and for it we have

Computing the deflection

$$\Delta \theta \approx \sin \Delta \theta = \frac{\gamma(n) \Delta t - \gamma(n + \Delta n) \Delta t}{\Delta n}.$$

Therefore, in the limit as $\Delta t \to 0$ and $\Delta n \to 0$, we get

$$\frac{d\theta}{dt} = -\frac{\partial \gamma}{\partial n}.$$

Now, the *curvature* of the ray is the derivative $d\theta/ds$, where s is arc length along the ray. But since the deflection is always extremely small, arc length is essentially $s = ct(1 + O(1/c))$. Therefore, $ct = s(1 + O(1/c))$, and so

$$\text{curvature} = \frac{d\theta}{ds} = \frac{d\theta}{dt}\frac{dt}{ds} = \frac{1}{c}\frac{d\theta}{dt} + O\left(\frac{1}{c^2}\right) = -\frac{1}{c}\frac{\partial \gamma}{\partial n} + O\left(\frac{1}{c^2}\right).$$

Chapter 8 Consequences

Rays turn toward the low-speed region

See the exercises. Thus, the ray will curve at a rate that is proportional to the rate at which the speed of light decreases in the direction normal to the ray.

The result is refraction, and we can think of it this way. If $\mathbf{n}$ is any unit vector, then the rate of change of γ in the direction of $\mathbf{n}$ is

$$\frac{\partial \gamma}{\partial n} = \mathbf{n} \cdot \nabla \gamma,$$

where $\nabla \gamma$ is the gradient of γ. In the figure below, we assume that the speed of light γ is constant on each line perpendicular to the plane and increases from left to right as shown. Consequently, the gradient $\nabla \gamma$ lies in the plane and points from left to right. If $\mathbf{n}$ is the normal to the ray that runs from front to back and lies in the plane, then $\mathbf{n} \cdot \nabla \gamma$ is positive, so the ray bends to the left. Along the other ray $\mathbf{n} \cdot \nabla \gamma = 0$, so *that* ray is undeflected.

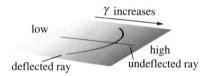

The 1911 Computation

The speed of light at any point depends on how fast a clock runs at that point. In Section 4.4 we saw that if two observers G and C were in a gravitational field at locations where the gravitational potential differed by $\Delta\Phi$, then when C measured a time interval of ΔT seconds, G would measure it as

$$\Delta\tau = \left(1 - \frac{\Delta\Phi}{c^2}\right) \Delta T \text{ seconds}.$$

While this was originally derived assuming a constant gravitational field, it holds quite generally. In particular, we can let Φ be the Newtonian gravitational potential of the sun,

$$\Phi = -\frac{GM}{r},$$

§8.3 The Bending of Light

where r is the distance in meters from the center of the sun and $M = 2 \times 10^{30}$ kg is the mass of the sun. We can also assume that the observer C is at infinity, where $\Phi = 0$ and the speed of light has its standard value $c = 3 \times 10^8$ m/sec in the vacuum. Then, at a distance r from the center of the sun, the speed of light is

$$\gamma(r) = \frac{c}{1 - (\Phi/c^2)} = c\left(1 + \frac{\Phi}{c^2} + O\left(\frac{1}{c^4}\right)\right) = c - \frac{GM}{cr}$$

if we ignore terms of order $O(1/c^3)$.

Now consider a light ray in the (x, y)-plane that if undeflected would be parallel to the y-axis at a distance X from the center of the sun. The total deflection it undergoes is the integral of its curvature:

$$\text{total deflection} = \int_{-\infty}^{\infty} \frac{d\theta}{ds}\, dy = -\frac{1}{c}\int_{-\infty}^{\infty} \frac{\partial \gamma}{\partial x}\, dy.$$

We have replaced $\partial \gamma/\partial n$ by $\partial \gamma/\partial x$ because the difference between the normal to the curve and the x-axis is of order $O(1/c)$. Now

$$\frac{\partial \gamma}{\partial x} = \frac{GMx}{cr^3} = \frac{GMX}{cr^3} + O\left(\frac{1}{c^2}\right)$$

along the light ray; therefore, to the precision we are considering,

$$\text{total deflection} = -\frac{GMX}{c^2}\int_{-\infty}^{\infty} \frac{dy}{(X^2 + y^2)^{3/2}}.$$

With the substitution $y = X \tan u$ the integral becomes

$$\int_{-\pi/2}^{\pi/2} \frac{\cos u}{X^2}\, du = \frac{2}{X^2},$$

so

$$\text{total deflection} = -\frac{2GM}{c^2 X}.$$

To make this as large as possible we choose a ray that nearly grazes the surface of the sun. Then

$$X = \text{radius of the sun} = 7 \times 10^8 \text{ meters},$$

and the magnitude of the deflection is

$$\frac{2GM}{c^2 X} = \frac{2(6.67 \times 10^{-11} \text{ m}^3/\text{kg-sec}^2)(2 \times 10^{30} \text{ kg})}{(3 \times 10^8 \text{ m/sec})^2 (7 \times 10^8 \text{ m})} = 4.2328 \times 10^{-6} \text{ radians}.$$

In degrees, this is

$$4.2328 \times 10^{-6} \text{ radians} \times \frac{180°}{\pi \text{ radian}} \times \frac{60'}{\text{degree}} \times \frac{60''}{\text{minute}} = 0.873''.$$

Einstein used the round number 4×10^{-6} radians and therefore got 0.83" for the deflection angle.

The 1916 Computation

Einstein's 1916 calculations (in [10]) are based on his relativistic model of the gravitational field of the sun that we found in the previous section. This is a weak field with spherical symmetry. In the frame $R: (x^0, x^1, x^2, x^3) = (ct, x, y, z)$ we write the metric for the field as

$$ds^2 = (dx^0)^2 - \sum_\alpha (dx^\alpha)^2 - \frac{2GM}{rc^2}\left((dx^0)^2 + \sum_{\alpha,\beta} \frac{x^\alpha x^\beta}{r^2} dx^\alpha dx^\beta\right),$$

ignoring terms of order $O(1/c^3)$. We must now see how this metric determines the speed of light γ at every point in spacetime.

Suppose $\mathbb{X}(t) = (ct, x^1(t), x^2(t), x^3(t))$ describes the worldcurve of an object in R, parametrized by the time t. Then, if we first use the Minkowski metric in R, we have

$$\|\mathbb{X}'\|^2 = c^2 - \left(\frac{dx^1}{dt}\right)^2 - \left(\frac{dx^2}{dt}\right)^2 - \left(\frac{dx^3}{dt}\right)^2 = c^2 - v^2,$$

where v is the ordinary velocity of the object. If the object is a photon, then $\|\mathbb{X}'\|^2 = 0$, implying $v = \gamma = c$ as we expect.

If we shift back to the metric that represents the gravitational field of the sun, then for the same worldcurve $\mathbb{X}(t)$ we have

$$0 = \left(\frac{ds}{dt}\right)^2 = c^2 - \underbrace{\sum_\alpha \left(\frac{dx^\alpha}{dt}\right)^2}_{\gamma^2} - \frac{2GM}{rc^2}\left(c^2 + \sum_{\alpha,\beta} \frac{x^\alpha x^\beta}{r^2} \frac{dx^\alpha}{dt}\frac{dx^\beta}{dt}\right)$$

§8.3 The Bending of Light

up to terms of order $O(1/c^2)$. Therefore, to the same order,

$$\gamma^2 = c^2 - \frac{2GM}{rc^2}\left(c^2 + \sum_{\alpha,\beta} \frac{x^\alpha x^\beta}{r^2} \frac{dx^\alpha}{dt} \frac{dx^\beta}{dt}\right)$$

$$= c^2\left[1 - \frac{2GM}{rc^2}\left(1 + \sum_{\alpha,\beta} \frac{x^\alpha x^\beta}{r^2 c^2} \frac{dx^\alpha}{dt} \frac{dx^\beta}{dt}\right)\right].$$

Taking square roots and using $\sqrt{1-2A} = 1 - A$ (ignoring terms of order A^2), we obtain

$$\gamma = c\left(1 - \frac{GM}{rc^2}\left(1 + \sum_{\alpha,\beta} \frac{x^\alpha x^\beta}{r^2 c^2} \frac{dx^\alpha}{dt} \frac{dx^\beta}{dt}\right)\right)$$

$$= c - \frac{GM}{rc} - \underbrace{\frac{GM}{rc} \sum_{\alpha,\beta} \frac{x^\alpha x^\beta}{r^2 c^2} \frac{dx^\alpha}{dt} \frac{dx^\beta}{dt}}_{\text{correction to 1911 calculation}}.$$

The first two terms are precisely the value of γ that Einstein obtained in 1911; we can therefore consider the rest to be a correction added to reflect a fuller account of the gravitational effect. As we shall see, one correction term has the same order of magnitude as the 1911 term GM/rc.

Let us determine γ along the worldcurve $\mathbb{X}$ of the photon that if undeflected would move in the (x, y)-plane at the fixed distance X from the y-axis. The *undeflected* worldcurve has the parametrization

$$\mathbb{X}(t) = (ct, X, ct, 0);$$

the deflection adds only lower-order terms:

$$\mathbb{X}(t) = \left(ct, X + O\left(\frac{1}{c}\right), ct + O(1), 0\right).$$

Note that we have left the $x^3 = z$-component as zero. In the exercises you are asked to show that if the x^3-components of $\mathbb{X}(t)$ and $\mathbb{X}'(t)$ are zero initially, they remain zero for all t. The

4-velocity of this photon is

$$\frac{d\mathbb{X}}{dt} = \left(c, O\left(\frac{1}{c}\right), c + O(1), 0\right).$$

That is,

$$\frac{dx^1}{dt} = O\left(\frac{1}{c}\right), \quad \frac{dx^2}{dt} = c + O(1), \quad \frac{dx^3}{dt} = 0,$$

and when we substitute these values into the formula for γ, ignoring terms of order $O(1/c^2)$, we obtain

$$\gamma = c - \frac{GM}{cr} - \frac{GM(x^2)^2}{cr^3}.$$

Exactly as in the 1911 calculations, total deflection is the integral of curvature along the worldcurve,

$$\text{total deflection} = -\frac{1}{c}\int_{-\infty}^{\infty} \frac{\partial \gamma}{\partial x^1} \, dx^2$$

to order $O(1/c^2)$. Now

$$\frac{\partial \gamma}{\partial x^1} = \frac{GMx^1}{cr^3} + \frac{3GMx^1(x^2)^2}{cr^5} = \frac{GMX}{c(X^2+y^2)^{3/2}} + \frac{3GMXy^2}{c(X^2+y^2)^{5/2}}$$

to order $O(1/c)$, and we have taken advantage of the fact that

$$x^1 = X + O\left(\frac{1}{c}\right), \quad r^2 = X^2 + (x^2)^2 + O\left(\frac{1}{c}\right) = X^2 + y^2 + O\left(\frac{1}{c}\right)$$

on the worldcurve. Therefore, the total deflection is

$$-\frac{GMX}{c^2}\int_{-\infty}^{\infty} \frac{dy}{(X^2+y^2)^{3/2}} - \frac{GMX}{c^2}\int_{-\infty}^{\infty} \frac{3y^2 \, dy}{(X^2+y^2)^{5/2}} + O\left(\frac{1}{c^3}\right).$$

The first term is just the 1911 calculation. The second term is new, but it has the same order of magnitude as the first. With the substitution $y = X \tan u$, the second integral becomes

$$\int_{-\pi/2}^{\pi/2} \frac{3\sin^2 u \cos u}{X^2} \, du = \frac{2}{X^2}.$$

This has the same value as the first integral, so the 1916 calculation equals twice the 1911 calculation:

$$\text{total deflection} = -\frac{4GM}{c^2 X} = 1.75''.$$

Finally, note that because Einstein's weak-field metric is asymptotically inertial, the deflections we measure here are taken with respect to a frame that is anchored in the background of fixed stars.

Exercises

1. Suppose $s = ct(1 + O(1/c))$. Show that $ct = s(1 + O(1/c))$ and hence that

$$\frac{dt}{ds} = \frac{1}{c}\left(1 + O\left(\frac{1}{c}\right)\right) = \frac{1}{c} + O\left(\frac{1}{c^2}\right).$$

Use this result to show that the curvature of a light ray is

$$\frac{d\theta}{ds} = \frac{1}{c}\frac{d\theta}{dt} + O\left(\frac{1}{c^2}\right) = -\frac{1}{c}\frac{\partial\gamma}{\partial n} + O\left(\frac{1}{c^2}\right).$$

2. Suppose $\mathbb{X}(t) = (ct, x(t), y(t), z(t))$ is a geodesic in the weak spherical metric and satisfies the conditions $z(0) = z'(0) = 0$. Show that $z(t) = 0$ for all t.

3. Verify that $\displaystyle\int_{-\infty}^{\infty} \frac{dy}{(X^2 + y^2)^{3/2}} = \int_{-\infty}^{\infty} \frac{3y^2\, dy}{(X^2 + y^2)^{5/2}} = \frac{2}{X^2}.$

8.4 Perihelion Drift

Classical Newtonian mechanics predicts that the orbit of each planet around the sun is an ellipse—at least if we ignore the influences of other planets. Of course, we *can't* ignore the other planets; they cause deviations in the orbit that can become quite significant over time. The innermost planet, Mercury, is especially susceptible; its orbital period is only 88 days, and its orbit is relatively elongated. One deviation in the orbit is the gradual drift in its perihelion; as early as 1859 it was noted that the observed drift exceeded the predicted value by about 43″ of arc per

Perturbations in planetary orbits

earth century. Now, a similar discrepancy between the predicted and observed orbits of the outer planet Uranus had led, a decade earlier, to the discovery of the planet Neptune, so it was natural to suppose that yet another undiscovered planet—given the name Vulcan in anticipation—would explain the unaccounted-for drift in Mercury's perihelion. But Vulcan has never been found, and the search for it was abandoned when Einstein showed in 1916 that his new theory of gravity explained the drift.

Ellipses

Cartesian equation

It turns out that the natural way to describe an elliptical orbit in celestial mechanics is by an equation involving polar coordinates:

$$r = \frac{k}{1 + e \cos\theta}.$$

We are, however, more familiar with the Cartesian equation

$$\frac{(x-p)^2}{a^2} + \frac{(y-q)^2}{b^2} = 1,$$

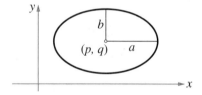

which describes an ellipse whose center is at $(x, y) = (p, q)$ and whose *semimajor* and *semiminor axes* are a and b, respectively, when $a \geq b$. We need to connect the Cartesian and polar equations and, at the same time, see how the geometric features of the ellipse can be extracted from them.

Foci and eccentricity

Geometrically, an ellipse is the locus of points whose distances to two fixed points, its *foci*, have a fixed sum. By taking a point on the major axis, we see that the sum must equal the length $2a$ of the major axis. By taking a point on the minor axis, we see that $a^2 = b^2 + f^2$, where f is the distance from the center of the ellipse to one focus. We define the **eccentricity** of the ellipse to be the ratio

$$\text{eccentricity: } e = \frac{f}{a} = \frac{\sqrt{a^2 - b^2}}{a} = \sqrt{1 - \left(\frac{b}{a}\right)^2}.$$

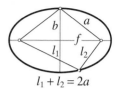

$l_1 + l_2 = 2a$

If $e = 0$, then $f = 0$; the two focal points coincide, and the ellipse is just a circle. If $e = 1$, then $b = 0$, so the ellipse has flattened to a line segment. Eccentricity therefore measures how much the ellipse differs from a circle.

Proposition 8.2 *If $0 < e < 1$, the polar equation $r = k/(1 + e\cos\theta)$ describes an ellipse whose major axis lies along the x-axis. It has eccentricity e and one focus at the origin. The semiaxes have lengths $a = k/(1 - e^2)$ and $b = k/\sqrt{1 - e^2}$. The least and greatest distances to the origin are $k/(1 + e)$ and $k/(1 - e)$.*

Polar equation

PROOF: Write the given equation as $r + er\cos\theta = r + ex = k$, or $r = k - ex$. Square this,

$$x^2 + y^2 = r^2 = k^2 - 2kex + e^2x^2,$$

and rewrite it, completing the square, as

$$(1 - e^2)\left(x^2 + \frac{2ke}{1 - e^2}x + \frac{k^2 e^2}{(1 - e^2)^2}\right) + y^2 = k^2 + \frac{k^2 e^2}{1 - e^2} = \frac{k^2}{1 - e^2}.$$

Now put it in the standard form

$$\frac{1}{k^2/(1 - e^2)^2}\left(x + \frac{ke}{1 - e^2}\right)^2 + \frac{1}{k^2/(1 - e^2)}y^2 = 1,$$

so the locus is an ellipse with center $(p, q) = (-ke/(1 - e^2), 0)$ and semiaxes $a = k/(1 - e^2)$ and $b = k/\sqrt{1 - e^2}$. Furthermore, since $b/a = \sqrt{1 - e^2}$,

$$\text{eccentricity} = \sqrt{1 - (1 - e^2)} = e.$$

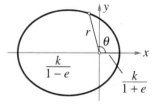

The center of the ellipse is at $p = -ke/(1 - e^2)$, and since the distance from the center to a focus is

$$f = ea = e\frac{k}{1 - e^2} = -p,$$

one focus is at the origin. The ellipse is closest to the origin when $\theta = 0$ and $r = k/(1+e)$; it is farthest when $\theta = \pi$ and $r = k/(1-e)$.

END OF PROOF

The Classical Orbit

We want to determine the motion of a planet according to Newtonian mechanics, under the assumption that the sun provides the sole gravitational force. If the center of the sun is at the origin and the planet is at position $\mathbf{x} = (x, y, z)$ at time t, then the equation of motion is

$$\ddot{\mathbf{x}} = -\frac{GM}{r^3}\mathbf{x}, \qquad r = \|\mathbf{x}\|;$$

see Section 4.3. As usual, dots denote derivatives with respect to t, M is the mass of the sun, and G is the gravitational constant. The function $\mathbf{x}(t)$ will give us the position of the planet at every moment, but this is more than we need. It will be enough to know is the *shape* of orbit; for that reason, we will develop arguments that will allow us to replace the vector function $\mathbf{x}(t)$ by the appropriate polar equation.

If $\mathbf{v} = \dot{\mathbf{x}}$ is the velocity vector, then

$$\frac{d}{dt}(\mathbf{x} \times \mathbf{v}) = \dot{\mathbf{x}} \times \mathbf{v} + \mathbf{x} \times \dot{\mathbf{v}} = \mathbf{v} \times \mathbf{v} - \frac{GM}{r^3}(\mathbf{x} \times \mathbf{x}) = \mathbf{0}.$$

Hence the vector $\mathbf{J} = \mathbf{x} \times \mathbf{v}$ is a constant, called the **angular momentum** of the planet. In particular, $\mathbf{x}$ lies in the plane orthogonal to $\mathbf{J}$, so the planet moves in that plane.

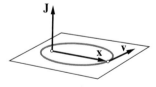

Rotating coordinates if necessary, we can assume that the planet moves in the (x, y)-plane. Let (r, θ) be polar coordinates in this plane:

$$x = r\cos\theta, \qquad y = r\sin\theta.$$

We want to find a description of the orbit as a function $r = r(\theta)$. As a first step in this direction, we find the components of the angular momentum vector

$$\mathbf{J} = \mathbf{x} \times \mathbf{v}$$
$$= (r\cos\theta, r\sin\theta, 0) \times (\dot r\cos\theta - r\dot\theta\sin\theta, \dot r\sin\theta + r\dot\theta\cos\theta, 0)$$
$$= (0, 0, r^2\dot\theta).$$

Therefore, the *magnitude* of angular momentum is $r^2\dot\theta = J = \|\mathbf{J}\|$, and this is a constant of the motion.

Next we write the equation of motion in terms of r. Since $r^2 = \mathbf{x} \cdot \mathbf{x}$, we have $2r\dot r = 2\mathbf{x} \cdot \dot{\mathbf{x}}$, or $r\dot r = \mathbf{x} \cdot \mathbf{v}$. Differentiating again (and letting $\mu = GM$), we get

The equation of motion in polar coordinates

$$r\ddot r + \dot r^2 = \mathbf{v} \cdot \mathbf{v} + \mathbf{x} \cdot \dot{\mathbf{v}} = v^2 - \frac{\mu}{r^3}\mathbf{x} \cdot \mathbf{x} = v^2 - \frac{\mu}{r}.$$

For any vectors $\mathbf{p}$ and $\mathbf{q}$ we have $(\mathbf{p} \cdot \mathbf{q})^2 + \|(\mathbf{p} \times \mathbf{q})\|^2 = p^2 q^2$; in particular,

$$(\mathbf{x} \cdot \mathbf{v})^2 + \|(\mathbf{x} \times \mathbf{v})\|^2 = (r\dot r)^2 + J^2 = r^2 v^2,$$

so $v^2 = \dot r^2 + J^2/r^2$, and the equation of motion becomes

$$r\ddot r + \dot r^2 = \dot r^2 + \frac{J^2}{r^2} - \frac{\mu}{r}, \quad \text{or} \quad r\ddot r = \frac{J^2}{r^2} - \frac{\mu}{r}.$$

The solution to this equation will be more transparent if we first make the change of variable $r = 1/u$. Then, keeping in mind that we want to convert r to a function of θ, we write

$$\dot r = -\frac{1}{u^2}\dot u = -\frac{1}{u^2}\frac{du}{d\theta}\dot\theta = -r^2\dot\theta\frac{du}{d\theta} = -J\frac{du}{d\theta},$$
$$\ddot r = -J\frac{d^2 u}{d\theta^2}\dot\theta = -J\frac{d^2 u}{d\theta^2}\frac{J}{r^2} = -J^2 u^2 \frac{d^2 u}{d\theta^2}.$$

Therefore, in terms of u and θ, the equation of motion becomes

$$\frac{1}{u}\left(-J^2 u^2 \frac{d^2 u}{d\theta^2}\right) = J^2 u^2 - \mu u,$$

or just

$$\frac{d^2u}{d\theta^2} + u = \frac{\mu}{J^2}.$$

The solution to this differential equation is

$$u(\theta) = \frac{\mu}{J^2} + A\cos(\theta - \alpha);$$

the cosine term has an arbitrary amplitude A and phase shift α. The phase shift simply alters the position of the planet in its orbit at time $t = 0$; since we are interested only in the orbit itself, we can take $\alpha = 0$. Then, if we let $e = AJ^2/\mu$, the solution is

$$u = \frac{\mu}{J^2}(1 + e\cos\theta), \quad \text{or} \quad r = \frac{J^2/\mu}{1 + e\cos\theta}.$$

This is the polar form of a conic section with a focus at the origin. When $0 < e < 1$, the conic is an ellipse with eccentricity e and perihelion distance $J^2/\mu(1 + e)$. Perihelion occurs when θ is an integer multiple of 2π.

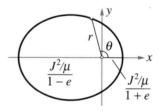

The Relativistic Orbit

Schwarzschild metric

We take as our relativistic model of the sun's gravitational field the Schwarzschild metric

$$ds^2 = c^2\left(1 - \frac{2\mu}{c^2r}\right)dt^2 - \frac{1}{1 - (2\mu/c^2r)}dr^2 - r^2(d\varphi^2 + \sin^2\varphi\, d\theta^2).$$

The worldcurve of the planetary orbit is a timelike geodesic $\mathbb{X}(\tau) = (t(\tau), r(\tau), \varphi(\tau), \theta(\tau))$ in which t, r, θ, and $\mu = GM$ have the same meaning they do in the classical model.

We look for a solution that lies in the (x, y)-plane; in spherical coordinates, this is $\varphi = \pi/2$. In the exercises you are asked to show that a geodesic for which $\varphi(0) = \pi/2$ and $\varphi'(0) = 0$ will have $\varphi(\tau) = \pi/2$ for *all* τ. (Primes denote differentiation with respect to the proper time τ.) Let us consider the geodesic equations for the remaining variables, beginning with θ:

Motions in the equatorial plane

$$\theta'' = -2\Gamma_{13}^3 r'\theta' = -\frac{2}{r}r'\theta'.$$

We can write this as $r^2\theta'' + 2rr'\theta' = (r^2\theta')' = 0$, implying that $r^2\theta'$ is a constant. In the classical model (where we replace τ by t), this constant is the angular momentum J. So we set $r^2\theta' = J$.

Angular momentum

The geodesic equation for t is

$$t'' = -2\Gamma_{01}^0 t'r' = -\frac{2\mu}{c^2 r^2}\frac{1}{1 - (2\mu/c^2 r)}t'r'.$$

$e^T \varepsilon \mid - \frac{2GM}{\rho c^2}$

Here again we express this as a derivative:

$$\left(1 - \frac{2\mu}{c^2 r}\right)t'' + \frac{2\mu}{c^2 r^2}t'r' = \left(\left(1 - \frac{2\mu}{c^2 r}\right)t'\right)' = 0.$$

Therefore,

$$\left(1 - \frac{2\mu}{c^2 r}\right)t' = A, \quad \text{or} \quad t' = \frac{A}{1 - (2\mu/c^2 r)}$$

for some constant A.

To get the equation of motion for r we use the metric condition itself instead of the geodesic equation. Since $ds^2 = c^2 d\tau^2$, we have

The equation for r

$$c^2 = c^2\left(1 - \frac{2\mu}{c^2 r}\right)(t')^2 - \frac{1}{1 - (2\mu/c^2 r)}(r')^2 - r^2(\theta')^2.$$

We have used the fact that $\varphi(\tau) \equiv \pi/2$, so $\varphi' = 0$ and $\sin^2 \varphi = 1$. Now substitute the expression we just obtained for t' and let

$$r' = \frac{dr}{d\theta}\theta'.$$

This gives

$$c^2 = c^2 \frac{A^2}{1-(2\mu/c^2 r)} - \frac{1}{1-(2\mu/c^2 r)} \left(\frac{dr}{d\theta}\right)^2 (\theta')^2 - r^2(\theta')^2$$

$$= c^2 \frac{A^2}{1-(2\mu/c^2 r)} - \frac{1}{1-(2\mu/c^2 r)} \left(\frac{dr}{d\theta}\right)^2 \frac{J^2}{r^4} - \frac{J^2}{r^2}.$$

Now multiply by $1 - (2\mu/c^2 r)$ and simplify the left-hand side:

$$c^2 - \frac{2\mu}{r} = c^2 A^2 - \left(\frac{dr}{d\theta}\right)^2 \frac{J^2}{r^4} - \frac{J^2}{r^2}\left(1 - \frac{2\mu}{c^2 r}\right).$$

If we let $u = 1/r$ as we did in the classical analysis, then this becomes

$$c^2 - 2\mu u = c^2 A^2 - J^2 \left(\frac{du}{d\theta}\right)^2 - J^2 u^2 \left(1 - \frac{2\mu}{c^2} u\right),$$

which can be rewritten as

$$\frac{1}{2}\left(\frac{du}{d\theta}\right)^2 + \frac{u^2}{2} = \frac{c^2(A^2-1)}{2J^2} + \frac{\mu}{J^2} u + \frac{\mu}{c^2} u^3.$$

We can now make this look like the classical equation; first differentiate with respect to θ,

$$\frac{du}{d\theta}\frac{d^2 u}{d\theta^2} + u\frac{du}{d\theta} = \frac{\mu}{J^2}\frac{du}{d\theta} + 3\frac{\mu}{c^2} u^2 \frac{du}{d\theta},$$

and then divide by $\frac{du}{d\theta}$:

$$\frac{d^2 u}{d\theta^2} + u = \frac{\mu}{J^2} + \underbrace{3\frac{\mu}{c^2} u^2}_{\text{correction}}.$$

Except that J has a slightly different value here, this is precisely the Newtonian equation of motion with a single correction term of order $O(1/c^2)$.

Perturb the classical solution

Since the relativistic equation differs from the Newtonian by a small term, we shall conjecture that the relativistic solution likewise differs from the Newtonian by a small term. Thus we suppose that the solution has the form

$$u(\theta) = \frac{\mu}{J^2}(1 + e\cos\theta) + \frac{v(\theta)}{c^2} + O\left(\frac{1}{c^4}\right).$$

Our goal is to confirm that $u(\theta)$ has this form and to determine $v(\theta)$. Substituting this expression into the differential equation, we obtain

$$\frac{1}{c^2}\left(\frac{d^2v}{d\theta^2} + v\right) = \frac{3\mu^3}{c^2 J^4}(1 + 2e\cos\theta + e^2\cos^2\theta) + O\left(\frac{1}{c^4}\right).$$

Using the identity $\cos 2\theta = 2\cos^2\theta - 1$ we can rewrite this as

$$\frac{d^2v}{d\theta^2} + v = \frac{3\mu^3}{J^4}\left(1 + \frac{e^2}{2}\right) + \frac{6e\mu^3}{J^4}\cos\theta + \frac{3\mu^3 e^2}{2J^4}\cos 2\theta + O\left(\frac{1}{c^2}\right).$$

If we ignore terms of order $O(1/c^2)$, this is a linear second-order differential equation for v of the form

$$L(v) = A + B + C,$$

where $L(v) = v'' + v$. Therefore, if v_A, v_B, v_C are particular solutions to the separate equations $L(v) = A$, $L(v) = B$, $L(v) = C$, then $v = v_A + v_B + v_C$ is a solution to the given equation $L(v) = A+B+C$. For the separate solutions you should check that we can take

$$v_A = \frac{3\mu^3}{J^4}\left(1 + \frac{e^2}{2}\right), \quad v_B = \frac{3e\mu^3}{J^4}\theta\sin\theta, \quad v_C = -\frac{\mu^3 e^2}{2J^4}\cos 2\theta,$$

so the general solution to the original differential equation becomes

$$u = \frac{\mu}{J^2}(1 + e\cos\theta) + \frac{3\mu^3}{c^2 J^4}\left(1 + \frac{e^2}{2} + e\theta\sin\theta - \frac{e^2}{6}\cos 2\theta\right) + O\left(\frac{1}{c^4}\right).$$

Of the new terms, only $e\theta\sin\theta$ is nonperiodic (because of the factor θ), so only this one has a potential cumulative effect on $u(\theta)$. To see that effect, let us first move the nonperiodic term to the Newtonian approximation:

One nonnegligible term

$$u = \frac{\mu}{J^2}(1 + e\cos\theta + eD\theta\sin\theta) + \frac{3\mu^3}{c^2 J^4}\left(1 + \frac{e^2}{2} - \frac{e^2}{6}\cos 2\theta\right) + O\left(\frac{1}{c^4}\right),$$

where $D = 3\mu^2/c^2 J^2$. Now,

$$e\cos(\theta - D\theta) = e\cos\theta\cos D\theta + e\sin\theta\sin D\theta,$$

and since $\cos D\theta = 1 + O(D^2)$ and $\sin D\theta = D\theta + O(D^3)$, we have

$$e\cos(\theta - D\theta) = e\cos\theta + eD\theta\sin\theta + O(D^2) = e\cos\theta + eD\theta\sin\theta + O\left(\frac{1}{c^4}\right).$$

This allows us, finally, to write our solution as

$$u(\theta) = \frac{\mu}{J^2}(1 + e\cos(\theta - D\theta)) + \underbrace{\frac{3\mu^3}{c^2 J^4}\left(1 + \frac{e^2}{2} - \frac{e^2}{6}\cos 2\theta\right)}_{\text{negligible cumulative effect}} + O\left(\frac{1}{c^4}\right);$$

the relativistic effect is concentrated in the term $\cos(\theta - D\theta)$.

Perihelion occurs when $\cos(\theta - D\theta) = 1$, and this happens when $\theta - D\theta = (1-D)\theta = 2n\pi$, or

$$\theta = \frac{2n\pi}{1-D} = 2n\pi(1 + D + O(D^2)) = 2n\pi + 2n\pi D + O\left(\frac{1}{c^4}\right).$$

Therefore, the angular position θ of perihelion *increases* by the amount

$$2\pi D = \frac{6\pi \mu^2}{c^2 J^2}$$

per revolution. Since the semimajor axis of the classical orbit is

$$a = \frac{J^2/\mu}{1-e^2}$$

(why?), we can express the increase in these readily measured terms:

$$2\pi D = \frac{6\pi \mu}{c^2 a(1-e^2)} = \frac{6\pi GM}{c^2 a(1-e^2)}.$$

For Mercury, $a = 57.9 \times 10^9$ meters, $e = 0.2056$, and $1 - e^2 = 0.9577$. Since there are about 36,525 earth days in an earth century and Mercury has a period of about 88 earth days, Mercury executes $36525/88 = 415$ orbits per century. The total increase in the perihelion angle in a century is therefore

$$415 \times \frac{6\pi GM}{c^2a(1-e^2)} = \frac{(2490\pi)(6.67 \times 10^{-11}\ \mathrm{m^3/kg\text{-}sec^2})(2 \times 10^{30}\ \mathrm{kg})}{(3 \times 10^8\ \mathrm{m/sec})^2(57.9 \times 10^9\ \mathrm{m})(0.9577)}$$
$$= 2.09 \times 10^{-4} \text{ radians}$$
$$= 43 \text{ seconds of arc.}$$

This is the last result cited in Einstein's 1916 paper on general relativity. Here is the final paragraph ([10], page 164):

> Calculation gives for the planet Mercury a rotation of the orbit of 43″ per century, corresponding exactly to astronomical observation (Leverrier); for the astronomers have discovered in the motion of the perihelion of this planet, after allowing for disturbances by other planets, an inexplicable remainder of this magnitude.

Visualizing the drift

INCORRECT, HE NOW SAYS. REWRITTEN. PAGE 434

The relativistic contribution is extremely small; even over 80 centuries it amounts to just under 1°, but it is enough to be visible in the figure on the next page. By contrast, the Newtonian calculation of the "disturbances by other planets" during the same time is enormous—a third of a complete revolution. The figure is drawn to scale and gives an accurate picture of the eccentricity of Mercury's orbit. The Newtonian drift (of about 5557″ per century) is shown by the dashed ellipse, the correct (relativistic) drift by the gray ellipse.

Exercises

1. Prove that for any vectors $\mathbf{p}$ and $\mathbf{q}$ in $\mathbf{R}^3$, we have $(\mathbf{p} \cdot \mathbf{q})^2 + (\mathbf{p} \times \mathbf{q})^2 = p^2q^2$, where $p = \|\mathbf{p}\|$, $q = \|\mathbf{q}\|$.
2. This exercise concerns the curve $r = \dfrac{k}{1 + e\cos\theta}$.
   (a) Give a complete description of the curve when $-1 < e < 0$.
   (b) Give a complete description of the curve when $e = \pm 1$. In particular, show that the curve is *not* an ellipse of eccentricity 1.

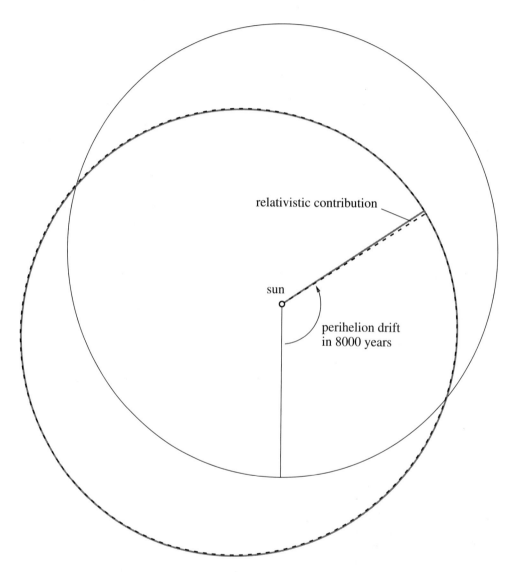

The orbit of Mercury

(c) Show that when $e > 1$ or $e < -1$, the curve is a hyperbola; determine the center of the hyperbola and the slope of its asymptotes.

3. Show that the period T of the classical planetary orbit can be obtained by integrating the equation $J = r^2\dot\theta$ to give

$$T = \frac{J^3}{\mu^2}\int_0^{2\pi}\frac{d\theta}{(1+e\cos\theta)^2} = Ba^{3/2},$$

where a is the length of the semimajor axis and B is a constant. This result, which can be restated as saying that a^3/T^2 has the same value for all orbits, is known as *Kepler's third law of planetary motion*.

4. Show that the classical planetary orbit can be a circle and, if it is, that the planet moves along it with constant angular velocity.

5. The differential equation for the relativistic orbit was obtained by dividing the equation

$$\frac{du}{d\theta}\frac{d^2u}{d\theta^2} + u\frac{du}{d\theta} = \frac{\mu}{J^2}\frac{du}{d\theta} + 3\frac{\mu}{c^2}u^2\frac{du}{d\theta}$$

by $du/d\theta$, tacitly assuming that $du/d\theta$ is not identically zero. But suppose $du/d\theta$ is identically zero; what orbits arise then?

6. (a) Show that the differential operator $L(v) = v'' + v$ is *linear*, that is, $L(v + w) = L(v) + L(w)$ and $L(av) = aL(v)$. Deduce that if v_A is a solution to the differential equation $L(v) = A$ and v_B is a solution to $L(v) = B$, then $v = v_A + v_B$ is a solution to $L(v) = A + B$.

(b) Verify that the functions v_A, v_B, and v_C given in the text (on page 429) satisfy the individual differential equations $L(v) = A$, $L(v) = B$, $L(v) = C$, where

$$A = \frac{3\mu^3}{J^4}\left(1 + \frac{e^2}{2}\right), \qquad B = \frac{6e\mu^3}{J^4}\cos\theta, \qquad C = \frac{3\mu^3 e^2}{2J^4}\cos 2\theta.$$

7. (a) For the relativistic orbit, write the geodesic equation for φ and show that $\varphi(\tau) \equiv \pi/2$ is a solution.

(b) Show that a relativistic orbit for which $\varphi(0) = \pi/2$ and $\varphi'(0) = 0$ has $\varphi(\tau) = \pi/2$ for all τ.

8. (a) Consider a planet whose classical orbit is an ellipse with semimajor axis a and fixed eccentricity e. Suppose its period is T earth centuries. Show that the perihelion drift of its orbit is
$$\frac{2\pi D}{T} = \frac{\text{constant}}{a^{5/2}}.$$

(b) The earth is about $2\frac{1}{2}$ times as far from the sun as Mercury. If the earth's orbit had the same eccentricity as Mercury's, what would its perihelion drift be per century? Ignore the effect of other planets so you determine only the "relativistic correction."

(c) Still ignoring the effect of other planets, what would the eccentricity of the earth's orbit have to be in order to make its perihelion drift equal to that of Mercury's orbit?

Further Reading for Chapter 8

Einstein devotes one section of his fundamental paper on general relativity to the weak-field approximation ([10], §21); the subject is covered in more detail in Dubrovnin et al. [7]. For the Schwarzschild solution and isotropic coordinates see Adler et al. [1]. The refraction of light is a standard topic in physics textbooks; see [13]. Pollard [27] has a concise and lucid introduction to the classical theory of planetary orbits, while Adler et al. [1] and Berry [2] discuss the relativistic theory.

Page 431, The paragraph "Visualizing the drift" states incorrectly the size of the drift due to the other planets. Replace the paragraph with the following.

In fact, most of the shift is due to the fact that the observations are made in a non-inertial frame; only about one-tenth is due to the Newtonian gravitational "disturbances by other planets." The relativistic contribution is extremely small; even over 80 centuries it amounts to just under 1°, but it is enough to be visible in the figure on the next page. By contrast, the observed shift during the same time is enormous—a third of a complete revolution. The figure is drawn to scale and gives an accurate picture of the eccentricity of Mercury's orbit. The non-relativistic contribution (of about 5557″ per century) is shown by the dashed ellipse, the correct (relativistic) drift by the gray ellipse.

Bibliography

[1] Ronald Adler, Maurice Bazin, and Menahem Schiffer. *Introduction to General Relativity*. McGraw-Hill, New York, second edition, 1975.

[2] Michael Berry. *Principles of Cosmology and Gravitation*. Cambridge University Press, Cambridge, 1976.

[3] Garrett Birkhoff and Saunders MacLane. *A Survey of Modern Algebra*. Macmillan, New York, second edition, 1953.

[4] Mary L. Boas. *Mathematical Methods in the Physical Sciences*. John Wiley & Sons, New York, second edition, 1983.

[5] M. Crampin and F. A. E. Pirani. *Applicable Differential Geometry*. Cambridge University Press, Cambridge, 1986.

[6] C. T. J. Dodson and T. Poston. *Tensor Geometry: The Geometric viewpoint and Its Uses*. Springer-Verlag, New York, 1991.

[7] B. A. Dubrovnin, A. T. Fomenko, and S. P. Novikov. *Modern Geometry—Methods and Applications*. Springer Verlag, New York, 1984.

[8] A. Einstein. Cosmological considerations on the general theory of relativity. In [20], *The Principle of Relativity*. Dover, 1952.

[9] A. Einstein. Does the inertia of a body depend on its energy-content. In [20], *The Principle of Relativity*. Dover, 1952.

[10] A. Einstein. The foundation of the general theory of relativity. In [20], *The Principle of Relativity*. Dover, 1952.

[11] A. Einstein. On the electrodynamics of moving bodies. In [20], *The Principle of Relativity*. Dover, 1952.

[12] A. Einstein. On the influence of gravitation on the propagation of light. In [20], *The Principle of Relativity*. Dover, 1952.

[13] Richard P. Feynman, Robert B. Leighton, and Matthew Sands. *The Feynman Lectures on Physics*. Addison-Wesley, Reading, MA, 1963.

[14] Theodore Frankel. *Gravitational Curvature: An Introduction to Einstein's Theory*. W. H. Freeman, San Francisco, 1979.

[15] Wilhelm Klingenberg. *A Course in Differential Geometry*. Springer-Verlag, New York, 1978.

[16] Erwin Kreyszig. *Introduction to Differential Geometry and Riemannian Geometry*. University of Toronto Press, Toronto, 1968.

[17] L. D. Landau and E. M. Lifshitz. *The Classical Theory of Fields*. Pergamon Press, Oxford, 1975.

[18] D. F. Lawden. *An Introduction to Tensor Calculus, Relativity and Cosmology*. John Wiley & Sons, New York, third edition, 1982.

[19] H. A. Lorentz. Michelson's interference experiment. In [20], *The Principle of Relativity*. Dover, 1952.

[20] H. A. Lorentz, A. Einstein, H. Minkowski, and H. Weyl. *The Principle of Relativity*. Dover, New York, 1952.

[21] F. K. Manasse and C. W. Misner. Fermi normal coordinates and some basic concepts in differential geometry. *J. Math. Phys.*, 4(6):735–745, June 1963.

[22] John McCleary. *Geometry from a Differentiable Viewpoint*. Cambridge University Press, Cambridge, 1994.

[23] H. Minkowski. Space and time. In [20], *The Principle of Relativity*. Dover, 1952.

[24] Charles W. Misner, Kip S. Thorne, and John Archibald Wheeler. *Gravitation*. W. H. Freeman, San Francisco, 1973.

[25] Gregory L. Naber. *Spacetime and Singularities*. Cambridge University Press, Cambridge, 1988.

[26] Abraham Pais. *'Subtle is the Lord...' The Science and the Life of Albert Einstein*. Oxford University Press, New York, 1982.

[27] Harry Pollard. *Mathematical Introduction to Celestial Mechanics*. Prentice-Hall, Englewood Cliffs, NJ, 1972.

[28] Edwin F. Taylor and John Archibald Wheeler. *Spacetime Physics*. W. H. Freeman, San Francisco, 1966.

[29] Edward R. Tufte. *The Visual Display of Quantitative Information*. Graphics Press, Cheshire, CT, 1983.

[30] I. M. Yaglom. *A Simple Non-Euclidean Geometry and Its Physical Basis*. Springer-Verlag, New York, 1979.

Index

c, speed of light, 7, 74
D/dt, covariant differential operator along a curve, 320
$E = mc^2$, equivalence of mass and energy, 138
$\mathbf{E}$, electric vector, 23
G, gravitation constant, 74, 168
$\Gamma_{jk,l}$, Γ^i_{jk}, Christoffel symbols of the first and second kinds, 263, 278
$\mathbf{H}$, magnetic vector, 23
h, Planck's constant, 139
$J_{p,q}$, Minkowski inner product matrix, 60, 102
$\mathbf{J}$, vector of
  angular momentum, 424
  electric current density, 23
K, Gaussian curvature, 246
K^h_j, tidal acceleration tensor, 350, 358–360

κ, curvature of a curve, 118, 121
$O(2, \mathbf{R})$, orthogonal group, 71
$O_+(2, \mathbf{R})$, group of rotations in $\mathbf{R}^2$, 51, 70
R_{hjkl}, R^i_{jkl}, covariant, mixed Riemann curvature tensors, 264, 278
R_{ik}, Ricci tensor, 266
ρ, density
  of electric charge, 23
  of matter, 179, 183
S^3, 3-sphere, 376
$T^{G_{i_1\cdots i_p}}_{j_1\cdots j_q}$, $T^{R_{i_1\cdots i_p}}_{j_1\cdots j_q}$, tensor of type (p, q) in the coordinate frames of G and R, respectively, 308
∇, gradient operator, 23, 182
$\nabla \cdot \mathbf{A}$, divergence of vector field $\mathbf{A}$, 182

439

∇^2, the Laplace operator, 24, 179, 182
$\wedge$, operator defining
   exterior product $dp \wedge dq$ of differentials, 188
   oriented parallelogram $\mathbf{v} \wedge \mathbf{w}$ spanned by $\mathbf{v}$ and $\mathbf{w}$, 210

acceleration, 124–136
   4-acceleration, 131, 133–136
   corresponds to curvature, 132, 134–136
   linear, 155–165
addition of velocities, 78–79
algebra of differential forms, *see* exterior product
angular momentum
   classical, 424
   relativistic, 427
arc length, 110–117
arc-length
   function, 115
   parametrization, *see also* curve, parametrized by arc length
   of great circles on sphere, 273–274
   of lines in hyperbolic plane, 283–284
area
   magnification factor, 217
   of a spherical cap, 155
   of region on a surface, 215–217
   oriented, 210–212
      in Gauss map, 248

Bianchi identity, 372
boost (velocity boost), 38, 46
   addition formula, 79

calibration
   of length, time, and mass, 74
   of two Euclidean frames, 51
   of two spacetime frames, 54
Cauchy–Schwarz inequality, 71
causal past and future of an event, 76
charge density ρ, 23
Christoffel symbols
   defined extrinsically, 263
   defined intrinsically, 278
   form gravitational field in relativity, 333
   not tensorial, 312–316
clock, *see* measurement
collision
   elastic, 98
   inelastic, 92, 99
commutative diagram, 35
congruence in Minkowski geometry, 64–70
conservation
   of 4-momentum, 98–101
   of energy–momentum, 370–372
   of momentum, 89, 91–93
contraction of a tensor, 265, 309
contravariant, *see* covariant
coordinate
   frame
      asymptotically inertial, 405
      Fermi, 351–362
      inertial, 90, 147–148, 155, 289, 330
      limited in size when not inertial, 146, 156–157
      linearly accelerating, 155–165

noninertial, 144
radar, 148, 156–162
rotating, 144–146
rulers and clocks, 148, 162–165, 289
transformation
arbitrary nonlinear, 292–293, 331
as dictionary, 13, 24, 147
Galilean, 12–14
induced on tangent plane, 294–297
Lorentz, 38, 46
coordinates
as measurements, 147–149
as names, 13, 97, 147
curvilinear, 205
dimensionally homogeneous, 137–138, 368, 369
compared to traditional, 102
in tangent plane, induced, 279–280, 293–296
isotropic, 406–409, 432
spherical, 181, 397
Coriolis force, 89, 344, 405
cosh (hyperbolic cosine), 43
cosmological constant, 376–378, 383
coth (hyperbolic cotangent), 43
covariance, 97, 300–303
general, 301
related to way a basis transforms, 301–303
special, 97
covariant
and contravariant quantities, 301–303

derivative
along a curve, 320
as limit of a difference quotient, 321–325
of a tensor, 316–317
csch (hyperbolic cosecant), 43
current density $\mathbf{J}$, 23
curvature
bending *versus* stretching, 196, 257
center of, 119, 121
Gaussian, 245–252, 278
manifested by distortions in flat charts, 196–197
negative, 250–252
on hyperbolic plane, 284
of a curve, 117–121, 242
of a surface, *see* curvature, Gaussian
of spacetime, 285–286
determines the way objects move, 331–333
evidence for, 152, 153, 165, 196–200
is determined by matter, 345, 374–376
radius of, 119, 121
scalar, 373
total, 255
vector, 118, 121
curve, parametrized, 108–121
by arc length, 115–121
nonsingular, 109
of constant curvature, 119–120
spacelike, 140
cylinder, generalized, 6

de Sitter spacetime, 230–240, 256, 289

deflection, *see* light, bending of light rays
determinant of a 2 × 2 matrix, 37
develop one surface on another, 258
developable, *see* tangent developable surface
diameter of a plane region, 246
dictionary, *see* coordinate transformation
differential
    expression for metric, 398, 401–409, 418, 426
    of a map, 295–297
differentials, rules for multiplying, *see* exterior product
Disquisitiones Generales (C. F. Gauss, "General Investigations of Curved Surfaces," 1827), 258
div, the divergence operator, 182
divergence, 185
    of a tensor, 372 313
    of a vector field, 182
    of Ricci tensor, 373–374
Doppler effect, 81–83, 85
    in a gravitational field, 188–191
    in a linearly accelerating frame, 163–165

eccentricity of an ellipse, 117, 422
eigenvalue and eigenvector of an $n \times n$ matrix, 37
Einstein, Albert
    "The happiest thought of my life", 361

clocks slow down in a gravitational field, 189–190
cosmological constant for spherical universe, 376–378
deduces Lorentz transformation from constancy of the speed of light, 31
devotes much of his fundamental paper on general relativity to developing tensors, 307
explains perihelion drift of Mercury's orbit, 422, 430–431
gravity causes light rays to bend, 413–421
introduces summation convention, 209
laws of physics must be generally covariant, 331
must use arbitrary frames to model gravity, 198–200
obtains a new theory of gravity by abandoning special status of inertial frames, 143
photoelectric effect, 139
proposes special relativity to solve troubling problems in physics, 1
treats Newtonian mechanics as a first approximation to general relativity, 385–386
weak field solution, 397–401

electric vector **E**, 23
electromagnetic potential, 333, 341
ellipse, 116, 422–423
elliptic integrals and functions, 117, 129
energy
　density, 368
　equivalent to mass, 100, 138
　kinetic, 100, 138
　　relativistic, 105, 138
　of light, 139
　potential, of gravitational field, 173–174
　rest, 105, 137
　total, 106, 138
energy–momentum
　tensor, 366–372, 383
　　has zero divergence, 372
　vector, 106, 138
equations of motion
　classical (Newtonian), 330, 424–426
　in a weak field, 390–394
　relativistic, 332, 426–429
equiangular spiral, 123
ether, medium for light waves, 17
event
　in spacetime, 2
　time-, space-, and lightlike, 53
evolute of a curve, 119
exterior product, 187

Fermi
　conditions, 356
　coordinates, 351–358, 383
Fermi, Enrico, 351

first fundamental form (metric), 208, 263, 277, 286
fission, converting mass to energy, 100
Fitzgerald contraction, 19–21, 84
　applied to the wave equation, 25
　implied by special relativity, 80
　in linearly accelerating frame, 160
　in rotating frame, 152
　of particle swarm, 367
Fitzgerald, B. F., explained results of Michelson–Morley experiment, 20
flip, as linear map, 35
fluid flow, incompressible, 183–185
focus (*plural* foci) of an ellipse, 422
force
　4-force, 132–133, 137–138
　and kinetic energy, 137
　in Newton's second law, 88–89
　in universal gravitation, 168
Foucault pendulum, 89
frame, *see* coordinate frame
freely falling body, 191–195, 330, 345–355
future
　ray, 58
　set, 55, 57
　　in de Sitter spacetime, 235
　vector, 57, 124

Galilean transformation, 12–14
  and conservation of momentum, 92
  applied to the wave equation, 24
  geometrically a shear, 13
Galileo's law of inertia, *see* Newton's laws of motion
Gauss map
  of a curve, 242
  of a surface, 245
  of cylinder or cone, 247
Gauss, Carl F., 257–258
Gaussian
  curvature, 245–252, 278
    and Riemann curvature tensor, 260–261
    taking sign into account, 248–252
  image, 245
generator
  of a cylinder, 6
  of a ruled surface, 239
geodesic, 267–275
  differential equations, 269
  in weak field, 392
  existence and uniqueness, 274
  has constant-speed parametrization, 270
  in hyperbolic plane, 281–285
  is covariant object, 313–315
  is worldcurve of freely falling particle, 331, 335–339
  not just "straightest path", 270–271
  on sphere, 272–274
  separation, 347–351, 383
    connection with Riemann curvature tensor, 349
    indication of tidal acceleration, 347
gravitational
  field, *see also* vacuum field equations; matter field equations
    can be produced by coordinate change, 199, 329
    classical (Newtonian) components, 171, 333
    constant, 333–341
    of a point source, 397–401, 432
    relativistic components, 332–333
    spherically symmetric, 396–409
    weak, 388–395
  potential
    and red shift, 190–191, 195–196
    classical (Newtonian), 171–174, 179–186, 333
    Laplacian of, 179, 181–186
    relativistic, 333
  radius, 404
gravity
  and curvature of spacetime, 285–286
  classical (Newtonian), 167–186
  comparing the theories of Newton and Einstein, 199
  in special relativity, 188–196

incompatible with
    Minkowski geometry,
    195–196, 329
great circle, arc-length
    parametrization,
    273–274

history, *see* worldcurve
Huygens' principle, 414
hyperbolic
    angle, 57–60
        preserved by hyperbolic
        rotations, 59
    functions, 43–48
        algebraic and calculus
        relations, 44
    plane, 281–285
hyperboloid (of one sheet), 230
    doubly-ruled surface, 239

index of a tensor
    covariant, contravariant, 264,
        303
    raising and lowering, 264
inertia, 88
inertial frame, 90, 147–148,
    155, 289, 330
inner product
    Euclidean, 49
    Minkowski, 60
interval
    between spacetime events, 56
    timelike, spacelike, and
        lightlike, 56
intrinsic geometry, 203, 257
    of an arbitrary 2-dimensional
        surface, 277–281
    of de Sitter spacetime,
        233–236

of the hyperbolic plane,
    281–285
of the sphere, 221–229
isometry
    of curves, 241
    of surfaces, 258
isotropic coordinates, 407

kinematics, 31
kinetics, 87
kosher, 48

Laplacian, 178, 182, *see also*
    gravitational field;
    Maxwell's equations
latitude and longitude
    on earth, 149–151
    on parametrized sphere, 222,
        272–273
Laue's scalar, 375
law of hyperbolic cosines, 62
length contraction, *see*
    Fitzgerald contraction
Leverrier, Urbain Jean Joseph,
    430
light
    and electromagnetism, 23
    bending of light rays,
        191–195, 413–421
        1911 computation,
            416–418
        1916 computation,
            418–421
    cone, 8, 32–35, 38
        determined by metric,
            288–289
        in de Sitter spacetime,
            233–240
    energy and frequency, 139

light (*cont.*)
  momentum and wavelength, 139
  speed of, c, 7
  travel time, in the Michelson–Morley experiment, 18
  waves and rays, 414
light-second, as unit of distance, 7
lightlike event, 53
linear map, geometric characterization (exercise), 40
local time, *see* Lorentz, H. A.
Lorentz
  group (orthochronous), 68
  transformation
    and hyperbolic reflection and rotation, 68
    as hyperbolic rotation, 46
    as velocity boost, 38, 46
    defined by Einstein to preserve the light cone, 31–39
    in conventional units, 103, 106–108
    preserves the wave equation, 26–27
Lorentz, H. A.
  local time, 27
  on the Michelson–Morley experiment, 18
  reluctant to accept Einstein's special relativity, 27
loxodrome, 230

Mach, Ernst, 198
magnetic vector **H**, 23
map, 292–303, *see also* coordinate transformation
  of the earth, 149–151, 153
mass
  gravitational, 168
  inertial, 88, 167
  proper, or rest, 91, 100
  relativistic, 91, 93–96, 104, 138
matter field equations
  classical (Newtonian), 182–186, 365
  in weak field, 394–395
  relativistic, 374–376
  with cosmological constant, 377–378
matter–energy, 101, 370
Maxwell's equations, 23
  covariant under Lorentz transformations but not Galilean, 97
  given by a 4-potential, 333, 341
Maxwell, James Clerk, 22
measurement
  grid of rulers and clocks, 148
  timings by radar and by clocks, 148–149
    disagree in noninertial frame, 148, 162
Mercury, 385, 421, 430–431
metric, or metric tensor
  asymptotically inertial, 405
  contravariant, 310–311
  determines lengths and areas on a surface, 213–217
  forms gravitational potential, 333

induced by "pull back", 354
inner product on tangent
    plane, 208–210, 278
  Minkowski, 299, 330, 356,
    388–391, 397–405, 418
  of weak field of a point
    source, 398–401
  on de Sitter spacetime, 232,
    233
  on hyperbolic plane, 282
  on intrinsically defined
    surface, 277
  on spacetime, 286, 298–300
  on the sphere, 222
  Schwarzschild, 401–405, 432
    in isotropic coordinates, 409
  written as a differential, 398,
    401–409, 418, 426
Michelson, Albert A., 17
Michelson–Morley experiment,
    15–19
Miller cylindrical projection,
    150
Minkowski
  geometry, 49–70
    congruence in, 64–70
    in tangent plane to de
      Sitter spacetime, 232
    incompatible with gravity,
      195–196
    reflection in, 65–69
    rigid motion in, 64–70
    length of a spacelike curve,
      140
  metric, 299, 330, 356,
    388–391, 397–405, 418
Minkowski, Hermann
  introduces a norm for
    spacetime events, 52

  places special relativity in a
    geometric context, 1
momentum, 88
  4-momentum, 94, 102, 105,
    131–133
    conservation of, 98–101
    covariant under Lorentz
      transformations, 95, 97
  conservation of, 89, 91–93
  of light, 139
  relativistic, 93–96
Morley, Edward W., 17
multi-index quantity, 262–266

Newton's laws
  of motion, 87–93
    first law (Galileo's law of
      inertia), 87–89
    second law, 88, 131–133
    third law, 88, 89, 91–93
  universal gravitation, 168
non-Euclidean geometry
  elliptic and hyperbolic, 154,
    281
  hyperbolic plane as model
    of, 281
  in rotating frame, 152
  on sphere, 149–155
nonsingular parametrization
  of a curve, 109
  of a surface, 206
norm
  Euclidean, 49
  Minkowski, 51–54
    in dimensionally
      homogeneous
      coordinates, 102
    in tangent plane, 288
normal to a surface, 207

objective past and future of an event, 75–76
objectivity
  and relativity, 10
  obtained by using generally covariant quantities, 331
observer
  Galilean, 10, 15–17, 143, 147, 155
  has subjective view of spacetime, 10
  in rotating frame, 144
  linearly accelerating, 155–165
  with arbitrary frame, 292
orthochronous, *see* Lorentz group
orthogonal
  group $O(2, R)$, 71
  vectors in spacetime, 63
orthonormal basis, 352
oscillatory motion, 127–130, 133–134

parallel
  transport, 317–325
    in Fermi coordinates, 352
    map, 321
    vectors or tensors along a curve, 321
parameter, 108, 204
particle density 4-vector, 368
past set in spacetime, 55
perihelion, 385
  drift, 421–431
perpendicular vectors in spacetime, 63
photo finish, as spacetime diagram, 4

photoelectric effect, 139
photon
  as particle of light, 7
  energy and frequency, 139, 190
  has zero rest mass, 141
  momentum and wave length, 139
Planck's constant, h, 139
planetary motion, 433
  classical orbit, 423–426
  Kepler's third law, 411, 431
  relativistic orbit, 426–431
potential, *see* gravitational potential; electromagnetic potential; energy
Preakness (horse race), 4
principal axes theorem, 290
Principia (Isaac Newton, *Philosophiæ Naturalis Principia Mathematica*, 1687), 87
principle
  of covariance, 97, 300
  of equivalence (of gravity and acceleration), 169
    implies gravity bends light rays, 191–195
    is only local, 169
  of general covariance, 331
  of relativity, *see* relativity
proper time, 79
  corresponds to arc length along a worldcurve, 125–127
  in rotating frame, 146
proper-time
  parameter, or function, 127

parametrization, *see* worldcurve, parametrized by proper time
Pythagorean theorem in Minkowski geometry, 63–64

quadratic form, Minkowski, 53–55, 240
    relation to Minkowski norm, 53

radar, *see* measurement
radian, hyperbolic, 58
radius of space
    in de Sitter spacetime, 232
    in Einstein's spherical universe, 376–378
rank of a tensor, 308
red shift, 81, 165
    as energy loss of a photon, 190–191
    gravitational, 190–191, 405–406
reflection, hyperbolic, 65–69
refraction of light, 413, 415–416
region, oriented, 210
relativity
    Galileo's principle, 10
    general, 143, 199
    special, 10, 90, 143, 298
Ricci tensor, 266
    alternative formula, 380
    in vacuum field equations, 361
Riemann curvature tensors
    and Gaussian curvature, 260–261

    defined extrinsically, 264
    defined intrinsically, 278
    in geodesic separation equations, 349–351
right triangle in spacetime, 63
rigid
    body, not physically meaningful in special relativity, 77
    motion in Minkowski geometry, 64–70
rotation
    Euclidean, 49–51
    hyperbolic, 46, 57–61, 64
    uniform, of an observer, 144–146
ruled surface, 239
rulers, *see* measurement

scalar curvature, 373
Schwarzschild
    metric, 401–405, 432
        in isotropic coordinates, 409
    radius, 404
Schwarzschild, Karl, 401
sech (hyperbolic secant), 43
second, *see also* light-second
    fundamental form, 263
semimajor, semiminor axes of an ellipse, 422
separation between spacetime events, 56
shear, as linear map, 13
shock wave, 77
simultaneity, not physically meaningful in special relativity, 75

singular
  point
    of a curve, 109
    of a surface, 205
    of sphere parametrization, 222
  solution of differential equations, 340
sinh (hyperbolic sine), 43
slice of space or spacetime, 8, 287
slow particle, 390
small quantity, 386–388
source of gravitational field
  continuous distribution, 179–186
  discrete, 172–173
spacelike event, 53
spacetime, 1
  as intrinsically defined surface, 277
  de Sitter, 230–240, 256
  diagram, 1–8
  sphere parametrization, 221–222, 377
  stress–energy tensor, *see* energy–momentum tensor
summation convention, 209
surface
  intrinsically defined, 277–278
  parametrized in $\mathbf{R}^3$, 204–207
    nonsingular, 206–207
    singular point, 205–206
swarm of noninteracting particles, 366–369, 375
Szechuan, 48

tangent
  developable (surface), 220
  plane, or space
    carries an inertial frame in spacetime, 289
    to embedded surface, 207
    to intrinsically defined surface, 278–281
    to spacetime, 286, 293–303
  vector, 108
    to an embedded surface, 205, 207–208
    to intrinsically defined surface, 279–281
    unit (unit speed), 118, 121
tanh (hyperbolic tangent), 43
  converts hyperbolic rotation to velocity boost, 46
tensor, 307–325
  covariant, contravariant, 308
  field, 308, 321, 325
  not preserved under differentiation, 311–312
  product, 326, 368
tensorial quantity, 308
theorema egregium of Gauss, 257–262
tidal acceleration, 174–179
  classical (Newtonian) equations, 345–348
  follows inverse cube law, so lunar tides are stronger than solar, 177
  matrix, 351
    classical, 178, 347–348
    relativistic, 350, 358–360
  relativistic equations, 350
  shown by geodesic separation, 347

tides
- distinguish between gravity and linear acceleration, 170
- effects revealed in orbiting spacecraft, 170
- Newtonian, 174–179
- on earth, 177

time dilation, 79–80, 103–105, 146, 164
- in gravitational field, 190, 405
- local, 132

timelike event, 53
trace of a 2 × 2 matrix, 37
train schedule (Paris–Lyon), as spacetime diagram, 3
trefoil knot, 220
triangle inequality, 62
Tufte, Edward R., 3
twin paradox, 130–131

units
- conventional, 74, 101–106
- converting between conventional and geometric, 101, 105
- geometric, 7, 74

vacuum field equations
- classical (Newtonian), 179, 182, 344–348
- for Schwarzschild solution, 402–404
- for weak field, 397–401
- relativistic, 358–362

vector field, 171–172, 183, 233, 317–325
- gradient, 171

velocity
- 4-velocity, 94, 102, 105
  - proper (corresponds to unit speed tangent vector), 104–105, 131
- boost, 46
- limitation, 74
  - implies mass is relative, 90–91
- in rotating frame, 146
- slope of worldcurve, 1–4, 11–13
- vector, 14–15, 88, *see also* tangent vector

Vulcan, 421

Watson, Bill, 48
wave equation, 24–27
work, gravitational, 173–174
worldcurve, or worldline, 2, 124–131
- nonsingular, 125
- of an observer, 11
- of constant acceleration, 134–136, 155–160, 163
- of freely falling particle is a geodesic, 331–332
- parametrized by proper time, 127–131

Undergraduate Texts in Mathematics

(continued from page ii)

Iooss/Joseph: Elementary Stability and Bifurcation Theory. Second edition.
Isaac: The Pleasures of Probability. *Readings in Mathematics.*
James: Topological and Uniform Spaces.
Jänich: Linear Algebra.
Jänich: Topology.
Kemeny/Snell: Finite Markov Chains.
Kinsey: Topology of Surfaces.
Klambauer: Aspects of Calculus.
Lang: A First Course in Calculus. Fifth edition.
Lang: Calculus of Several Variables. Third edition.
Lang: Introduction to Linear Algebra. Second edition.
Lang: Linear Algebra. Third edition.
Lang: Undergraduate Algebra. Second edition.
Lang: Undergraduate Analysis.
Lax/Burstein/Lax: Calculus with Applications and Computing. Volume 1.
LeCuyer: College Mathematics with APL.
Lidl/Pilz: Applied Abstract Algebra. Second edition.
Logan: Applied Partial Differential Equations.
Macki-Strauss: Introduction to Optimal Control Theory.
Malitz: Introduction to Mathematical Logic.
Marsden/Weinstein: Calculus I, II, III. Second edition.
Martin: The Foundations of Geometry and the Non-Euclidean Plane.
Martin: Geometric Constructions.
Martin: Transformation Geometry: An Introduction to Symmetry.
Millman/Parker: Geometry: A Metric Approach with Models. Second edition.
Moschovakis: Notes on Set Theory.
Owen: A First Course in the Mathematical Foundations of Thermodynamics.
Palka: An Introduction to Complex Function Theory.
Pedrick: A First Course in Analysis.
Peressini/Sullivan/Uhl: The Mathematics of Nonlinear Programming.
Prenowitz/Jantosciak: Join Geometries.
Priestley: Calculus: A Liberal Art. Second edition.
Protter/Morrey: A First Course in Real Analysis. Second edition.
Protter/Morrey: Intermediate Calculus. Second edition.
Roman: An Introduction to Coding and Information Theory.
Ross: Elementary Analysis: The Theory of Calculus.
Samuel: Projective Geometry. *Readings in Mathematics.*
Scharlau/Opolka: From Fermat to Minkowski.
Schiff: The Laplace Transform: Theory and Applications.
Sethuraman: Rings, Fields, and Vector Spaces: An Approach to Geometric Constructability.
Sigler: Algebra.
Silverman/Tate: Rational Points on Elliptic Curves.
Simmonds: A Brief on Tensor Analysis. Second edition.
Singer: Geometry: Plane and Fancy.
Singer/Thorpe: Lecture Notes on Elementary Topology and Geometry.
Smith: Linear Algebra. Third edition.
Smith: Primer of Modern Analysis. Second edition.
Stanton/White: Constructive Combinatorics.
Stillwell: Elements of Algebra: Geometry, Numbers, Equations.
Stillwell: Mathematics and Its History.
Stillwell: Numbers and Geometry. *Readings in Mathematics.*
Strayer: Linear Programming and Its Applications.
Thorpe: Elementary Topics in Differential Geometry.
Toth: Glimpses of Algebra and Geometry. *Readings in Mathematics.*

Troutman: Variational Calculus and Optimal Control. Second edition.
Valenza: Linear Algebra: An Introduction to Abstract Mathematics.

Whyburn/Duda: Dynamic Topology.
Wilson: Much Ado About Calculus.